COMPUTATIONAL GENOMICS

COMPUTATIONAL GENOMICS

By

Dr. M. Prakash

Dept. of Zoology
M.M.H. Post Graduate College
Ghaziabad (U.P.)
(India)

DISCOVERY PUBLISHING HOUSE PVT. LTD.
NEW DELHI-110 002

First Published-2009

ISBN: 978-81-8356-466-3

Published by:
DISCOVERY PUBLISHING HOUSE PVT. LTD.
4831/24, Ansari Road, Prahlad Street
Darya Ganj, New Delhi-110002 (India)
Phone: 23279245 • Fax: 91-11-23253475
E-mail: dphbooks@rediffmail.com • dphtemp@indiatimes.com
web: www.discoverypublishinghouse.com

Printed at:
Sachin Printers
Delhi

Preface

The present title *"Computational Genomics"* is an exciting new field bringing together life scientists, mathematicians, computer scientists and engineers to explore a new and deeper understanding of biological systems. The present text has been compiled for computer science and mathematics graduate and level undergraduate students. It will also be of interest to molecular biologists interested in bioinformatics.

The rationable of the book is to present algorithmic ideas in computational biology and to show how they are connected to molecular biology and to biotechnology. To achieve this goal, the book has a substantial "computating biology without formulas" component that presents biological motivation and computational ideas in a simple way. This simplified presentation of biology and computing aims to make the book accessible to computer scientists entering this new areas and to biologists who do not have sufficient background for more involved computational techniques. Every chapter has an introductory section that describes both computational and biological ideas without any formulas. Thus book concentrate on computational ideas rather than details of the algorithms and makes special efforts to prevent there ideas in a simple way.

To make the work more comprehensive and informative, the author has consulted many authoritative books, research journals, abstracts, monographs etc. He is grateful to all those great scholars whose work are cited or substantially reproduced.

There can be no claim to originality except in the manner of treatment and much of the information has been obtained from the books and scientific journals available in the different libraries.

The author expresses his thanks to his friends and colleagues whose continue inspirations have initiated him to bring out this book.

The author expresses his gratitude to Mr. Wasan and staff of M/s Discovery Publishing House Pvt. Ltd. for their whole hearted cooperation in the publication of this book.

In the mean time, the author will remain sincerely responsible for any shortcomings of the book and be grateful to the readers for their suggestions and constructive criticism for the continuous betterment of the book. He takes this opportunity to appeal to the readers to send their suggestions straightaway to his Publisher.

Author

Contents

Introduction

Automatic programming—i.e., having computer programs automatically write computer programs—has a long history in the field of artificial intelligence. Many different approaches have been tried, but as yet no general method has been found for automatically producing the complex and robust programs needed for real applications. Some early evolutionary computation techniques were aimed at automatic programming. The evolutionary programming approach of Fogel, *Owens*, and *Walsh* (1966) evolved simple programs in the form of finite-state machines. Early applications of genetic algorithms to simple automatic-programming tasks were performed by Cramer (1985) and by Fujiki and Dickinson (1987), among others. The recent resurgence of interest in automatic programming with genetic algorithms has been, in part, spurred by John Koza's work on evolving Lisp programs via *"genetic programming."* The idea of evolving computer programs rather than writing them is very appealing to many. This is particularly true in the case of programs for massively parallel computers, as the difficulty of programming such computers is a major obstacle to their widespread use. *Hillis's* work on evolving efficient sorting networks is one example of automatic programming for parallel computers. My own work with *Crutchfield*, *Das*, and *Hraber* on evolving cellular automata to perform computations is an example of automatic programming for a very different type of parallel architecture.

LISP PROGRAMS

John Koza (1992,1994) has used a form of the genetic algorithm to evolve Lisp programs to perform various tasks. *Koza* claims that his method— *"genetic programming"* (GP)—has the potential to produce programs of the necessary complexity and robustness for general automatic programming. Programs in Lisp can easily be expressed in the form of a *"parse tree,"* the object the GA will work on.

As a simple example, consider a program to compute the orbital period P of a planet given its average distance A from the Sun. Kepler's Third Law states that $P^2 = cA^3$, where c is a constant. Assume that P is expressed in units of earth years and A is expressed in units of the Earth's average distance from the Sun, so $c = 1$. In FORTRAN such a program might be written as

```
PROGRAM ORBITAL_PERIOD
C   # Mars #
    A = 1.52
    P = SQRT(A * A * A)
    PRINT P
END ORBITAL_PERIOD
```

where * is the multiplication operator and SQRT is the square-root operator. (The value for A for Mars. In Lisp, this program could be written as

```
(defun orbital_period ()
    ; Mars ;
    (setf A 1.52)
    (sqrt (* A (* A A))))
```

In Lisp, operators precede their arguments: *e.g.*, $X * Y$ is written $(* X Y)$. The operator "setf assigns its second argument (a value) to its first argument (a variable). The value of the last expression in the program is printed automatically.

Assuming we know A, the important statement here is (SQRT $(* A (* A A))$). A simple task for automatic programming might be to automatically discover this expression, given only observed data for P and A.

Expressions such as (SQRT $(* A (* A A))$ can be expressed as parse trees, as shown in figure 1.1. In Koza's GP algorithm, a candidate solution is expressed as such a tree rather than as a bit string. Each tree consists of *functions* and *terminals*. In the tree shown in figure 1.1, SQRT is a function that takes one argument, * is a function that takes two arguments, and A is a terminal. Notice that the argument to a function can be the result of another function—*e.g.*, in the expression above one of the arguments to the top-level * is $(* A A)$.

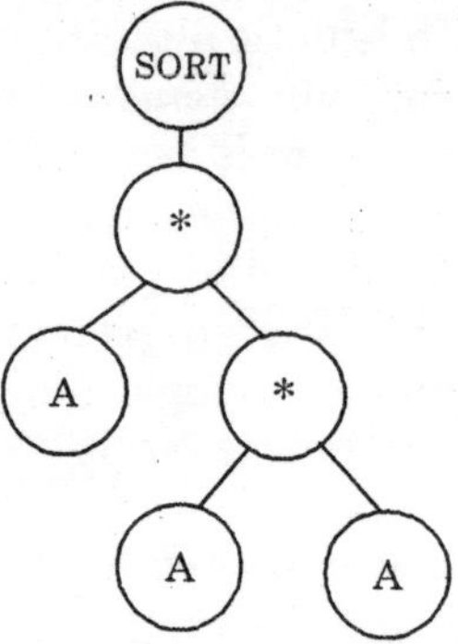

*Fig. 1.1. Parse tree for the Lisp expression (SQRT (*A (*A *A A))).*

Koza's algorithm is as follows :

A. Choose a set of possible functions and terminals for the program. The idea behind GP is, of course, to evolve programs that are difficult to write, and in general one does not know ahead of time precisely which functions and terminals will be needed in a successful program. Thus, the user of GP has to make an intelligent guess as to a reasonable set of functions and terminals for the problem at hand. For the orbital-period problem, the function set might be {+, , *, /, ,} and the terminal set might simply consist of {A}, assuming the user knows that the expression will be an arithmetic function of A.

B. Generate an initial population of random trees (programs) using the set of possible functions and terminals. These random trees must be syntactically correct programs—the number of branches extending from each function node must equal the number of arguments taken by that function. Three programs from a possible randomly generated initial population are displayed in figure 1.2. Notice that the randomly generated programs can be of different sizes (*i.e.*, can have different numbers of nodes and levels in the trees). In principle a randomly generated tree can be any size, but in practice Koza restricts the maximum size of the initially generated trees.

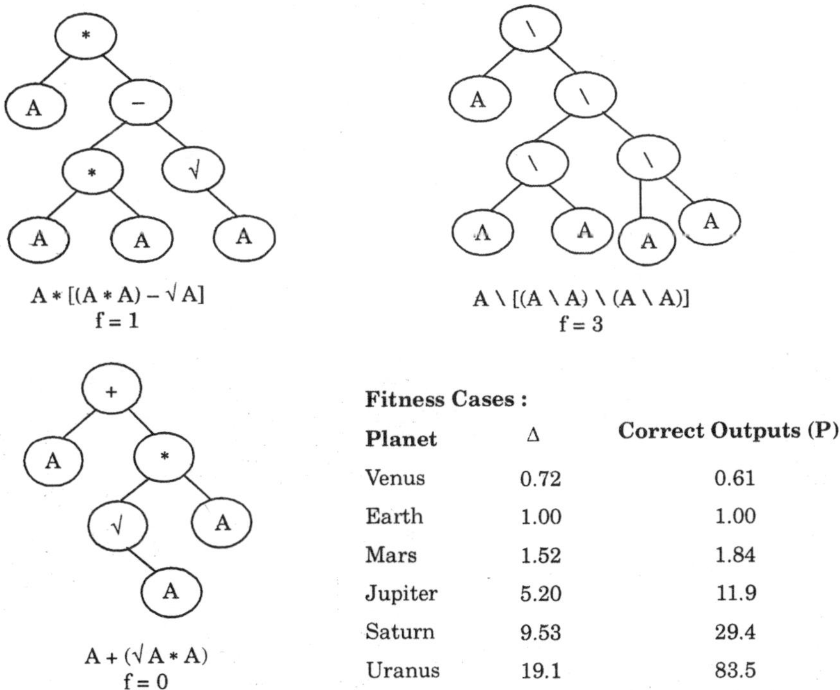

Fitness Cases :

Planet	Δ	**Correct Outputs (P)**
Venus	0.72	0.61
Earth	1.00	1.00
Mars	1.52	1.84
Jupiter	5.20	11.9
Saturn	9.53	29.4
Uranus	19.1	83.5

Fig. 1.2. Three programs from a possible randomly generated initial population for the orbital-period task. The expression represented by each tree is printed beneath the tree. Also printed is the fitness f (number of outputs within 20% of correct output) of each tree on the given set of fitness cases. A is given in units of Earth's semi major axis of orbit; P is given in units of Earth years.

C. Calculate the fitness of each program in the population by running it on a set of "fitness cases". For the orbital-period example, the fitness cases might be a set of empirical measurements of *P* and *A*. The fitness of a program is a function of the number of fitness cases on which it performs correctly. Some fitness functions might give partial credit to a program for getting close to the correct output. For example, in the orbital-period task, we could define the fitness of a program to be the number of outputs that are within 20% of the correct value. Figure 1.2 displays the fitnesses of the three sample programs according to this fitness function on the given set of fitness cases. The randomly generated programs in the initial population are not likely to do very well; however, with a large enough population some of them will do better than others by chance. This initial fitness differential provides a basis for *"natural selection."*

Apply selection, crossover, and mutation to the population to form a new population. In Koza's method, 10% of the trees in the population (chosen probabilistically in proportion to fitness) are copied without modification into the new population. The remaining 90% of the new population is formed by crossovers between parents selected (again probabilistically in proportion to fitness) from the current population. Crossover consists of choosing a random point in each parent and exchanging the subtrees beneath those points to produce two offspring. Figure 1.3 displays one possible crossover event. Notice that, in contrast to the simple *GA*, crossover here allows the size of a program to increase or decrease. Mutation might performed by choosing a random point in a tree and replacing the subtree beneath that point by a randomly generated subtree. Koza (1992) typically does not use a mutation operator in his applications; instead he uses initial populations that are presumably large enough to contain a sufficient diversity of building blocks so that crossover will be sufficient to put together a working program.

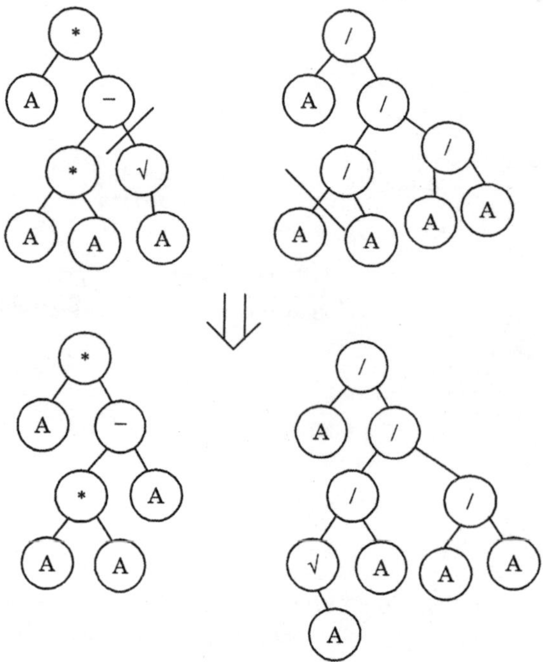

Fig. 1.3. An example of crossover in the genetic programming algorithm. The two parents are shown at the top of the figure, the two offspring below. The crossover points are indicated by slashes in the parent trees.

It may seem difficult to believe that this procedure would ever result in a correct program—the famous example of a monkey randomly hitting the keys on a typewriter and producing the works of Shakespeare comes to mind. But, surprising as it might seem, the GP technique has succeeded in evolving correct programs to solve a large number of simple (and some not-so-simple) problems in optimal control, planning, sequence induction, symbolic regression, image compression, robotics, and many other domains. One example is the block-stacking problem illustrated in figure 1.4. The goal was to find a program that takes any initial configuration of blocks—some on a table, some in a stack—and places them in the stack in the correct order. Here the correct order spells out the word "universal." The functions and terminals Koza used for this problem were a set of sensors and actions defined by Nilsson (1989). The terminals consisted of three sensors (available to a hypothetical robot to be controlled by the resulting program), each of which returns (*i.e.*, provides the controlling Lisp program with) a piece of information:

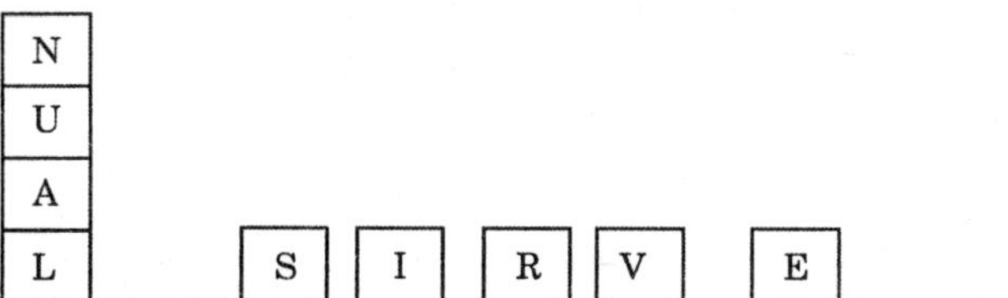

Fig. 1.4. One initial state for the block-stacking problem. The goal is to find a plan that will stack the blocks correctly (spelling "universal") from any initial state.

CS (*"current stack"*) returns the name of the top block of the stack. If the stack is empty, CS returns NIL (which means *"false"* in Lisp).

TB (*"top correct block"*) returns the name of the topmost block on the stack such that it and all blocks below it are in the correct order. If there is no such block, TB returns NIL.

NN (*"next needed"*) returns the name of the block needed immediately above TB in the goal "universal." If no more blocks are needed, this sensor returns NIL.

In addition to these terminals, there were five functions available to GP:

MS(*x*) (*"move to stack"*) moves block *x* to the top of the stack if *x* is on the table, and returns *x*. (In Lisp, every function returns a value. The returned value is often ignored.)

MT(*x*) (*"move to table"*) moves the block at the top of the stack to the table if block *x* is anywhere in the stack, and returns *x*.

DU (*expression*1, *expression* 2) ("do until") evaluates *expression*1 until *expression* 2 (a predicate) becomes TRUE.

NOT (*expression* 1) returns TRUE if *expression*1 is NIL; otherwise it returns NIL.

EQ (*expression*1, *expression* 2) returns TRUE if *expression*1 and *expression*2 are equal (*i.e.*, return the same value).

The programs in the population were generated from these two sets. The fitness of a given program was the number of sample fitness cases (initial configurations of blocks) for which the stack was correct after the program was run. Koza used 166 different fitness cases, carefully constructed to cover the various classes of possible initial configurations.

The initial population contained 300 randomly generated programs. Some examples (written in Lisp style rather than tree style) follow:

(EQ (MT CS) NN)

"Move the current top of stack to the table, and see if it is equal to the next needed." This clearly does not make any progress in stacking the blocks, and the program's fitness was 0.

(MS TB)

"Move the top correct block on the stack to the stack." This program does nothing, but doing nothing allowed it to get one fitness case correct: the case where all the blocks were already in the stack in the correct order.

(EQ (MS NN) (EQ (MS NN) (MS NN)))

"Move the next needed block to the stack three times." This program made some progress and got four fitness cases right, giving it fitness 4. (Here EQ serves merely as a control structure. Lisp evaluates the first expression, then evaluates the second expression, and then compares their value. EQ thus performs the desired task of executing the two expressions in sequence — we do not actually care whether their values are equal.)

By generation 5, the population contained some much more successful programs. The best one was (DU (MS NN) (NOT NN)) (*i.e.*, "Move the next needed block to the stack until no more blocks are needed"). Here we have the basics of a reasonable plan. This program works in all cases in which the blocks in the stack are already in the correct order: the program moves the remaining blocks on the table into the stack in the correct order. There were ten such cases in the total set of 166, so this program's fitness was 10. Notice that this program uses a building block—(MS NN)—that was discovered in the first generation and found to be useful there.

In generation 10 a completely correct program (fitness 166) was discovered:

(EQ (DU (MT CS) (NOT CS)) (DU (MS NN) (NOT NN))).

This is an extension of the best program of generation 5. The program empties the stack onto the table and then moves the next needed block to the stack until no more blocks are needed. GP thus discovered a plan that works in all cases, although it is not very efficient. Koza (1992) discusses how to amend the fitness function to produce a more efficient program to do this task. The block stacking example is typical of those found in Koza's books in that it is a relatively simple sample problem from a broad domain (planning).

A correct program need not be very long. In addition, the necessary functions and terminals are given to the program at a fairly high level. For example, in the block stacking problem GP was given the high-level actions MS, MT, and so on; it did not have to discover them on its own. Could GP succeed at the block stacking task if it had to start out with lower-level primitives? O'Reilly and Oppacher (1992), using GP to evolve a sorting program, performed an experiment in which relatively low-level primitives (*e.g.*, "if-less-*than*" and "*swap*") were defined separately rather than combined *a priori* into "if-less-than-then-swap" Under these conditions, GP achieved only limited success.

This indicates a possible serious weakness of GP, since in most realistic applications the user will not know in advance what the appropriate high-level primitives should be; he or she is more likely to be able to define a larger set of lower-level primitives. Genetic programming, as originally defined, includes no mechanism for automatically chunking parts of a program so they will not be split up under crossover, and no mechanism for automatically generating hierarchical structures (*e.g.*, a main program with subroutines) that would facilitate the creation of new high-level primitives from built-in low-level primitives. These concerns are being addressed in more recent research. Koza (1992, 1994) has developed methods for encapsulation and automatic definition of functions. Angeline and Pollack (1992) and O'Reilly and Oppacher (1992) have proposed other methods for the encapsulation of useful subtrees. Koza's GP technique is particularly interesting from the standpoint of evolutionary computation because it allows the size (and therefore the complexity) of candidate solutions to increase over evolution, rather than keeping it fixed in the standard GA.

The lack of sophisticated encapsulation mechanisms has so far limited the degree to which programs can usefully grow. In addition, there are other open questions about the capabilities of GP. Does it work well because the space of Lisp expressions is in some sense *"dense"* with correct programs for the relatively simple tasks Koza and other GP researchers have tried? This was given as one reason for the success of the artificial intelligence program AM, which evolved Lisp expressions to discover *"interesting"* conjectures in mathematics, such as the Goldbach conjecture (every even number is the sum of two primes). Koza refuted this hypothesis about GP by demonstrating how difficult it is to randomly generate a successful program to perform some of the tasks for which GP evolves successful programs.

One could speculate that the space of Lisp expressions (with a given set of functions and terminals) is dense with useful intermediate-size building blocks for the tasks on which GP has been successful. GP's ability to find solutions quickly (*e.g.*, within 10 generations using a population of 300) lends credence to this speculation. GP also has not been compared systematically with other techniques that could search in the space of parse trees. For example, it would be interesting to know if a hill climbing technique could do as well as GP on the examples Koza gives. One test of this was reported by O'Reilly and Oppacher (1994 *a*, *b*), who defined a mutation operator for parse trees and used it to compare GP with a simple hill-climbing technique similar to random-mutation hill climbing and with simulated annealing.

Comparisons were made on five problems, including the block stacking problem described above. On each of the five, simulated annealing either equaled or significantly outperformed GP in terms of the number of runs on which a correct solution was found and the average number of fitness-function evaluations needed to find a correct program. On two out of the five, the simple hill climber either equaled or exceeded the performance of GP. Though five problems is not many for such a comparison in view of the number of problems on which GP has been tried, these results bring into question the claim that the crossover operator is a major contributor to GP's success. O'Reilly and Oppacher (1994a) speculate from their results that the parse-tree representation "may be a more fundamental asset to program induction than any particular search technique," and that "perhaps the concept of building blocks is irrelevant to GP." These speculations are well worth further investigation, and it is imperative to characterize the types of problems for which crossover is a useful operator and for which a GA will be likely to outperform gradient-ascent strategies such as hill climbing and simulated annealing.

Some other questions about GP :

Will the technique scale up to more complex problems for which larger programs are needed? Will the technique work if the function and terminal sets are large?

How well do the evolved programs generalize to cases not in the set of fitness cases? In most of Koza's examples, the cases used to compute fitness are samples from a much larger set of possible fitness cases. GP very often finds a program that is correct on all the given fitness cases, but not enough has been reported on how well these programs do on the "*out*-of-*sample*" cases. We need to know the extent to which GP produces programs that generalize well after seeing only a small fraction of the possible fitness cases. To what extent can programs be optimized for correctness, size, and efficiency at the same time?

Genetic programming's success on a wide range of problems should encourage future research addressing these questions.

CELLULAR AUTOMATA

A quite different example of automatic programming by genetic algorithms is found in work done by James Crutchfield, Rajarshi Das, Peter Hraber, and myself on evolving cellular automata to perform computations. This project has elements of both problem solving and scientific modeling. One motivation is to understand how natural evolution creates systems in which "emergent computation" takes place—that is, in which the actions of simple components with limited information and communication give rise to coordinated global information processing. Insect colonies, economic systems, the immune system, and the brain have all been cited as examples of systems in which such emergent computation occurs. However, it is not well understood *how* these natural systems perform computations. Another motivation is to find ways to engineer sophisticated emergent computation in decentralized multi-processor systems, using ideas from how natural decentralized systems compute. Such systems have many of the desirable properties for computer systems they are sophisticated, robust, fast, and adaptable information processors.

Using ideas from such systems to design new types of parallel computers might yield great progress in computer science. One of the simplest systems in which emergent computation can be studied is a one-dimensional binary-state cellular automation (CA)—a one-dimensional lattice of N two-state machines ("cells"), each of which changes its state as a function only of the current states in a local neighbourhood. (The well-known "game of Life" is an example of a two-dimensional CA.) A one-dimensional CA is illustrated in figure 1.5. The lattice starts out with an initial configuration of cell states (zeros and ones) and this configuration changes in discrete time steps in which all cells are updated simultaneously according to the CA "rule" Æ. (Here *I* use the term "state" to refer to a local state s_i—the value of the single cell at site *i*. The term "configuration" will refer to the pattern of local states over the entire lattice.)

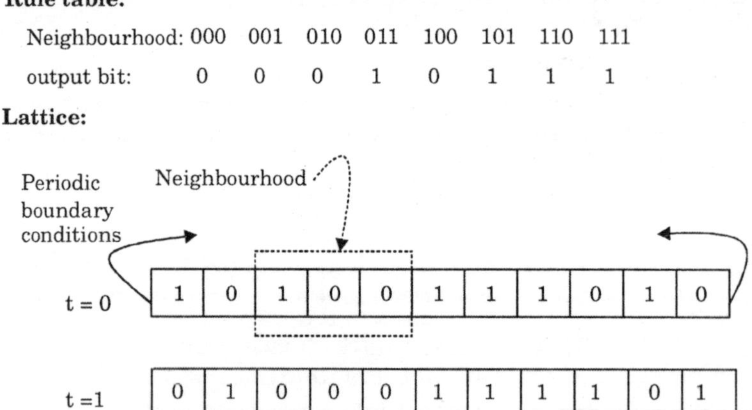

Fig. 1.5. Illustration of a one-dimensional, binary-state, nearest-neighbour (r = 1) cellular automation with N = 11. Both the lattice and the rule table for updating the lattice are illustrated. The lattice configuration is shown over one time step. The cellular automaton has periodic boundary conditions: the lattice is viewed as a circle, with the leftmost cell the right neighbour of the rightmost cell, and vice versa.

A CA rule Æ can be expressed as a lookup table (*"rule table"*) that lists, for each local neighbourhood, the update state for the neighbourhood's central cell. For a binary-state CA, the update states are referred to as the "output bits" of the rule table. In a one-dimensional CA, a neighbourhood consists of a cell and its *r* (*"radius"*), neighbours on either side. The CA illustrated in figure 1.5 has $r = 1$. It illustrates the "majority" rule: for each neighbourhood of three adjacent cells, the new state is decided by a majority vote among the three cells. The CA illustrated in figure 1.5. like all those I will discuss here, has periodic boundary conditions:

$s_i = s_{i+N}$. In figure 1.5 the lattice configuration is shown iterated over one time step. Cellular automata have been studied extensively as mathematical objects, as models of natural systems, and as architectures for fast, reliable parallel computation. However, the difficulty of understanding the emergent behaviour of CAs or of designing CAs to have desired behaviour has up to now severely limited their use in science and engineering and for general computation. Our goal is to use GAs as a method for engineering CAs to perform computations. Typically, a CA performing a computation means that the input to the computation is encoded as an initial configuration, the output is read off the configuration after some time step, and the intermediate steps that transform the input to the output are taken as the steps in the computation. The *"program"* emerges from the CA rule being obeyed by each cell. (Note that this use of CAs as computers differs from the impractical though theoretically interesting method of constructing a universal turing machine in a CA; see Mitchell, Crutchfield, and Hraber 1994 *b* for a comparison of these two approaches.)

The behaviour of one-dimensional CAs is often illustrated by a "*space*-time *diagram*"—a plot of lattice configurations over a range of time steps, with ones given as black cells and zeros given as white cells and with time increasing down the page. Figure 1.6 shows such a diagram for a binary-state $r = 3$ CA in which the rule table's output bits were filled in at random. It is shown iterating on a randomly generated initial configuration. Random-looking patterns, such as the one shown, are typical for the vast majority of CAs. To produce CAs that can perform sophisticated parallel computations, the genetic algorithm must evolve CAs in which the actions of the cells are not random-looking but are coordinated with one another so as to produce the desired result. This coordination must, of course, happen in the absence of any central processor or memory directing the coordination.

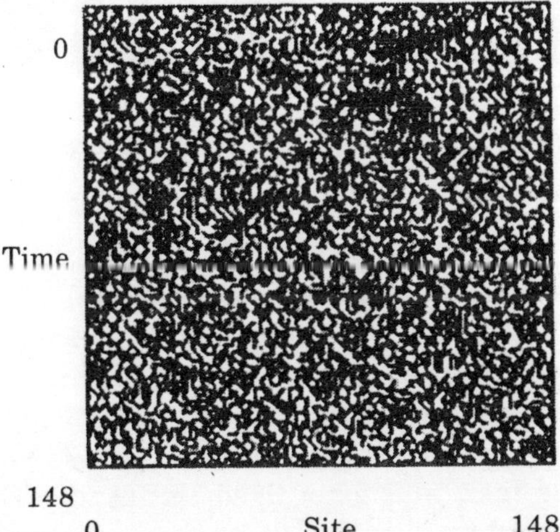

Fig. 1.6. Space-time diagram for a randomly generated $r = 3$ cellular automation, iterating on a randomly generated initial configuration. $N = 149$ sites are shown, with time increasing down the page. Here cells with state 0 are white and cells with state 1 are black.

Some early work on evolving CAs with genetic algorithms was done by Norman Packard and his colleagues. John Koza (1992) also applied the GP paradigm to evolve CAs for simple random-number generation. Our work builds on that of Packard (1988). As a preliminary project, we used a form of the GA to evolve one-dimensional, binary-state $r = 3$ CAs to perform a density-classification task. The goal is to find a CA that decides whether or not the initial configuration contains a majority of ones (*i.e.*, has high density). If it does, the whole lattice should eventually go to an unchanging configuration of all ones; all zeros otherwise. More formally, we call this task the $p_c = \frac{1}{2}$ task. Here Á denotes the density of ones in a binary-state CA configuration and *Ac* denotes a "critical" or threshold density for classification. Let $Á_0$ denote the density of ones in the initial configuration (IC). If $Á_0 > Á_c$, then within M time steps the CA should go to the fixed-point configuration of all ones (*i.e.*, all cells in state 1 for all subsequent *t*); otherwise, within M time steps it should go to the fixed-point configuration of all zeros. M is a parameter of the task that depends on the lattice size *N*. It may occur to the reader that the majority rule mentioned above might be a good candidate for solving this task.

Figure 1.7 gives space-time diagrams for the $r = 3$ majority rule (the output bit is decided by a majority vote of the bits in each seven-bit neighbourhood) on two ICs, one with $p < \frac{1}{2}$ and one with $p > \frac{1}{2}$ As can be seen, local neighbourhoods with majority ones map to regions of all ones and similarly for zeros, but when an all-ones region and an all-zeros region border each other, there is no way to decide between them, and both persist. Thus, the majority rule does not perform the $p_c = \frac{1}{2}$ task.

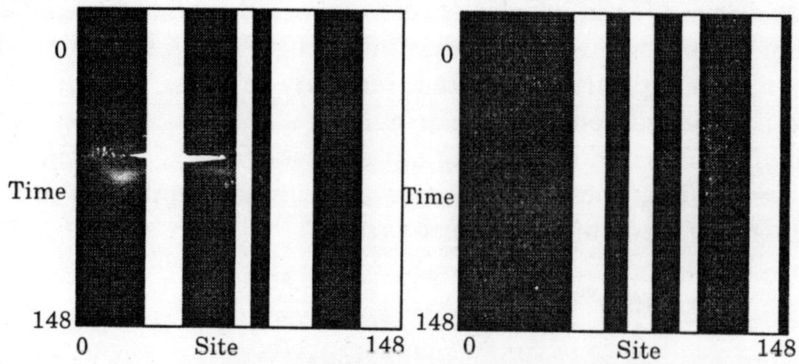

Fig. 1.7. Space-time diagrams for the r = 3 majority rule. In the left diagram, $p_0 < \frac{1}{2}$ *in the right diagram* $p_0 > \frac{1}{2}$.

Designing an algorithm to perform the $p_c = \frac{1}{2}$ task is trivial for a system with a central controller or central storage of some kind, such as a standard computer with a counter register or a neural network in which all input units are connected to a central hidden unit. However, the task is nontrivial for a small-radius ($r << N$) CA, since a small-radius CA relies only on local interactions mediated by the cell neighbourhoods. In fact, it can be proved that no finite-radius CA with periodic boundary conditions can perform this task perfectly across all lattice sizes, but even to perform this task well for a fixed lattice size requires more powerful computation than can be performed by a single cell or any linear combination of cells. Since the ones can be distributed throughout the CA lattice, the CA must transfer information over large distances (H *N*). To do this requires the global coordination of cells that are separated by large distances and that cannot communicate directly. How can this be done? Our interest was to see if the GA could devise one or more methods. The chromosomes evolved by the GA were bit strings representing CA rule tables. Each chromosome consisted of the output bits of a rule table, listed in lexicographic order of neighbourhood. The chromosomes representing rules were thus of length $2^{2r+1} = 128$ (for binary $r = 3$ rules).

The size of the rule space the GA searched was thus 2^{128}—far too large for any kind of exhaustive search. In our main set of experiments, we set $N = 149$ (chosen to be reasonably large but not computationally intractable). The GA began with a population of 100 randomly generated chromosomes. The fitness of a rule in the population was calculated by (*i*) randomly choosing 100 ICs (initial configurations) that are uniformly distributed over $\acute{A}$ $\hat{I}$ [0.0, 1.0], with exactly half with $\acute{A}\acute{A}_c$ and (*ii*) running the rule on each IC either until it arrives at a fixed point or for a maximum of approximately 2B time steps, and (iii) determining whether the final pattern is correct—*i.e.*, N zeros for $\acute{A}_0\acute{A}_c$ and N ones for $\acute{A}_0\acute{A}_c$. The initial density, $\acute{A}_0$, was never exactly $\frac{1}{2}$ since N was chosen to be odd. The rule's fitness, f_{100}, was the fraction of the 100 ICs on which the rule produced the correct final pattern. No partial credit was given for partially correct final configurations. A few comments about the fitness function are in order. First, as was the case in Hillis's sorting-networks project, the number of possible input cases (2^{149} for $N = 149$) was far too large to test exhaustively.

Instead, the GA sampled a different set of 100 ICs at each generation. In addition, the ICs were not sampled from an unbiased distribution (*i.e.*, equal probability of a one or a zero at each site in the IC), but rather from a flat distribution across $\acute{A}$ $\hat{I}$ [0, 1] (*i.e.*, ICs of each density from $\acute{A} = 0$ to $\acute{A} = 1$ were approximately equally represented). This flat distribution was used because the unbiased distribution is binomially distributed and thus very strongly peaked at $p = \frac{1}{2}$. The ICs selected from such a distribution will likely all have $p \approx \frac{1}{2}$ the hardest cases to classify. Using an unbiased sample made it too difficult for the GA to ever find any high-fitness CAs. Our version of the GA worked as follows. In each generation, (*i*) a new set of 100 ICs was generated, (*ii*) f_{100} was calculated for each rule in the population, (*iii*) the population was ranked in order of fitness, (*iv*) the 20 highest-fitness (*"elite"*) rules were copied to the next generation without modification, and (*v*) the remaining 80 rules for the next generation were formed by single-point crossovers between randomly chosen pairs of elite rules. The parent rules were chosen from the elite with replacement—that is, an elite rule was permitted to be chosen any number of times. The offspring from each crossover were each mutated twice. This process was repeated for 100 generations for a single run of the GA. Note that this version of the GA differs from the simple GA in several ways. First, rather than selecting parents with probability proportional to fitness, the rules are ranked and selection is done at random from the top 20% of the population. Moreover, all of the top 20% are copied without modification to the next generation, and only the bottom 80% are replaced. This is similar to the selection method — called "(¼ + ≫)"—used in some evolution strategiest. This version of the GA was the one used by Packard (1988), so we used it in our experiments attempting to replicate his work and in our subsequent experiments.

Selecting parents by rank rather than by absolute fitness prevents initially stronger individuals from quickly dominating the population and driving the genetic diversity down too early. Also, since testing a rule on 100 ICs provides only an approximate gauge of the true fitness, saving the top 20% of the rules was a good way of making a "first cut" and allowing rules that survive to be tested over more ICs. Since a new set of ICs was produced every generation, rules that were copied without modification were always retested on this new set. If a rule performed well and thus survived over a large number of generations, then it was likely to be a genuinely better rule than those that were not selected, since it was tested with a large set of ICs. An alternative method would be to test every rule in each generation on a much larger set of ICs, but this would waste computation time.

Too much effort, for example, would go into testing very weak rules, which can safely be weeded out early using our method. As in most applications, evaluating the fitness function (here, iterating each CA) takes up most of the computation time. Three hundred different runs were performed, each starting with a different random-number seed. On most runs the GA evolved a nonobvious but rather unsophisticated class of strategies. One example, a rule here called $Æ_{a'}$, is illustrated in figure 1.8*a*. This rule had f_{100} H0.9 in the generation in which it was discovered (*i.e.*, $Æ_a$ correctly classified 90% of the ICs in that generation). Its "strategy" is the following: Go to the fixed point of all zeros unless there is a sufficiently large block of adjacent (or almost adjacent) ones in the 1C. If so, expand that block. This strategy does a fairly good job of classifying low and high density under f_{100}: it relies on the appearance or absence of blocks of ones to be good predictors of $Á_0$, since high-density ICs are statistically more likely to have blocks of adjacent ones than low density ICs. Similar strategies were evolved

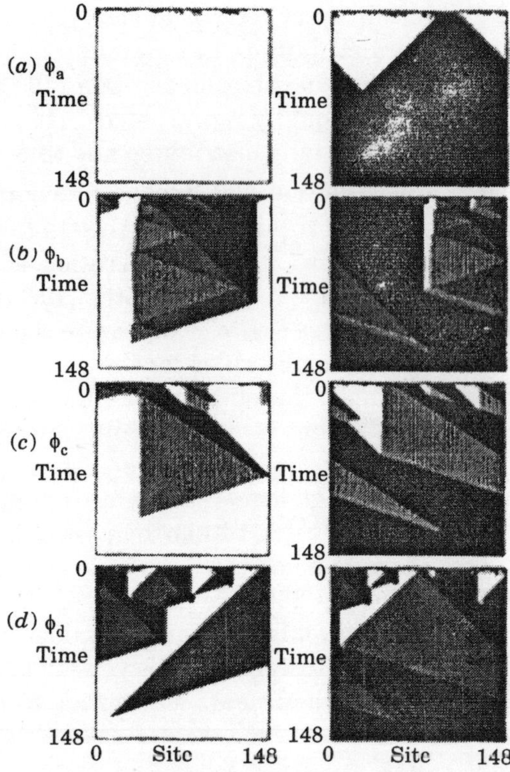

Fig. 1.8. Space-time diagrams from four different rules discovered by the GA (adopted from Das, Mitchell, and Crutch field 1994 by permission of the authors.. The left diagrams have $\infty > \frac{1}{2}$; the right diagrams have $\infty < \frac{1}{2}$. All are correctly classified. Fitness increases from (a) to (d). The "gray" area in (d) is actually a checkerboard pattern of alternating zeros and ones.

in most runs. On approximately half the runs, *"expand ones"* strategies were evolved, and on approximately half the runs, the opposite *"expand zeros"* strategies were evolved.

These block-expanding strategies were initially surprising to us and even seemed clever, but they do not count as sophisticated examples of computation in CAs: all the computation is done locally in identifying and then expanding a "sufficiently large" block. There is no notion of global coordination or interesting information flow between distant cells —two things we claimed were necessary to perform well on the task. In Mitchell, Crutchfield, and Hraber 1994a we analyzed the detailed mechanisms by which the GA evolved such block-expanding strategies.

This analysis uncovered some quite interesting aspects of the GA, including a number of impediments that, on most runs, kept the GA from discovering better-performing rules. These included the GA's breaking the $P_c = \frac{1}{2}$ task's symmetries for short-term gains in fitness, as well as an "overfitting" to the fixed lattice size and the unchallenging nature of the samples of ICs. The biased, flat distribution of ICs over *ÁI* [0, 1] helped the GA get a leg up in the early generations.

We found that calculating fitness on an unbiased distribution of ICs made the problem too difficult for the GA early on — it was unable to find improvements to the rules in the initial population. However, the biased distribution became too easy for the improved CAs later in a run, and these ICs did not push the GA hard enough to find better solutions. Recall that the same problem plagued Hillis's GA until he introduced host-parasite co-evolution. We are currently exploring a similar co-evolution scheme to improve the GA's performance on this problem. The weakness of $Æ_a$ and similar rules is clearly seen when they are tested using an unbiased distribution of ICs. We defined a rule *Æ's* "unbiased performance" $P_N(\phi)$ as the fraction of correct classifications produced by *Æ* within approximately $2N$ time steps on 10,000 ICs on a lattice of length *N,* chosen from an unbiased distribution over *Á*. As mentioned above, since the distribution is unbiased, the ICs are very likely to have *Á* H 0.5. These are the very hardest cases to classify, so $P_N(\phi)$ gives a lower bound on *Æ's* overall performance.

Table 1.1 gives $P_N(\phi)$ values for several different rules each for three values of *N*. The majority rule, unsurprisingly, has $P_N = 0$ for all three values of *N*. The performance of $Æ_a$ (the block-expanding rule of figure 1.8*a*) decreases significantly as *N* is increased. This was true for all the block-expanding rules: the performance of these rules decreased dramatically.

Table 1.1. Measured values of P_N at various values of *N* for six different $r = 3$ rules: the majority rule, four rules discovered by the GA in different runs ($Æ_a$ $Æ_d$), and the GKL rule . The subscripts for the rules discovered by the GA indicate the pair of space-time diagrams illustrating their behaviour in figure 1.8. The standard deviation of *p* 149, when calculated 100 times for the same rule, is approximately 0.004. The standard deviations for P_N; for larger *N* are higher.

CA	*Symbol*	*N = 149*	*N = 599*	*N = 999*
Majority	Æmaj	0.000	0.000	0.000
Expand 1-blocks	$Æ_a$	0.652	0.515	0.503
Particle-based	$Æ_b$	0.697	0.580	0.522
Particle-based	$Æ_c$	0.742	0.718	0.701
Particle-based	$Æ_d$	0.769	0.725	0.714
GKL	ÆGKL	0.816	0.766	0.757

for larger *N,* since the size of block to expand was tuned by the GA for $N = 149$.

Despite these various impediments and the unsophisticated rules evolved on most runs, on several different runs in our initial experiment the GA discovered rules with significantly higher performance and rather sophisticated strategies. The typical space-time behaviour of three such rules (each from a different run) are illustrated in figure 1.8*b*-1.8*d* SomeP_N values for these three "particle-based" rules are given in table 1.1. As can be seen, P_N is significantly higher for these rules than for the typical block-expanding rule $Æ_a$. In addition, the performances of the most highly fit rules remain relatively constant as N is increased, meaning that these rules can generalize better than can $Æ_a$. Why does *Æd,* for example, perform relatively well on the A = $\frac{1}{2}$ task? In figure 1.8*d* it can be seen that, although the patterns eventually converge to fixed points, there is a transient phase during which spatial and temporal transfer of information about the density in local regions takes place. This local information interacts with other local information to produce the desired final state. Roughly, $Æ_d$ successively classifies *"local"* densities with a locality range that increases with time. In regions where there is some ambiguity, a *"signal"* is propagated. This is seen either as a checkerboard pattern propagated in both spatial directions or as a vertical black-to-white boundary. These signals indicate that the classification is to be made at a larger scale. Note that regions centered about each signal locally have A = $\frac{1}{2}$. The consequence is that the signal patterns can propagate, since the density of patterns with A = $\frac{1}{2}$ is neither increased nor decreased under the rule.

The creation and interactions of these signals can be interpreted as the locus of the computation being performed by the CA—they form its emergent program. The above explanation of how $Æ_d$ performs the A = $\frac{1}{2}$ task is an informal one obtained by careful scrutiny of many space-time patterns. Can we understand more rigorously how the rules evolved by the GA perform the desired computation? Understanding the results of GA evolution is a general problem—typically the GA is asked to find individuals that achieve high fitness but is not told how that high fitness is to be attained. One could say that this is analogous to the difficulty biologists have in understanding the products of natural evolution (*e.g.*, us).

We computational evolutionists have similar problems, since we do not specify what ' solution evolution is supposed to create; we ask only that it find some solution. In many cases, particularly in automatic-programming applications, it is difficult to understand exactly how an evolved high-fitness individual works. In genetic programming, for example, the evolved programs are often very long and complicated, with many irrelevant components attached to the core program performing the desired computation. It is usually a lot of work—and sometimes almost impossible—to figure out by hand what that core program is. The problem is even more difficult in the case of cellular automata, since the emergent *"program"* performed by a given CA is almost always impossible to extract from the bits of the rule table. A more promising approach is to examine the space-time patterns created by the CA and to *"reconstruct"* from those patterns what the algorithm is. Crutchfield and Hanson have developed a general method for reconstructing and understanding the "intrinsic" computation embedded in space-time patterns in terms of *"regular domains,"* "particles" and "particle interactions". This method is part of their "computational mechanics" framework for understanding computation in physical systems. A detailed discussion of computational mechanics and particle-based computation is beyond the scope of this chapter. Very briefly, for those familiar with formal language theory, regular domains are regions of spacetime consisting of words in the same regular language—that is, they are regions that are computationally simple. Particles are localized boundaries between regular domains. In computational mechanics, particles are identified as information carriers, and collisions between particles are identified as the loci of important information processing. Particles and particle interactions form a high-level language for describing computation in spatially extended systems such as CAs. Figure 1.9 hints at this higher level of description: to produce it we filtered the regular domains from the space-time behaviour of a GA-evolved CA to leave only the particles and their interactions, in terms of which the emergent algorithm of the CA can be understood.

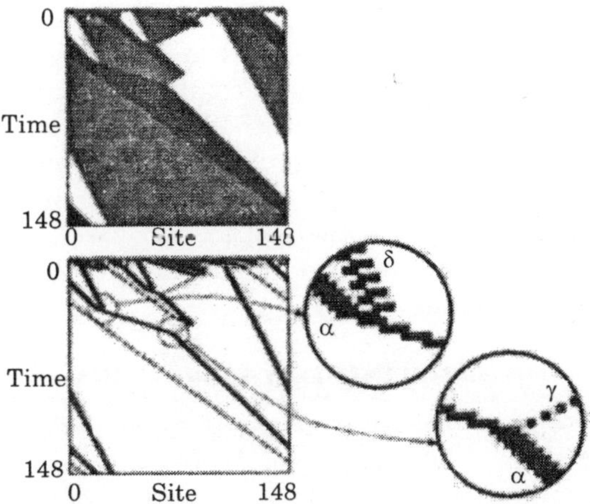

Fig. 1.9. A space-time diagram of a GA-evolved rule for the $p_c = \frac{1}{2}$ *task, and the same diagram with the regular domains filtered out, leaving only the particles and particle interactions.*

Interestingly, it turns out that the behaviour of the best rules discovered by the GA (such as $Æ_d$) is very similar to the behaviour of the well-known Gacs-Kurdyumov-Levin (GKL) rule Figure 1.10 is a space-time diagram illustrating its typical behaviour. The GKL rule ($Æ_{GKL}$) was designed by hand to study reliable computation and phase transitions in one-dimensional spatially extended systems, but before we started our project it was also the rule with the best-known performance (for CAs with periodic boundary conditions) on the $p_c = \frac{1}{2}$ task. Its

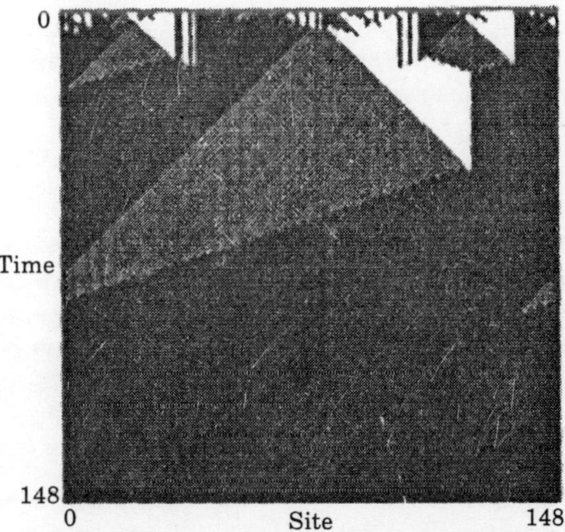

Fig. 1.10. Space-time diagram for the GKL rule with $\infty > \frac{1}{2}$.

unbiased performance is given in the last row of table. 1.1. The difference in performance betw een $Æ_d$ and $Æ_{GKL}$ is due to asymmetries in $Æ_d$ that are not present in $Æ_{GKL}$. Further GA evolution of $Æ_d$ (using an increased number of ICs) has produced an improved version that approximately equals the performance of the $Æ_{GKL}$.

Rajarshi Das has gone further and, using the aforementioned particle analysis, has designed by hand a rule that slightly outperforms $Æ_{GKL}$. The discovery of rules such as *Æb*, *Æd* is significant, since it is the first example of a GA's producing sophisticated emergent computation in decentralized, distributed systems such as CAs. It is encouraging for the prospect of using GAs to automatically evolve computation in more complex systems. Moreover, evolving CAs with GAs also gives us a tractable framework in which to study the mechanisms by which an evolutionary process might create complex coordinated behaviour in natural decentralized distributed systems. For example, by studying the GA's behaviour, we have already learned how evolution's breaking of symmetries can lead to suboptimal computational strategies; eventually we may be able to use such computer models to test ways in which such symmetry breaking might occur in natural evolution.

PREDICTING DATA ANALYSIS

A major impediment to scientific progress in many fields is the inability to make sense of the huge amounts of data that have been collected via experiment or computer simulation. In the fields of statistics and machine learning there have been major efforts to develop automatic methods for finding significant and interesting patterns in complex data, and for forecasting the future from such data; in general, however, the success of such efforts has been limited, and the automatic analysis of complex data remains an open problem. Data analysis and prediction can often be formulated as search problems—for example, a search for a model explaining the data, a search for prediction rules, or a search for a particular structure or scenario well predicted by the data. In this section I describe two projects in which a genetic algorithm is used to solve such search problems—one of predicting dynamical systems, and the other of predicting the structure of proteins.

Predicting Dynamical Systems

Norman Packard (1990) has developed a form of the GA to address this problem and has applied his method to several data analysis and prediction problems. The general problem can be stated as follows: A series of observations from some process (*e.g.*, a physical system or a formal dynamical system) take the form of a set of pairs,

$$[(\vec{x}^1 \cdot y^1),...,(\vec{x}^N \cdot y^N)].$$

where $\vec{x}^i = (x_1^i,...,x_n^i)$ are independent variables and y^i is a dependent variable ($1 d i d N$). For example, in a weather prediction task, the independent variables might be some set of features of today's weather (*e.g.*, average humidity, average barometric pressure, low and high temperature, whether or not it rained), and the dependent variable might be a feature of tomorrow's weather (*e.g.*, rain). In a stock market prediction task, the independent variables might be $\vec{x} = (x(t_1), x(t_2), ..., x(t_n))$, representing the values of the value of a particular stock (the "state variable") at successive time steps, and the dependent variable might be $y = x(t_n + k)$, representing the value of the stock at some time in the future. (In these examples there is only one dependent variable y for each vector of independent variables $\vec{x}$; a more general form of the problem would allow a vector of dependent variables for each vector of independent variables.)

Packard used a GA to search through the space of sets of conditions on the independent variables for those sets of conditions that give good predictions for the dependent variable. For example, in the stock market prediction task, an individual in the GA population might be a set of conditions such as

$$c = [(\$20 \le \text{Price of Xerox stock on day 1}) \wedge (\$25 \le \text{Price of Xerox stock on day 2} \le \$27) \wedge (\$22 \le \text{Price of Xerox stock on day 3} \le \$25)]$$

where "^" is the logical operator "AND" This individual represents all the sets of three days in which the given conditions were met (possibly the empty set if the conditions are never met). Such a condition set C thus specifies a particular subset of the data points (here, the set of all 3-day periods). Packard's goal was to use a GA to search for condition sets that are good predictors of *something*—in other words, to search for condition sets that specify subsets of data points whose dependent-variable values are close to being uniform. In the stock market example, if the GA found a condition set such that all the days satisfying that set were followed by days on which the price of Xerox stock rose to approximately \$30, then we might be confident to predict that, if those conditions were satisfied today, Xerox stock will go up.

The fitness of each individual C is calculated by running all the data points ($\vec{x}y$) in the training set through C and, for each $\vec{x}$ that satisfies C, collecting the corresponding y. After this has been done, a measurement is made of the uniformity of the resulting values of y. If the y values are all close to a particular value Å, then C is a candidate for a good predictor for y—that is, one can hope that a new $\vec{x}$ that satisfies C will also correspond to a y value close to Å. On the other hand, if the y values are very different from one another, then $\vec{x}$ satisfying C does not seem to predict anything about the corresponding y value. As an illustration of this approach, I will describe the work done by Thomas Meyer and Norman Packard (1992) on finding "regions of predictability" in time series generated by the Mackey-Glass equation, a chaotic dynamical system created as a model for blood flow.

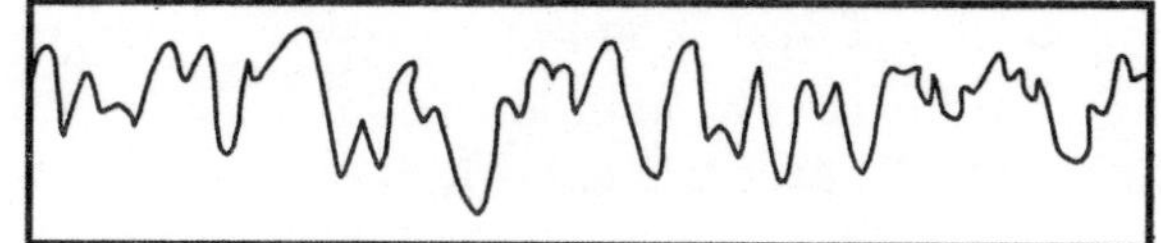

Fig. 1.11. Plot of a time series from Mackey-Glass equation with Ä = 150. Time is plotted on the horizontal axis; x(t)s> is plotted on the vertical axis.

$$\frac{dx}{dt} = \frac{ax(t-\tau)}{1+[x(t-\tau)]^c} - bx(t).$$

Here $x(t)$ is the state variable, t is time in seconds, and a, b, c, and Ä are constants. A time series from this system (with Ä set to 150) is plotted in figure 1.11.

To form the data set, Meyer and Packard did the following: For each data point i, the independent variables $\vec{x}^i$ are 50 consecutive values $x(t)$ (one per second):

$$\vec{x}^i = (x_1^i, x_2^i, \ldots, x_{50}^i).$$

The dependent variable for data point i, y^i, is the state variable t' time steps in the future: $y^i = x^I{}_{50} + t'$. Each data point $(\vec{x}^i, y^i)$ is formed by iterating the Mackey-Glass equation with a different initial condition, where an initial condition consists of values for $\{x_1$ Ä,..., $x_0\}$.

Meyer and Packard used the following as a fitness function:

$$f(C) = -\log_2\left(\frac{\sigma}{\sigma_0}\right) - \frac{\alpha}{N_C},$$

where Ã is the standard deviation of the set of *y's* for data points satisfying *C*, $\tilde{A}_0$ is the standard deviation of the distribution of *y's* over the entire data set, N_C is the number of data points satisfying condition C, and ± is a constant. The first term of the fitness function measures the amount of information in the distribution of *y's* for points satisfying *C*, and the second term is a penalty term for poor statistics—if the number of points satisfying *C* is small, then the first term is less reliable, so *C* should have lower fitness. The constant ± can be adjusted for each particular application.

Meyer and Packard used the following version of the GA:

1. Initialize the population with a random set of *Cs*.
2. Calculate the fitness of each *C*.
3. Rank the population by fitness.
4. Discard some fraction of the lower-fitness individuals and replace them by new *Cs* obtained by applying crossover and mutation to the remaining *Cs*.
5. Go to step 2.

(Their selection method was, like that used in the cellular-automata project described above, similar to the "(¼ + »)" method of evolution strategies.) Meyer and Packard used a form of crossover known in the GA literature as *"uniform crossover"*. This operator takes two Cs and exchanges approximately half the "genes" (conditions). That is, at each gene position in parent A and parent B, a random decision is made whether that gene should go into offspring A or offspring B. An example follows:

Parent A: $[(3.2 \le x_6 \le 5.5) \wedge (0.2 \le x_8 \le 4.8) \wedge (3.4 \le x_9 \le 9.9)]$ }

Parent B: $[(6.5 \le x_2 \le 6.8) \wedge (1.4 \le x_4 \le 4.8) \wedge (1.2 \le x_9 \le 1.7)$
$\wedge (4.8 \le x_{16} \le 5.1)]$

Offspring A: $[(3.2 \le x_6 \le 5.5) \wedge (1.4 \le x_4 \le 4.8) \wedge (3.4 \le x_9 \le 9.9)]$

Offspring B: $[(6.5 \le x_2 \le 6.8) \wedge (0.2 \le x_8 \le 4.8) \wedge (1.2 \le x_9 \le 1.7)]$
$\wedge (4.8 \le x_{16} \le 5.1)]$

Here offspring *A* has two genes from parent *A* and one gene from parent *B*. Offspring *B* has one gene from parent *A* and three genes from parent *B*.

In addition to crossover, four different mutation operators were used:

Add a new condition:

$[(3.2 \le x_6 \le 5.5) \wedge (0.2 \le x_8 \le 4.8)]$
$\rightarrow [(3.2 \le x_6 \le 5.5) \wedge (0.2 \le x_8 \le 4.8) \wedge (3.4 \le x_9 \le 9.9)]$

$[(3.2 \le x_6 \le 5.5) \wedge (0.2 \le x_8 \le 4.8) \wedge (3.4 \le x_9 \le 9.9)]$
$\rightarrow [(3.2 \le x_6 \le 5.5) \wedge (3.4 \le x_9 \le 9.9)]$

Broaden or shrink a range:

$[(3.2 \le x_6 \le 5.5) \wedge (0.2 \le x_8 \le 4.8)]$
$\rightarrow [(3.9 \le x_6 \le 4.8) \wedge (0.2 \le x_8 \le 4.8)]$

Shift a range up or drawn:

$[(3.2 \le x_6 \le 5.5) \wedge (0.2 \le x_8 \le 4.8)]$
$\rightarrow [(3.2 \le x_6 \le 5.5) \wedge (1.2 \le x_8 \le 5.8)]$

The results of running the GA using these data from the Ä = 150 time series with $t' = 150$ are illustrated in Figure 1.12 and Figure 1.13. Figure 1.12 gives the four highest-fitness condition sets found by the GA, and figure 1.13 shows the four results of those condition sets. Each of the four plots in figure 1.13 shows the trajectories corresponding to data points $(\vec{x}^i, y^i)$ that satisfied the condition set. The leftmost white region is the initial 50 time steps during which the data were taken. The vertical lines in that region represent the various conditions on $\vec{x}$.

$\vec{x}$ given in the condition set. For example, in plot *a* the leftmost vertical line represents a condition on x_{20} (this set of trajectories is plotted starting at time step 20), and the rightmost vertical line in that region represents a condition on x_{49}. The shaded region represents the period of time between time steps 50 and 200, and the rightmost vertical line marks time step 200 (the point at which the y^i observation was made). Noticc that in cach of thcsc plots thc values of y^i fall into a very narrow range, which means that the GA was successful in finding subsets of the data for which it is possible to make highly accurate predictions.

These results are very striking, but some questions immediately arise. First and most important, do the discovered conditions yield correct predictions for data points outside the training set (*i.e.*, the set of data points used to calculate fitness), or do they merely describe chance statistical fluctuations in the data that were learned by the GA? Meyer and Packard performed a number of *"out of sample"* tests with data points outside the training set that satisfied the evolved condition sets and found that the results were robust—the y^i values for these data points also tended to be in the narrow range.

Exactly how is the GA solving the problem? What are the schemas that are being processed? What is the role of crossover in finding a good solution? Uniform crossover of the type used here has very different properties than single-point crossover, and its use makes it harder to figure out what schemas are being recombined. Meyer found that turning crossover off and relying solely on the four mutation operators did not make a big difference in the GA's performance; as in the case of genetic programming, this raises the question of whether the GA is the best method for this task. An interesting extension of this work would be to perform control experiments comparing the performance of the GA with that of other search methods such as hill climbing.

To what extent are the results restricted by the fact that only certain conditions are allowed (*i.e.*, conditions that are conjunctions of ranges on independent variables)? Packard (1990) proposed a more general form for conditions that also allows disjunctions (('s); an example might be

$$[(3.2 \le x_6 \le 5.5) \vee (1.1 \le x_6 \le 2.5)] \wedge [0.2 \le x_8 \le 4.8)]$$

where we are given two nonoverlapping choices for the conditions on x_6. A further generalization proposed by Packard would be to allow disjunctions between sets of conditions.

To what extent will this method succeed on other types of prediction tasks? Packard (1990) proposes applying this method to tasks such as weather prediction, financial market prediction, speech recognition, and visual pattern recognition. Interestingly, in 1991 Packard left the Physics Department at the University of Illinois to help form a company to predict financial markets. As i write this the company has not yet gone public with their results, but stay tuned.

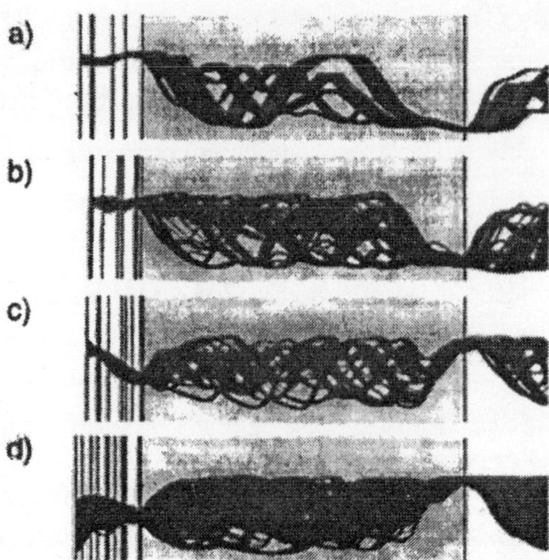

Fig. 1.12. Results of the four highest-fitness condition sets found by the GA. Each plot shows trajectories of data points that satisfied that condition set. The leftmost white region is the initial 50 time steps during which data were taken. The vertical lines in that region represent the various conditions $\vec{x}$ on given in the condition set. The vertical line on the right-hand side represents the time at which the prediction is to be made. Note how the trajectories narrow at that region, indicating that the GA has found conditions for good predictability.

PREDICTING MOLECULAR PROTEIN

One of the most promising and rapidly growing areas of GA application is data analysis and prediction in molecular biology. GAs have been used for, among other things, interpreting nuclear magnetic resonance data to determine the structure of DNA, finding the correct ordering for an unordered group of DNA fragments and predicting protein structure. Proteins are the fundamental functional building blocks of all biological cells. The main purpose of DNA in a cell is to encode instructions for building up proteins out of amino acids; the proteins in turn carry out most of the structural and metabolic functions of the cell. A protein is made up of a sequence of amino acids connected by peptide bonds. The length of the sequence varies from protein to protein but is typically on the order of 100 amino acids.

Owing to electrostatic and other physical forces, the sequence "folds up" to a particular three-dimensional structure. It is this three-dimensional structure that primarily determines the protein's function. The three-dimensional structure of a Crambin protein (a plant-seed protein consisting of 46 amino acids) is illustrated in figure 1.13. The three-dimensional structure of a protein is determined by the particular sequence of its amino acids, but it is not currently known precisely how a given sequence leads to a given structure. In fact, being able to predict a protein's structure from its amino acid

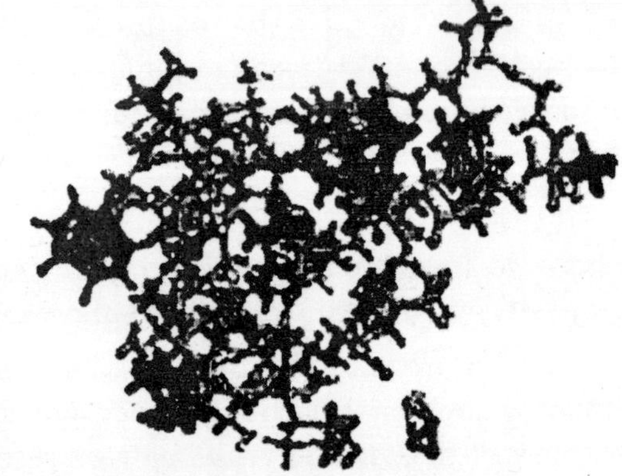

Fig. 1.13. A representation of the three-dimensional structure of a Crambin protein.

sequence is one of the most important unsolved problems of molecular biology and biophysics. Not only would a successful prediction algorithm be a tremendous advance in the understanding of the biochemical mechanisms of proteins, but, since such an algorithm could conceivably be used to *design* proteins to carry out specific functions, it would have profound, far-reaching effects on biotechnology and the treatment of disease. Recently there has been considerable effort toward developing methods such as GAs and neural networks for automatically predicting protein structures. The relatively simple GA prediction project of Steffen Schulze-Kremer (1992) illustrates one way in which GAs can be used on this task; it also illustrates some potential pitfalls.

Schulze-Kremer took the amino acid sequence of the Crambin protein and used a GA to search in the space of possible structures for one that would fit well with Crambin's amino acid sequence. The most straight-forward way to describe the structure of a protein is to list the three-dimensional coordinates of each amino acid, or even each atom. In principle, a GA could use such a representation, evolving vectors of coordinates to find one that resulted in a plausible structure. But, because of a number of difficulties with that representation (*e.g.*, the usual crossover and mutation operators would be too likely to create physically impossible structures), Schulze-Kremer instead described protein structures using "torsion angles"—roughly, the angles made by the peptide bonds connecting amino acids and the angles made by bonds in an amino acid's "side chain." (See Dickerson and Geis 1969 for an overview of how three-dimensional protein structure is measured.) Schulze-Kremer used 10 torsion angles to describe each of the N (46 in the case of Crambin) amino acids in the sequence for a given protein. This collection of N sets of 10 torsion angles completely defines the three-dimensional structure of the protein. A chromosome, representing a candidate structure with N amino acids, thus contains N sets of ten real numbers. This representation is illustrated in figure 1.14.

torsion angles

amino acid 1	**amino acid 2**	...	**amino acid 46**
$\varphi: 66.3°$	$\varphi: -72.2°$		.
$\nu: 45.2°$	$\psi: 23.1°$		.
$\propto: 180.0°$	$\omega: 180.0°$		.
$\chi^1: -22.7°$	$\chi^1: 111.4°$		
$\chi^1: 127.1°$	$\chi^2: 120.2°$		
$\chi^3: -100.0°$	$\chi^3: -22.1°$		
$\chi^4: 33.2°$	$\chi^4: 32.2°$		
$\chi^5: -125.9°$	$\chi^5: -87.3°$		
$\chi^6: 55.4°$	$\chi^6: -95.2°$		
$\chi^7: 76.6°$	$\chi^7: -54.1°$		

chromosome:

[66.3 45.2 180.0 22.7 127.1 –100.0 32.2 –125.9 55.4 76.6] [–27.2 23.1 ...1...1...]

Fig. 1.14 : An illustration of the representation for protein structure used in Schulze-Kremer's experiments.

The next step is to define a fitness function over the space of chromosomes. The goal is to find a structure that has low potential energy for the given sequence of amino acids. This goal is based on the assumption that a sequence of amino acids will fold to a minimal-energy state, where energy is a function of physical and chemical properties of the individual amino acids and their spatial interactions (*e.g.*, electrostatic pair interactions between atoms in two spatially adjacent amino acids). If a complete description of the relevant forces were known and solvable, then in principle the minimum-energy structure could be calculated. However, in practice this problem is intractable, and biologists instead develop approximate models to describe the potential energy of a structure. These models are essentially intelligent guesses as to what the most relevant forces will be. Schulze-Kremer's initial experiments used a highly simplified model in which the potential energy of a structure was assumed to be a function of only the torsion angles, electrostatic pair interactions between atoms, and van der Waals pair interactions between atoms. The goal was for the GA to find a structure (defined in terms of torsion angles) that minimized this simplified potential-energy function for the amino acid sequence of Crambin.

Each of the N amino acids in the sequence is represented by 10 torsion angles: *Æ È É* and x^1 x^7. A chromosome is a list of these N sets of 10 angles. Crossover points are chosen only at amino acid boundaries.

In Schulze-Kremer's GA, crossover was either two-point (*i.e.*, performed at two points along the chromosome rather than at one point) or uniform (*i.e.*, rather than taking contiguous segments from each parent to form the offspring, each *"gene"* is chosen from one or the other parent, with a 50% probability for each parent). Here a *"gene"* consisted of a group of 10 torsion angles; crossover points were chosen only at amino acid boundaries. Two mutation operators designed to work on real numbers rather than on bits were used: the first replaced a randomly chosen torsion angle with a new value randomly chosen from the 10 most frequently occurring angle values for that particular bond, and the second incremented or decremented a randomly chosen torsion angle by a small amount. The GA started on a randomly generated initial population often structures and ran for 1000 generations. At each generation the fitness was calculated (here, high fitness means low potential energy), the population was sorted by fitness, and a number of the highest-fitness individuals were selected to be parents for the next generation (this is, again, a form of rank selection). Offspring were created via crossover and mutation. A scheme was used in which the probabilities of the different mutation and crossover operators increased or decreased over the course of the run.

In designing this scheme, Schulze-Kremer relied on his intuitions about which operators were likely to be most useful at which stages of the run. The GA's search produced a number of structures with quite low potential energy—in fact, much lower than that of the actual structure for Crambin! Unfortunately, however, none of the generated individuals was structurally similar to Crambin. The snag was that it was too easy for the GA to find low-energy structures under the simplified potential energy function; that is, the fitness function was not sufficiently constrained to force the GA to find the actual target structure. The fact that Schulze-Kremer's initial experiments were not very successful demonstrates how important it is to get the fitness function right—here, by getting the potential-energy model right, or at least getting a good enough approximation to lead the GA in the right direction.

NEURAL NETWORKS

Neural networks are biologically motivated approaches to machine learning, inspired by ideas from neuroscience. Recently some efforts have been made to use genetic algorithms to evolve aspects of neural networks. In its simplest "feedforward" form (figure 2.16), a neural network is a collection of connected activatable units ("neurons") in which the connections are weighted, usually with real-valued weights. The network is presented with an activation pattern on its input units, such a set of numbers representing features of an image to be classified (*e.g.*, the pixels in an image of a handwritten letter of the alphabet). Activation spreads in a forward direction from the input units through one or more layers of middle (*"hidden"*) units to the output units over the weighted connections. Typically, the activation coming into a unit from other units is multiplied by the weights on the links over which it spreads, and then is added together with other incoming activation. The result is typically thresholded (*i.e.*, the unit *"turns on"* if the resulting activation is above that unit's threshold). This process is meant to roughly mimic the way activation spreads through networks of neurons in the brain.

In a feedforward network, activation spreads only in a forward direction, from the input layer through the hidden layers to the output layer. Many people have also experimented with *"recurrent"* networks, in which there are feedback connections as well as feedforward connections between layers. After activation has spread through a feedforward network, the resulting activation pattern on the output units encodes the network's *"answer"* to the input (*e.g.*, a classification of the input pattern as the letter A). In most applications, the network learns a correct mapping between input and output patterns via a learning algorithm. Typically the weights are initially set to small random values. Then a set of training inputs is presented sequentially to the network. In back-propagation learning procedure after each input has propagated through the network and an output has been produced, a *"teacher"* compares the activation value at each output unit with the correct values, and the weights in the network are adjusted in order to reduce the difference between the network's output and the correct output. Each iteration of this procedure is called a *"training cycle,"* and a complete pass of training cycles through the set of training inputs is called a *"tarining epoch."* (Typically many training epochs are needed for a network to learn to successfully classify a given set of training inputs.) This type of procedure is known as *"supervised learning,"* since a teacher supervises the learning by providing correct output values to guide the learning process. In *"unsupervised training"* there is no teacher, and the learning system must learn on its own using less detailed environmental feedback on its performance. There are many ways to apply GAs to neural networks. Some aspects that can be evolved are the weights in a fixed network, the network architecture (*i.e.*, the number of units and their interconnections can change), and the learning rule used by the network. Here I will describe four different projects, each of which was a genetic algorithm to evalve one of these aspects.

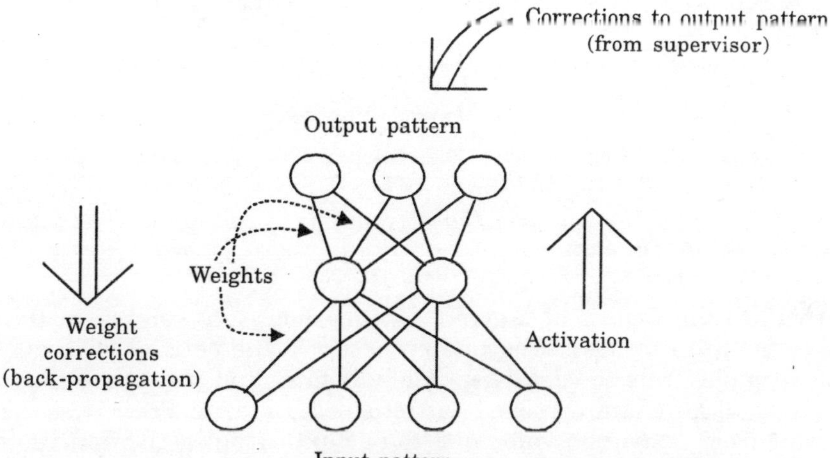

Fig. 1.15W. A schematic diagram of a simple feedforward neural network and the backpropagation process by which weight values are adjusted.

Fixed Network

David Montana and Lawrence Davis (1989) took the first approach—evolving the weights in a fixed network. That is, Montana and Davis were using the GA *instead* of back-propagation as a way of finding a good set of weights for a fixed set of connections. Several problems associated with the back-propagation algorithm (*e.g.*, the tendency to get stuck at local optima in weight space, or the unavailability of a "teacher" to supervise learning in some tasks) often make it desirable to find alternative weight training schemes. Montana and Davis were interested in using neural networks to classify underwater sonic *"lofargrams"* (similar to spectrograms) into two classes: *"interesting"* and *"not interesting."*

The overall goal was to "detect and reason about interesting signals in the midst of the wide variety of acoustic noise and interference which exist in the ocean." The networks were to be trained from a database containing lofargrams and classifications made by experts as to whether or not a given lofargram is *"interesting."* Each network had four input units, representing four parameters used by an expert system that performed the same classification. Each network had one output unit and two layers of hidden units (the first with seven units and the second with ten units). The networks were fully connected feedforward networks — that is, each unit was connected to every unit in the next higher layer. In total there were 108 weighted connections between units. In addition, there were 18 weighted connections between the noninput units and a *"threshold unit"* whose outgoing links implemented the thresholding for each of the non-input units, for a total of 126 weights to evolve. The GA was used as follows. Each chromosome was a list (or *"vector"*) of 126 weights. Figure 1.16 shows (for a much smaller network) how the encoding was done: the weights were read off the network in a fixed order (from left to right and from top to bottom) and placed in a list. Notice that each *"gene"* in the chromosome is a real number rather than a bit.

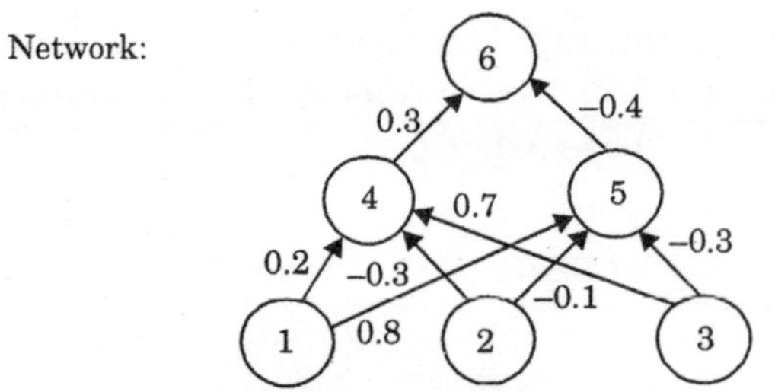

Fig. 1.16. Illustration of Montana and Davis's encoding of network weights into a list that serves as a chromosome for the GA. The units in the network are numbered for later reference. The real-valued numbers on the links are the weights.

To calculate the fitness of a given chromosome, the weights in the chromosome were assigned to the links in the corresponding network, the network was run on the training set (here 236 examples from the database of lofargrams), and the sum of the squares of the errors (collected over all the training cycles) was returned. Here, an *"error"* was the difference between the desired output activation value and the actual output activation value. Low error meant high fitness. An initial population of 50 weight vectors was chosen randomly, with each weight being between '.0 and + 1.0. Montana and Davis tried a number of different genetic operators in various experiments.

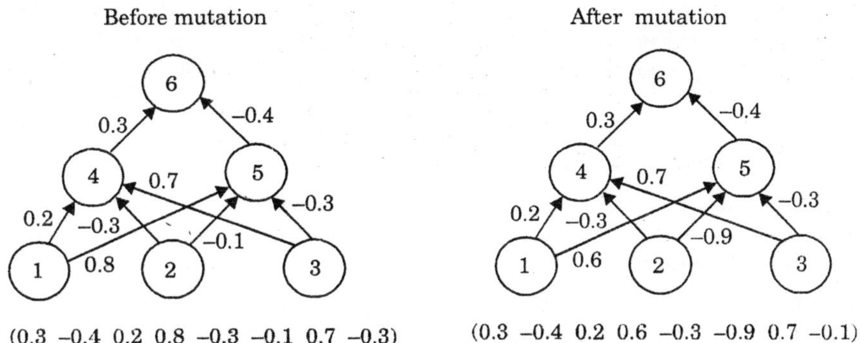

Fig. 1.17 : Illustration of Montana and Davis's mutation method. Here the weights on incoming links to unit 5 are mutated.

The mutation and crossover operators they used for their comparison of the GA with back-propagation are illustrated in Figure. 1.18 and Figure 1.18. The mutation operator selects *n* non-input units and, for each incoming link to those units, adds random value between '.0 and + 1.0 to the weight on the link. The crossover operator takes two parent weight vectors and, for each non-input unit in the offspring vector, selects one of the parents at random and copies the weights on the incoming links from that parent to the offspring. Notice that only one offspring is created. The performance of a GA using these operators was compared with the performance of a back-propagation algorithm. The GA had a population of 50 weight vectors, and a rank-selection method was used. The GA was allowed to run for 200 generations (*i.e.*, ı0,000 network evaluations). The back-propagation algorithm was allowed to run for 5000 iterations, where one iteration is a complete epoch (a complete pass through the training data).

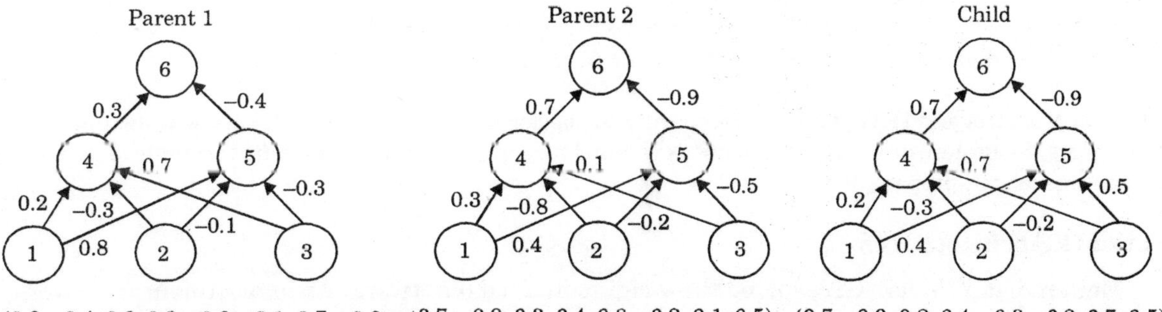

Fig. 1.18. Illustration of Montana and Davis's crossover method. The offspring is created as follows: for each non-input unit, a parent is chosen at random and the weights on the incoming links to that unit are copied from the chosen parent. In the child network shown here, the incoming links to unit 4 come from parent 1 and the incoming links to units 5 and 6 come from parent 2.

Montana and Davis reasoned that two network evaluations under the GA are equivalent to one back-propagation iteration, since back-propagation on a given training example consists of two parts — the forward propagation of activation (and the calculation of errors at the output units) and the backward error propagation (and adjusting of the weights). The GA performs only the first part. Since the second part requires more computation, two GA evaluations takes less than half the computation of a single back-propagation iteration. The results of the comparison are displayed in Fig. 1.19. Here one back-propagation iteration is plotted for every two GA evaluations. The x axis gives the number of iterations, and the y axis gives the best evaluation (lowest sum of squares of errors) found by that time. It can be seen that the GA significantly outperforms back-propagation on this task, obtaining better weight vectors more quickly. This experiment shows that in some situations the GA is a better training method for networks than simple back-propagation. This does not mean that the GA will outperform back-propagation in all cases. It is also possible that enhancements of back-propagation might help it overcome some of the problems that prevented it from performing as well as the GA in this experiment. Schaffer, Whitley, and Eshelman (1992) point out that the GA has not been found to outperform the best weight-adjustment methods (*e.g.*, *"quickprop"*) on supervised learning tasks, but they predict that the GA will be most useful in finding weights in tasks where back-propagation and its relatives cannot be used, such as in unsupervised learning tasks, in which the error at each output unit is not available to the learning system, or in situations in which only sparse reinforcement is available. This is often the case for *"neurocontrol"* tasks, in which neural networks are used to control complicated systems such as robots navigating in unfamiliar environments.

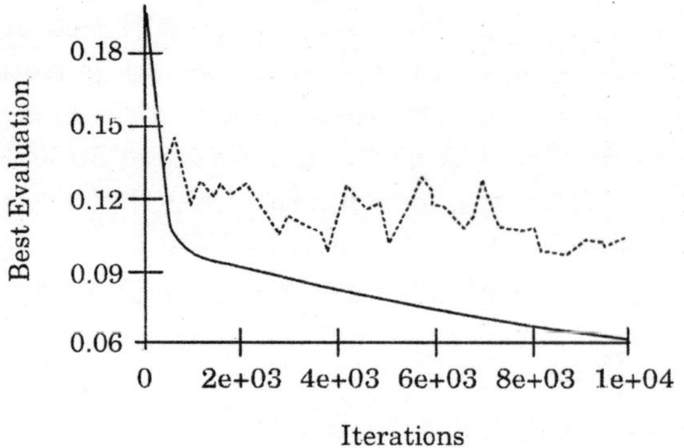

Fig. 1.19. Montana and Davis's results comparing the performance of the GA with back-propagation. The figure plots the best evaluation (lower is better) found by a given iteration. Solid line: genetic algorithm. Broken line: back—propagation.

Network Architectures

Montana and Davis's GA evolved the weights in a fixed network. As in most neural network applications, the architecture of the network—the number of units and their interconnections—is decided ahead of time by the programmer by guesswork, often aided by some heuristics (*e.g.*, "more hidden units are required for more difficult problems") and by trial and error. Neural network researchers know all too well that the particular architecture chosen can determine the success or failure of the application, so they would like very much to be able to automatically optimize the procedure of designing an architecture for a particular application. Many believe that GAs are well suited for this task. There have been several efforts along these lines, most of which fall into one of two categories: direct encoding and grammatical encoding. Under direct encoding, a network architecture is directly encoded into a GA chromosome. Under grammatical encoding, the GA does not evolve network architectures: rather, it evolves grammars that can be used to develop network architectures.

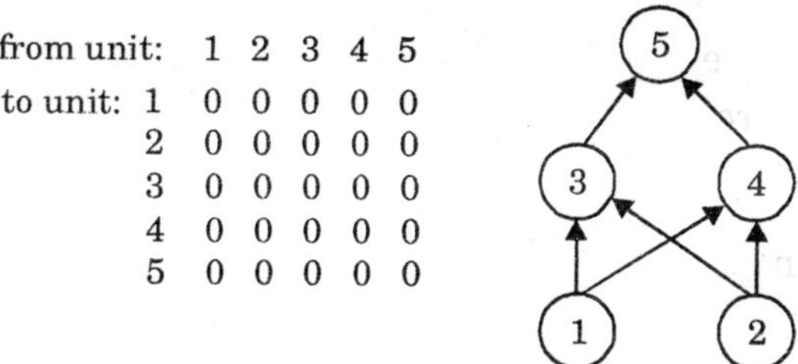

chromosome: 0 0 0 0 0 0 0 0 0 0 1 1 0 0 0 1 1 0 0 0 0 0 1 1 0

Fig. 1.20. An illustration of Miller, Todd, and Hegde's representation scheme. Each entry in the matrix represents the type of connection on the link between the "from unit" (column) and the "to unit" (row). The rows of the matrix are strung together to make the bit-string encoding of the network, given at the bottom of the figure. The resulting network is shown at the right.

Direct Encoding

The method of direct encoding is illustrated in work done by Geoffrey Miller, Peter Todd, and Shailesh Hegde (1989), who restricted their initial project to feedforward networks with a fixed number of units for which the GA was to evolve the connection topology. As is shown in figure 1.21. the connection topology was represented by an *N x N* matrix (5 × 5 in figure 1.21) in which each entry encodes the type of connection from the *"from unit"* to the *"to unit."* The entries in the connectivity matrix were either "0" (meaning no connection) or *"L"* (meaning a "learnable" connection—*i.e.*, one for which the weight can be changed through learning). Figure 1.21 also shows how the connectivity matrix was transformed into a chromosome for the GA (*"O"* corresponds to 0 and *"L"* to 1) and how the bit string was decoded into a network. Connections that were specified to be learnable were initialized with small random weights. Since Miller, Todd, and Hegde restricted these networks to be feedforward, any connections to input units or feedback connections specified in the chromosome were ignored. Miller, Todd, and Hegde used a simple fitness-proportionate selection method and mutation (bits in the string were flipped with some low probability). Their crossover operator randomly chose a row index and swapped the corresponding rows between the two parents to create two offspring.

The intuition behind that operator was similar to that behind Montana and Davis's crossover operator—each row represented all the incoming connections to a single unit, and this set was thought to be a functional building block of the network. The fitness of a chromosome was calculated in the same way as in Montana and Davis's project: for a given problem, the network was trained on a training set for a certain number of epochs, using back-propagation to modify the weights. The fitness of the chromosome was the sum of the squares of the errors on the training set at the last epoch. Again, low error translated to high fitness.

Miller, Todd, and Hegde tried their GA on three tasks:

XOR :

The single output unit should turn on (*i.e.*, its activation should be above a set threshold) if the exclusive-or of the initial values (1 = on and 0 = off) of the two input units is 1.

Four Quadrant

The real-valued activations (between 0.0 and 1.0) of the two input units represent the coordinates of a point in a unit square. All inputs representing points in the lower left and upper right quadrants of the square should produce an activation of 0.0 on the single output unit, and all other points should produce an output activation of 1.0.

Encoder/Decoder

The output units (equal in number to the input units) should copy the initial pattern on the input units. This would be trivial, except that the number of hidden units is smaller than the number of input units, so some encoding and decoding must be done. These are all relatively easy problems for multi-layer neural networks to learn to solve under back-propagation. The networks had different numbers of units for different tasks (ranging from 5 units for the XOR task to 20 units for the encoder/decoder task); the goal was to see if the GA could discover a good connection topology for each task. For each run the population size was 50, the crossover rate was 0.6, and the mutation rate was 0.005. In all three tasks, the GA was easily able to find networks that readily learned to map inputs to outputs over the training set with little error. However, the three tasks were too easy to be a rigorous test of this method—it remains to be seen if this method can scale up to more complex tasks that require much larger networks with many-more interconnections.

Grammatical Encoding

The method of grammatical encoding can be illustrated by the work of Hiroaki Kitano (1990), who points out that direct-encoding approaches become increasingly difficult to use as the size of the desired network increases. As the network's size grows, the size of the required chromosome increases quickly, which leads to problems both in performance (how high a fitness can be obtained) and in efficiency (how long it takes to obtain high fitness). In addition, since direct-encoding methods explicitly represent each connection in the network, repeated or nested structures cannot be represented efficiently, even though these are common for some problems. The solution pursued by Kitano and others is to encode networks as grammars; the GA evolves the grammars, but the fitness is tested only after a *"development"* step in which a network develops from the grammar. That is, the *"genotype"* is a grammar, and the *"phenotype"* is a network derived from that grammar.

A grammar is a set of rules that can be applied to produce a set of structures (*e.g.*, sentences in a natural language, programs in a computer language, neural network architectures). A simple example is the following grammar:

$$S \rightarrow aSb,$$

$$S \rightarrow \varepsilon,$$

Here S is the start symbol and a nonterminal, a and b are terminals, and μ is the empty-string terminal. (S' μ means that S can be replaced by the empty string.) To construct a structure from this grammar, start with S, and replace it by one of the allowed replacements given by the righthand sides (*e.g.*, *S' aSb)*. Now take the resulting structure and replace any nonterminal (here *S)* by one of its allowed replacements (*e.g.*, *aSb ' aaSbb)*. Continue in this way until no nonterminals are left (*e.g.*, *aaSbb' aabb,* using *S*' μ). It can easily be shown that the set of structures that can be produced by this grammar are exactly the strings $a^n b^n$ consisting of the same number of *as* and *bs* with all the *as* on the left and all the *bs* on the right.

Kitano applied this general idea to the development of neural networks using a type of grammar called a *"graph-generation grammar,"* a simple example of which is given in figure

1.22a Here the right-hand side of each rule is a 2 × 2 matrix rather than a one-dimensional string. Capital letters are nonterminals, and lower-case letters are terminals. Each lower-case letter from a through p represents one of the 16 possible 2 × 2 arrays of ones and zeros. In contrast to the grammar for $a^n b^n$ given above, each nonterminal in this particular grammar has exactly one right-hand side, so there is only one structure that can be formed from this grammar: the 8 × 8 matrix shown in figure 1.21b this matrix can be interpreted as a connection matrix for a neural network: a 1 in row i and column j, $i' j$ means that unit i is present in the network and a 1 in row i and column $i; i$ means that there is a connection from unit i to unit j. The result is the network shown in figure 1.21c. which, with appropriate weights, computes the Boolean function XOR. Kitano's goal was to have a GA evolve such grammars. Figure 1.23 illustrates a chromosome encoding the grammar given in figure 1.21a. The chromosome is divided up into separate rules, each of which consists of five loci.

The first locus is the left-hand side of the rule; the second through fifth loci are the four symbols in the matrix on the right-hand side of the rule. The possible alleles at each locus are the symbols $A - Z$ and $a - p$. The first locus of the chromosome is fixed to be the start symbol, S; at least one rule taking S into a 2 × 2 matrix is necessary to get started in building a network from a grammar. All other symbols are chosen at random. A network is built applying the grammar rules encoded in the chromosome for a predetermined number of iterations. (The rules that take $a - p$ to the 16 2 × 2 matrices of zeros and ones are fixed and are not represented in the chromosome.

$$S \to \begin{matrix} A & B \\ C & D \end{matrix} \quad A \to \begin{matrix} c & p \\ a & c \end{matrix} \quad B \to \begin{matrix} a & a \\ a & c \end{matrix} \quad C \to \begin{matrix} a & a \\ a & a \end{matrix} \quad D \to \begin{matrix} a & a \\ a & b \end{matrix}$$

$$a \to \begin{matrix} 0 & 0 \\ 0 & 0 \end{matrix} \quad b \to \begin{matrix} 0 & 0 \\ 0 & 1 \end{matrix} \quad c \to \begin{matrix} 1 & 0 \\ 0 & 1 \end{matrix} \quad C \to \begin{matrix} 0 & 1 \\ 0 & 1 \end{matrix} \quad p \to \begin{matrix} 1 & 1 \\ 1 & 1 \end{matrix}$$

(a)

$$S \Rightarrow \begin{matrix} A & B \\ C & D \end{matrix} \Rightarrow \begin{matrix} c & p & a & a \\ a & c & a & c \\ a & a & a & a \\ a & a & a & b \end{matrix} \Rightarrow \begin{matrix} 1 & 0 & 1 & 1 & 0 & 0 & 0 & 0 \\ 0 & 1 & 1 & 1 & 0 & 0 & 0 & 0 \\ 0 & 0 & 1 & 0 & 0 & 0 & 0 & 1 \\ 0 & 0 & 0 & 1 & 0 & 0 & 0 & 1 \\ 0 & 0 & 0 & 0 & 0 & 0 & 0 & 0 \\ 0 & 0 & 0 & 0 & 0 & 0 & 0 & 0 \\ 0 & 0 & 0 & 0 & 0 & 0 & 0 & 0 \\ 0 & 0 & 0 & 0 & 0 & 0 & 0 & 1 \end{matrix}$$

(b)

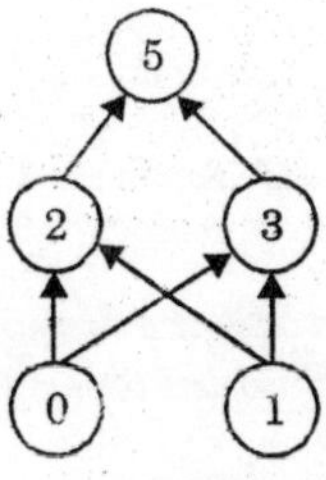

(c)

Fig. 1.21. Illustration of the use of Kitano's "graph generation grammar" to produce a network to solve the XOR problem, (a) Grammatical rules, (b) A connection matrix is produced from the grammar. (c) The resulting network.

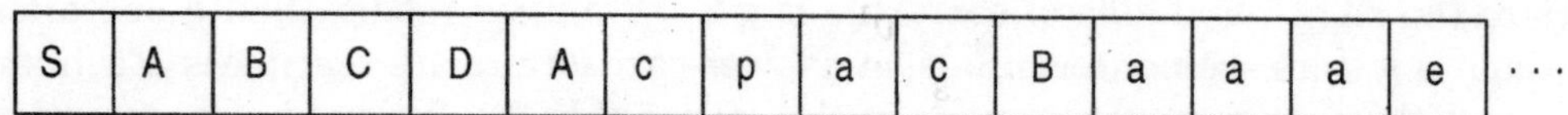

Fig. 1.22. Illustration of a chromosome encoding a grammar.

In the simple version used by Kitano, if a nonterminal (*e.g.*, *A*) appears on the left-hand side in two or more different rules, only the first such rule is included in the grammar. The fitness of a grammar was calculated by constructing a network from the grammar, using back-propagation with a set of training inputs to train the resulting network to perform a simple task, and then, after training, measuring the sum of the squares of the errors made by the network on either the training set or a separate test set.

The GA used fitness-proportionate selection, multi-point crossover (crossover was performed at one or more points along the chromosome), and mutation. A mutation consisted of replacing one symbol in the chromosome with a randomly chosen symbol from the $A - Z$ and $a - p$ alphabets. Kitano used what he called *"adaptive mutation"*: the probability of mutation of an offspring depended on the Hamming distance (number of mismatches) between the two parents.

High distance resulted in low mutation, and vice versa. In this way, the GA tended to respond to loss of diversity in the population by selectively raising the mutation rate. Kitano (1990) performed a series of experiments on evolving networks for simple "encoder/decoder" problems to compare the grammatical and direct encoding approaches. He found that, on these relatively simple problems, the performance of a GA using the grammatical encoding method consistently surpassed that of a GA using the direct encoding method, both in the correctness of the resulting neural networks and in the speed with which they were found by the GA. An example of Kitano's results is given in figure 1.23, which plots the error rate of the best network in the population (averaged over 20 runs) versus generation.

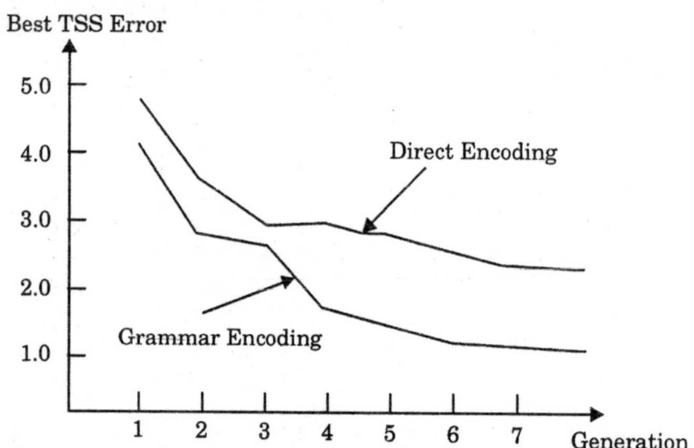

Fig. 1.23. Results from Kitano's experiment comparing the direct and grammatical encoding methods. Total sum squared (TSS) error for the average best individual (over 20 runs) is plotted against generation. (Low TSS is desired).

In the grammatical encoding runs, the GA found networks with lower error rate, and found the best networks more quickly, than in the direct encoding runs. Kitano also discovered that the performance of the GA scaled much better with network size when grammatical encoding was used—performance decreased very quickly with network size when direct encoding was used, but stayed much more constant with grammatical encoding what accounts for the grammatical encoding method's apparent superiority? Kitano argues that the grammatical encoding method can easily create *"regular"*, repeated patterns of connectivity, and that this is a result of the repeated patterns that naturally come from repeatedly applying grammatical rules.

We would expect grammatical encoding approaches to perform well on problems requiring this kind of regularity. Grammatical encoding also has the advantage of requiring shorter chromosomes, since the GA works on the instructions for building the network (the grammar) rather than on the network structure itself. For complex networks, the latter could be huge and intractable for any search algorithm. Although these attributes might lend an advantage in general to the grammatical encoding method, it is not clear that they accounted for the grammatical encoding method's superiority in the experiments reported by Kitano (1990). The encoder/decoder problem is one of the simplest problems for neural networks; moreover, it is interesting only if the number of hidden units is smaller than the number of input units. This was enforced in Kitano's experiments with direct encoding but not in his experiments with grammatical encoding. It is possible that the advantage of grammatical encoding in these experiments was simply due to the GA's finding network topologies that make the problem trivial; the comparison is thus unfair, since this route was not available to the particular direct encoding approach being compared. Kitano's idea of evolving grammars is intriguing, and his informal arguments are plausible reasons to believe that the grammatical encoding method (or extensions of it) will work well on the kinds of problems on which complex neural networks could be needed. However, the particular experiments used to support the arguments are not convincing, since the problems may have been too simple. An extension of Kitano's initial work, in which the evolution of network architecture and the setting of weights are integrated, is reported in Kitano 1994.

Learning Rule

David Chalmers (1990) took the idea of applying genetic algorithms to neural networks in a different direction: he used GAs to evolve a good learning rule for neural networks. Chalmers limited his initial study to fully connected feedforward networks with input and output layers only, no hidden layers. In general a learning rule is used during the training procedure for modifying network weights in response to the network's performance on the training data. At each training cycle, one training pair is given to the network, which then produces an output. At this point the learning rule is invoked to modify weights. A learning rule for a single-layer, fully connected feedforward network might use the following local information for a given training cycle to modify the weight on the link from input unit i to output unit j:

a_i : the activation of input unit i

o_j : the activation of output unit j

t_j : the training signal (*i.e.*, correct activation, provided by a teacher) on output unit j

w_{ij}: the current weight on the link from i to j.

The change to make in weight w_{ij}, "w_{ij}, is a function of these values:

$$\Delta w_{ij} = \eta(t_j o_j - a_j o_j)$$

The chromosomes in the GA population encoded such functions.

Chalmers made the assumption that the learning rule should be a linear function of these variables and all their pairwise products. That is, the general form of the learning rule was

$$\Delta w_{ij} = k_0(k_1 w_{ij} + k_2 a_j + k_3 a_j + k_4 t_j + k_5 wja_j + k_6 w_{ij} o_j$$
$$+ k_7 w_{ij} t_j + k_8 a_i o_j + k_9 a_i t_j + k_{10} o_j t_j).$$

The k_m (1dmd10) are constant coefficients, and k_0 is a scale parameter that affects how much the weights can change on any one cycle. (k_0 is called the *"learning rate"*). Chalmers's assumption about the form of the learning rule came in part from the fact that a known good learning rule for such networks — the "Widrow-Hoff " or "delta" rule— has the form where n is a constant representing the learning rate. One goal of Chalmers's work was to see if the GA could evolve a rule that performs as well as the delta rule.

The task of the GA was to evolve values for the k_m*'s*. The chromosome encoding for the set of k_m*'s* is illustrated in figure 1.25. The scale parameter k^0 is encoded as five bits, with the zeroth bit encoding the sign (1 encoding + and 0 encoding) and the first through fourth bits encoding an integer n: k_0, = 0 if $n = 0$; otherwise $|k_0| = 2^{n-9}$. Thus k_0 can take on the values 0, ± 1/256, ±1/128, ..., ±32, ±64. The other coefficients k_m are encoded by three bits each, with the zeroth bit encoding the sign and the first and second bits encoding an integer n. For $i = 1$ ... 10, $k_m = 0$ if $n = 0$; otherwise $|k_m| = 2^{n-1}$.

Genome encoding:

k_0	k_1	k_2	k_3
1 0 0 1 0	0 0 1	0 0 0	1 1 0

k_0 encoded by 5 bits:

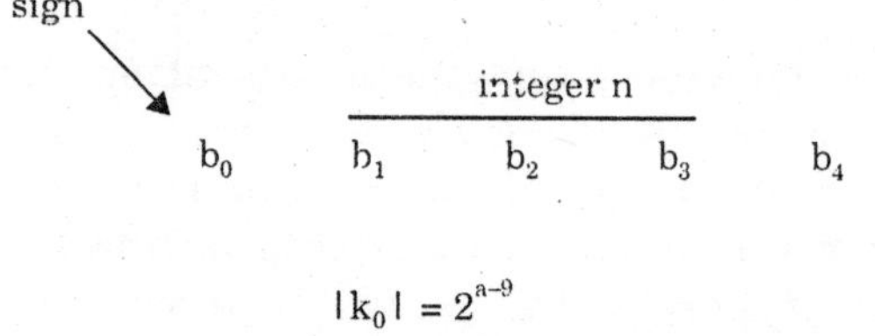

Other k's encoded by 3 bits each:

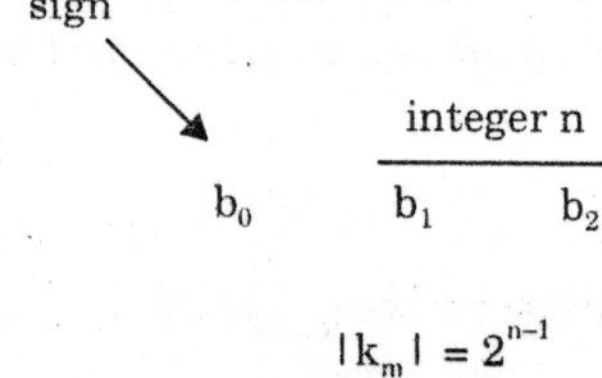

Fig. 1.24. Illustration of the method for encoding the k_m*s in Chalmers's system.*

It is known that single-layer networks can learn only those classes of input-output mappings that are *"linearly separable"*. As an "environment" for the evolving learning rules, Chalmers used 30 different linearly separable mappings to be learned via the learning rules. The mappings always had a single output unit and between two and seven input units. The fitness of each chromosome (*learning rule*) was determined as follows. A subset of 20 mappings was selected from the full set of 30 mappings. For each mapping, 12 training examples were selected. For each of these mappings, a network was created with the appropriate number of input units for the given mapping (each network had one output unit). The network's weights were initialized randomly. The network was run on the training set for some number of epochs (typically 10), using the learning rule specified by the chromosome. The performance of the learning rule on a given mapping was a function of the network's error on the training set, with low error meaning high performance. The overall fitness of the learning rule was a function of the average error of the 20 networks over the chosen subset of 20 mappings—low average error translated to high fitness. This fitness was then transformed to be a percentage, where a high percentage meant high fitness. Using this fitness measure, the GA was run on a population of 40 learning rules, with two-point crossover and standard mutation. The crossover rate was 0.8 and the mutation rate was 0.01. Typically, over 1000 generations, the fitness of the best learning rules in the population rose from between 40% and 60% in the initial generation (indicating no significant learning ability) to between 80% and 98%, with a mean (over several runs) of about 92%. The fitness of the delta rule is around 98%, and on one out of a total of ten runs the GA discovered this rule).

On three of the ten runs, the GA discovered slight variations of this rule with lower fitness. These results show that, given a somewhat constrained representation, the GA was able to evolve a successful learning rule for simple single-layer networks. The extent to which this method can find learning rules for more complex networks (including networks with hidden units) remains an open question, but these results are a first step in that direction. Chalmers suggested that it is unlikely that evolutionary methods will discover learning methods that are more powerful than back-propagation, but he speculated that the GA might be a powerful method for discovering learning rules for unsupervised learning paradigms (*e.g.*, reinforcement learning) or for new classes of network architectures (*e.g.*, recurrent networks). Chalmers also performed a study of the generality of the evolved learning rules. He tested each of the best evolved rules on the ten mappings that had not been used in the fitness calculation for that rule (the "test set"). The mean fitness of the best rules on the original mappings was 92%, and Chalmers found that the mean fitness of these rules on the test set was 91.9%. In short, the evolved rules were quite general. Chalmers then looked at the question of how diverse the environment has to be to produce general rules. He repeated the original experiment, varying the number of mappings in each original environment between 1 and 20. A rule's *evolutionary fitness* is the fitness obtained by testing a rule on its original environment. A rule's *test fitness* is the fitness obtained by testing a rule on ten additional tasks not in the original environment. Chalmers then measured these two quantities as a function of the number of tasks in the original environment. The results are shown in figure 1.26. The two curves are the mean evolutionary fitness and the mean test fitness for rules that were tested in an environment with the given number of tasks. This plot shows that while the evolutionary fitness stays roughly constant for different numbers of environmental tasks, the test fitness increases sharply with the number of tasks, leveling off somewhere between 10 and 20 tasks. The conclusion is that the evolution of a general learning rule requires a diverse environment of tasks. (In this case of simple single-layer networks, the necessary degree of diversity is fairly small).

2

Genomic Informatics

What are these areas of intense research labeled bioinformatics and functional genomics? If we take literally much of the recently published "news and views," it seems that the often stated claim that the last century was the century of physics, whereas the twenty-first will be the century of biology, rests significantly on these new research areas. We might therefore ask: What is new about them? After all, compu- tational or mathematical biology has been around for a long time. Surely much of bioinformatics, particularly that associated with evolution and genetic analyses, does not appear very new. In fact, the related work of researchers like R. A. Fisher, J. B. S. Haldane, and Sewell Wright dates nearly to the beginning of the 1900s. The modern analytical approaches to genetics, evolution, and ecology rest directly on their and similar work. Even genetic mapping easily dates to the 1930s, with the work of T. S. Painter and his students of *Drosophila* (still earlier if you include T. H. Morgan's work on X-linked markers in the fly).

A short historical-evie might provide ' a useful perspective on this anticipated century of biology and allow us to view the future from a firmer foundation. First of all, it should be helpful to recognize that it was very early in the so-called century of physics that modern biology began, with a paper read by Hermann Miiller at a 1921 meeting in Toronto. Miller, a student of Morgan's, stated that although of submicroscopic size, the gene was clearly a physical particle of complex structure, not just a working construct! Miiller noted that the gene is unique from its product, and that it is normally duplicated unchanged, but once mutated, the new form is in turn duplicated faithfully. The next 30 years, from the early 1920*s* to the early 1950s, were some of the most revolutionary in the science of biology. In my original field of physics, the great insights of relativity and quantum mechanics were already being taught to undergraduates; in biology, the new one-gene-one-enzyme concept was leading researchers to new understandings in biochemistry, genetics, and evolution.

The detailed physical nature of the gene and its product were soon obtained. By midcentury, the unique linear nature of the protein and the gene were essentially known from the work of Frederick Sanger and Erwin Chargraff. All that remained was John Kendrew's structural analysis of sperm whale myoglobin and Jamaes Watson and Francis Crick's double helical model for DNA. Thus by the mid-1950s, we had seen the physical gene and one of its products, and the motivation was in place to find them all. Of course, the genetic code needed to be determined and restriction enzymes discovered, but the beginning of modern molecular biology was on its way.

We might say that much of the last century was the century of applied physics, and the last half of the century was applied molecular biochemistry, generally called molecular biology! So what happened to create bioinformatics and functional genomics? It was, of course, the wealth of sequence data, first protein and then genomic. Both are based on some very clever chemistry and the late 1940s molecular sizing by chromatography. Frederick Sanger's sequencing of insulin and Wally Gilbert and Allan Maxam's sequence of the Lactose operator from *E. coli* showed that it could be done. In principle, all genetic sequences, including the human genome, were determinable; and, if determinable, they were surely able to be engineered, suggesting that the economics and even the ethics of biological research was about to change. The revolution was already visible to some by the 1970s. The science or discipline of analyzing and organizing sequence data defines for many the bioinformatics realm. It had two somewhat independent beginnings. The older was the attempt to related amino acid sequences to the three dimensional structure and function of proteins.

The primary focus was the understanding of the sequence's encoding of structure and, in turn, the structure's encoding of biochemical function. Beginning with the early work of Sanger and Kendrew, progress continued such that, by the mid-1960s, Margaret Dayhoff had formally created the first major database of protein sequences. By 1973, we had the start of the database of X-ray crystallographic determined protein atomic coordinates under Tom Koetzle at the Brookhaven National Laboratory. From early on, Dayhoff seemed to understand that there was other very fundamental information available in sequence data, as shown in her many phylogenetic trees. This was articulated most clearly by Emile Zuckerkandl and Linus Pauling as early as 1965, that within the sequences lay their evolutionary history.

There was a second fossil record to be deciphered. It was that recognition that forms the true second beginning of what is so often thought of as the heart of bioinformatics, comparative sequence analyses. The seminal paper was by Walter Fitch and Emanuel Margoliash, in which they constructed a phylogenetic tree from a set of cytochrome sequences. With the advent of more formal analysis methods and larger datasets the marriage between sequence analysis and computer science emerged as naturally as it had with the analysis of tens of thousands of diffraction spots in protein structure determination a decade before. As if proof was needed that comparative sequence analysis was of more than academic interest, Russell Doolittle demonstrated that we could explain the onc gene v-sis's properties as an aberrant growth factor by assuming that related functions are carried out by sequence similar proteins. By 1990, nearly all of the comparative sequence analysis methods had been refined and applied many times.

The result was a wealth of new functional and evolutionary hpotheses. Many of these led directly to new insights and experimental validation. This in turn made the 40 years between 1950 and 1990 the years that brought reality to the dreams seeded in those wondrous previous 40 years of genetics and biochemistry. It is interesting to note that during this same 40 years, computers developed from the wartime monsters through the university mainframes and the lab bench workstation to the powerful personal computer. In fact, Doolittle's early successful comparative analysis was done on one of the first personal computers, an Apple II.

The link between computers and molecular biology is further seen in the justification of initially placing GenBank at the Los Alamos National Laboratory rather than at an academic institution. This was due in large part to the laboratory's then immense computer resources, which in the year 2000 can be found in a top-of he-line laptop! What was new to computational biology was the data and the anticipated amount of it. The human genome project was being formally initiated by 1990. Within the century's final decade, the genomes of more than two dozen microorganisms, along with yeast and *C. elegans,* the worm, would be completely sequenced. By the summer of the new century's very first year, the fruit fly genome would be sequenced, as well as 85 percent of the entire human genome. Although envisioned as possible by the late 1970s, no one foresaw the wealth of full genomic sequences that would be available at the start of the new millennium. What challenges remained at the informatics level? Major database problems and some additional algorithm development will still surely come about. And, even though we still cannot predict a protein's structure or function directly from its sequence, *de novo,* straightforward sequence comparisons with such a wealth of data can generally infer both function and structure from the identification of close homologues previously analyzed.

It has slowly become obvious that there are at least four major problems here: first, most *"previously analyzed"* sequences obtained their annotation via sequence comparative inheritance, and not by any direct experimentation; second, many proteins carry out very different cellular roles even when their biochemical functions are similar; third, there are even proteins that have evolved to carry out functions distinct from those carried out by their close homologues; and, finally, many proteins are multidomained and thus multifunctional, but identified by only one function. When we compound these facts with the lack of any universal vocabulary throughout much of molecular biology, there is great confusion, even with interpreting standard sequence similarity analysis. Even more to the point of the future of bioinformatics is knowing that the function of a protein or even the role in the cell played by that function is only the starting point for asking real biological questions.

Asking questions beyond what biochemistry is encoded in a single protein or protein domain is still challenging. However, asking what role biochemistry plays in the life of the cell, which many refer to as functional genomics, is clearly even more challenging from the computational side.

The analysis of genes and gene networks and their regulation may be even more complicated. Here we have to deal with alternate spliced gene products with potentially distinct functions and highly degenerate short DNA regulatory words. So far, sequence comparative methods have had limited success in these cases. What will be the future role of computation in biology in the first few decades of this century? Surely many of the traditional comparative sequence analyses, including homologous extension protein structure modeling and DNA signal recognition, will continue to play major roles. As already demonstrated, standard statistical and clustering methods will be used on gene expression data. It is obvious, however, that the challenge for the biological sciences is to begin to understand how the genome parts list encodes cellular function—not the function of the individual parts, but that of the whole cell and organism. This, of course, has been the motivation underlying most of molecular biology over the last 20 years.

The difference now is that we have the parts lists for multiple cellular organisms. These are complete parts lists rather than just a couple of genes identified by their mutational or other effects on a single pathway or cellular function. The past logic is now reversible: rather than starting with a pathway or physiological function, we can start with the parts list either to generate testable models or to carry out large-scale exploratory experimental tests. The latter, of course, is the logic behind the mRNA expression chips, whereas the former leads to experiments to test new regulatory network or metabolic pathway models. The design, analysis, and refinement of such complex models will surely require new computational approaches. The analysis of the RNA expression data requires the identification of various correlations between individual gene expression profiles and between those profiles and different cellular environments or types. These, in turn, require some model concepts as to how the behavior of one gene may effect that of others, both temporally and spatially. Some straightforward analyses of RNA expression data have identified many differences in gene expression in cancer versus noncancer cells and for different growth conditions. Such data have also been used in an attempt to identify common or shared regulatory signals in bacteria.

Yet expression data's full potential is not close to being realized. In particular, when gene expression data can be fully coupled to protein expression, modification, and activity, the very complex genetic networks should begin to come into view. In higher animals, for example, proteins can be complex products of genes through alternate exon splicing. We can anticipate that mRNA-based microarray expression analysis will be replaced by exon expression analysis. Here again, modeling will surely play a critical role, and the type of computational biology envisioned by population and evolutionary geneticists such as Wright may finally become a reality.

The extraction of how the organism's range of behavior or environment responses is encoded in the genome, is the ultimate aim of functional genomics. Many people in what is now called bioinformatics will recall that much of the wondrous mathematical modeling and analysis associated with population and evolutionary biology was at best suspect and at worst ignored by molecular biologists over the last 30 years or so. At the beginning of the new millennium, perhaps those thinkers should be viewed as being ahead of their time. Note, it was not that serious mathematics is not necessary to understand anything as complex as interacting populations, but only that the early biomodelers did not have the needed data.

Today we are rapidly approaching the point where we can measure not only a population's genetic variation, but nearly all the genes that might be associated with a particular environmental response.

It is the data that has created the latest aspect of the biological revolution. Just imagine what we will be able to do with a dataset composed of distributions of genetic variation among different subpopulations of fruit fly living in distinctly different environments, or what might we learn about our own evolution by having access to the full range of human and other primate genetic variation for all 40,000 to 100,000 human genes? It is perhaps best for those anticipating the challenges of bioinformatics and computational genomics to think about how biology is likely to be taught by the end of the second decade of this century. Will the complex mammalian immune system be presented as a logical evolutionary adaptation of an early system for cell-cell communication that developed into a cell-cell recognition system, and then self-nonself recognition? Will it become obvious that the use by yeast of the G-protein couple receptors to recognize matting types would become one of the main components of nearly all higher organisms sensor systems? Like physics, where general rules and laws are taught at the start and the details are left for the computer, biology will surely be presented to future generations of students as a set of basic systems that have been duplicated and adapted to a very wide range of cellular and organismic functions following basic evolutionary principles constrained by Earth's geological history.

3

Computational Genomics

Once a complete genome sequence is known, it should in principle be possible to identify all the genes and uncover all their functions by computational methods. In reality, this is not possible. Is it because the current computational methods are imperfect or because the information in the genome is insufficient? Whichever to believe, anyone would agree that additional data and knowledge will help to interpret the complete genome sequence information. Post-genomics is an emerging field for developing new experimental and computational technologies, such as DNA chips and protein chips, for generating and analyzing different types of systematic data, such as gene expressions and polymorphisms, and for expanding our biological knowledge based on the genomic information. Here again, differences arise depending on the views or directions taken for post-genome analyses.

In the traditional view of molecular biology, after the genome is the transcriptome, and then the proteome. The transcriptome represents a whole set of mRNAs expressed in the cell of a given tissue under a given condition. The proteome usually represents a whole set of proteins expressed in the cell and how they interact with each other, but it may also mean structural genomics to systematically determine a catalog of protein 3D structures. Computational molecular biology has been the discipline of choice to analyze sequence and 3D structural information of DNAs, RNAs, and proteins in order to understand molecular functions. However, the analysis of individual molecules would never be sufficient for understanding higher order functions of cells and organisms, represented by another axis. Furthermore, although the biological macromolecules of DNAs, RNAs, and proteins may play major roles, there are other substances that together make up the entire chemical complement of the cell. The third axis emphasizes the roles of chemical compounds and metal ions in biological functions. The two additional axes represent an extension of the traditional molecular biology they are in fact the conceptual basis of KEGG, the Kyoto Encyclopedia of Genes and Genomes. In our view, the genome is simply an information storage of how to make individual molecular building blocks of life.

The genome does not contain much information about the wiring of building blocks—for example, how they interact to make up a cell or to exert cellular functions. The wiring information is likely to be distributed in the cell and more dynamic in nature. Although the molecular wiring diagram of the cell may not be computable from the information in the genome alone, it may still be predictable, at least to some extent, if we have sufficient knowledge of actual wiring in living cells and if empirical relations to genomes can be found. Thus, we have been computerizing current knowledge on molecular pathways and complexes in the *PATH WAY* database, and analyzing possible relations to the gene catalogs of all the completely sequenced genomes and some partial genomes that are stored in the *GENES* database in *KEGG*. We have also been collecting information about chemical compounds and chemical reactions in the *LIGAND* database. Such information is essential for understanding the dynamic interactions of the cell with its environment. In traditional computational molecular biology, the data objects to be analyzed consist of elements that are abstracted to symbols at the atomic level, such as *C* for carbon in the protein 3D structure, or to other symbols at the molecular level, such as *C* for cysteine in the amino acid sequence.

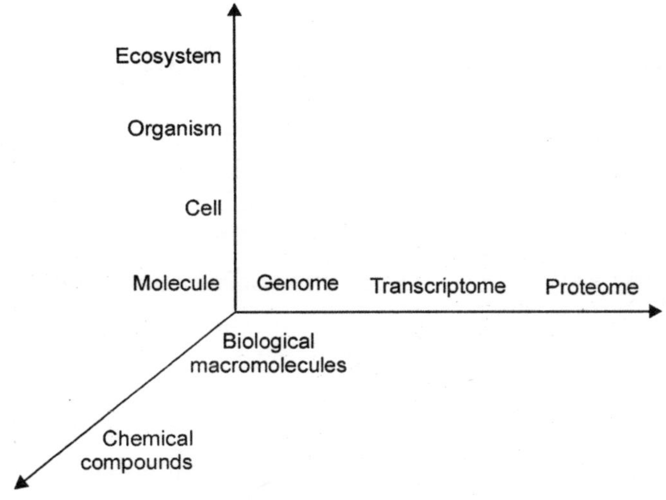

Fig. 3.1. Post-genomics in three directions.

In computational genomics as we define here, the data objects to be analyzed are the genome, which is a sequence of genes, the pathway that is a network of interacting proteins, and other types of relations among genes or gene products. Thus, symbols are used for an abstract representation of data elements at a higher level, such as polA for a gene in the genome and Ras for a protein in the pathway. *KEGG* is a computational resource for analyzing network of such higher level symbols. It is highly integrated with the existing molecular biology resources for analyzing sequences of molecular symbols and networks of atomic symbols.

Table 3.1. Examples of biological complex systems at different levels.

Complex system	*Node*	*Edge (interaction)*
Protein	Atom	Atomic interaction
Cell	Molecule	Molecular interaction
Brain	Cell	Cellular interaction
Ecosystem	Organism	Organism interaction
Civilization	Human	Human interaction

COMPLEX SYSTEMS

Life is a manifestation of biological complex systems at different levels, as exemplified in table 3.1. A complex system consists of nodes and edges, namely, building blocks and their interactions, and it is interacting with the environment. The protein is a complex system consisting of atoms and atomic interactions. Under the physiological environment, the protein assumes the native 3D structure, which makes it possible to perform a specific biological function. When the environment is perturbed, the native structure is disrupted and the protein loses its function. The structural change occurs in a narrow range of environmental conditions, which is like a phase transition in physical phenomena and which represents a systemic behavior of the complex system. At a higher level of abstraction, the cell may be viewed as a complex system consisting of molecules and molecular interactions.

When there is a proper network of molecules, such as a series of enzymes catalyzing successive reaction steps in a metabolic pathway or a set of proteins that forms a signal transduction pathway, then the cell is able to perform its specific function, such as biosynthesis of amino acids or response to environmental stresses. Thus, the specific network of molecules in the cell can be related to a higher order cellular function, which is like relating the specific 3D structure of the protein to its molecular function. When the cell is perturbed, for example, by a foreign substance in the environment or a mutation in the genome, a dynamic change may be observed in the global network of interacting molecules. Such a systemic response to a perturbation is again a common feature of complex systems.

Table 3.2. Graph representation of KEGG data objects.

Graph (data object)	*Node*	*Edge (interaction or relation)*
Genome	Gene	Ajacency
Transcriptome	Gene	Expression similarity
Proteome	Protein	Direct interaction
Protein universe	Protein	Sequence similarity or 3D structural similarity
Gene universe	Gene	Orthology, paralogy, or xenology
Complex	Gene product (protein or RNA)	Direct interaction
Pathway	Gene product or complex	Generalized protein-protein interaction (direct interaction, gene expression relation, or enzyme-enzyme relation)

Network: = Pathway | Complex

Graph Representation

Although there are other systemic phenomena at still higher levels, we focus our analysis on the level of molecular interactions because this is the level where the information in the genome can be directly correlated. We also extend the concept of edges to other types of relations. Thus, many data objects in *KEGG* are represented by a graph, which is a set of nodes and edges, as summarized in table 3.2. The genome is a graph consisting of one-dimensionally connected nodes (genes). The transcriptome generated by systematic gene expression profile analyses can be interpreted as a graph of expression similarity from which clusters of coregulated genes can be identified.

The proteome obtained by yeast two hybrid system experiments or mass spectroscopy experiments suggest a graph of possible protein-protein interactions in complexes and pathways.

In addition to experimental data on genomes, transcriptomes, and proteomes, the result of computational analyses can also be represented by a graph, such as the protein universe viewed as a hierarchy of structurally similar proteins and the gene universe representing evolutionary relations of genes and organisms. One of the major objectives of *KEGG* is to computerize data and knowledge on molecular pathways and complexes that are involved in various cellular processes.

KEGG contains a unique data object termed the generalized protein-protein interaction network, or simply the network, which is an abstract network of gene products. Although there may be different ways of representing a network of interacting molecules in the cell, the representation in KEGG focuses on proteins and RNAs that are directly linked to genes in the genome. The generalized protein-protein interaction includes: (*a*) a direct interaction such as binding, modification, or cleavage; (*b*) an indirect interaction involving gene expression, namely, the relation between a transcription factor and a target gene product; and (*c*) another indirect interaction representing the relation of two enzymes that catalyze two successive reaction steps. It must be noted that DNAs and chemical compounds are not considered as the nodes of the network, but rather they are part of the edges. Of course, details of protein-DNA interactions in gene expressions and protein-ligand interactions in enzymatic reactions are useful information in actual data analysis. It must also be noted that the term protein used here actually includes an RNA or a complex of proteins and/or RNAs.

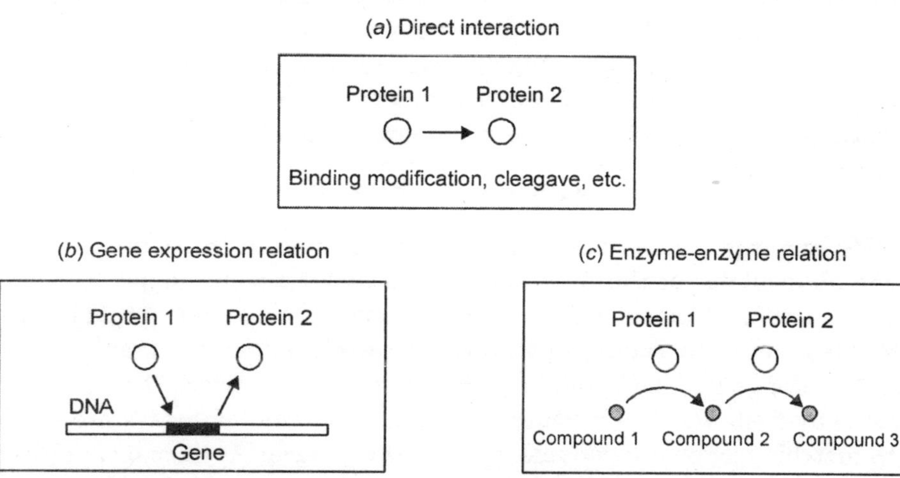

Fig. 3.2. The network of interacting molecules in the cell, such as a pathway or a complex, is represented in KEGG by the three types of generalized protein-protein interactions.

Functional Hierarchy

The graph representation shown in table 3.2 is the most basic content of the *KEGG* ontology, which is a formal specification of entities and their relations in *KEGG*. In addition, the network (pathway or complex) is hierarchically structured, as shown in Fig. 3.3. The three categories in the top hierarchy, metabolism, genetic information processing, and environmental information processing, are the three essential aspects of life in any organism; the fourth category of cellular processes contains divergent aspects of cellular functions in various organisms. As of October 2000, the KEGG network hierarchy has been further subdivided into 21 subcategories and over 120 sub-subcategories.

KEGG Network Hierarchy
As of October 2000

Metabolism
- Carbohydrate Metabolism
- Energy Metabolism
- Lipid Metabolism
- Nucleotide Metabolism
- Amino Acid Metabolism
- Metabolism of Other Amino Acids
- Metabolism of Complex Carbohydrates
- Metabolism of Complex Lipids
- Metabolism of Cofactors and Vitamins
- Metabolism of Other Substances

Genetic Information Processing
- Transcription
- Translation
- Sorting and Degradation
- Replication and Repair

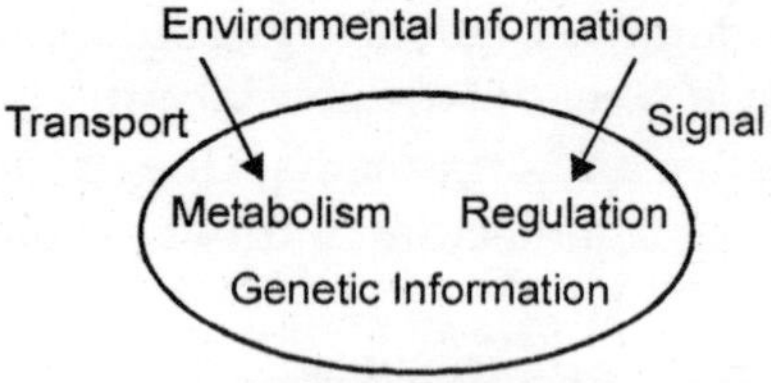

Environmental Information
Transport
Signal
Metabolism
Regulation
Genetic Information
KEGG Virtual Cell

Environmental Information Processing
- Membrane Transport
- Signal Transduction
- Ligand-Receptor Interaction

Cellular Processes
- Cell Motility
- Cell Cycle and Cell Division
- Cell Death
- Development

Fig. 3.3. The functional hierarchy of the KEGG network data. The top two levels are shown here.

Once the complete genome sequence is determined, it has become customary to present a hierarchical classification of gene functions. In contrast to various existing classification schemes, the KEGG functional hierarchy is assigned to the network, rather than to individual genes, because any higher order function involving a cell or an organism is an attribute of the network. KEGG does provide a hierarchical classification of genes for each genome, which is automatically generated in the process of matching genes in the genome and gene products in the network.

PATHWAY DATABASE

The *KEGG*, ontology summarized above is implemented in the databases shown in table 3.3, which are all available at the GenomeNet (http://www.genome.ad.jp/). The main database *PATHWAY* is a collection of known pathways (and complexes) that are involved in various cellular processes. Mostly from the literature, a pathway is drawn manually as a graphical diagram based on the concept of the generalized protein-protein interaction network. This is, what is called a reference pathway from which a number of organism-specific pathways are computationally generated by matching against individual genes in the genome. At the moment, there are about two hundred reference pathways; each pathway contains, on the average, about 30 proteins.

Table 3.3. KEGG databases

Data object	*Database*	*Data type*	*Content*
Network	PATHWAY	Graph	Generalized protein-protein interaction networks for various cellular processes
Genome	GENES	Node	Gene catalogs for completely sequenced genomes and some partial genomes
	GENOME	Graph	Genome maps and information about organisms
Transcriptome	EXPRESSION	Graph	Microarray gene expression profiles
Proteome	BRITE	Graph	Protein-protein interactions and relations
Environment	LIGAND	Edge	Chemical compounds and chemical reactions

The *KEGG* pathways are .divided into metabolic pathways and regulatory pathways, which correspond to metabolism and the rest, respectively, in the functional hierarchy shown in figure 3.3. A metabolic pathway involving enzymatic reactions on chemical substances consists of enzyme-enzyme relations, whereas a regulatory pathway involving macromolecular reactions and interactions mostly consists of direct interactions and gene expression relations. For computational purposes of using KEGG metabolic pathways, an auxiliary file is provided containing an entire list of enzyme-enzyme relations. Such binary relation files are not yet available for regulatory pathways.

Genes Database

The *GENES* database provides the gene catalog information for all the completely sequenced genomes and some partial genomes, including human and mouse. An entry in *GENES* contains sequence information and functional annotation, together with links to the *PATHWAY* and *GENOME* databases in *KEGG* and to other outside databases. When the complete genome sequence is publicly made available in Gen-Bank, it is incorporated in the *GENES* database within a few days, and the assignment of EC numbers and ortholog identifiers is performed within a few weeks. Here the ortholog identifier is an extension of the EC numbering system for enzymes. It is applicable to all proteins and RNAs, and it can distinguish subunits or genes with the same EC number. Once the assignment of ortholog identifiers is manually performed, organism-specific pathways are automatically generated by matching genes in the genome and gene products in the pathway. They are represented by the coloring of gene product nodes (boxes) in the reference pathways. This is possible because each node in the pathway is associated either with an ortholog identifier or with the combination of an organism name and a gene name. The matching process also generates a gene catalog for each organism, which is a hierarchical classification of genes according to the functional hierarchy of *KEGG* pathways.

The gene catalog is manipulated by what is called the hierarchical text browser. For most genomes, the hierarchical classification of gene functions provided by the original authors is also made compatible with the hierarchical text browser. After the initial annotation of ortholog identifiers, efforts are continuously made to standardize the terminology across organisms and to provide the most up-to-date information according to new experimental evidence reported in the literature, convincing results of our pathway analysis, and functional annotations of the *SWISSPROT* and other databases. The gene annotations are maintained by the Web-based *KEGG* annotation tool, which is linked to a relational database and which is integrated with *GFIT* and other computational tools.

Genome Database

The *GENOME database* is a collection of genome maps containing information about chromosomal locations of genes for completely sequenced genomes. The genome map is manipulated by the Java-based genome map browser. There are again two versions of genome maps, original and *KEGG*, corresponding to the two versions of the gene catalogs. They differ in the coloring of genes that represent functional hierarchy and also the links made to individual gene entries. The *GENOME* database is associated with the taxonomy and text information about each organism.

Ortholog Group Table

The *KEGG ortholog group table is* a condensation of the results obtained by the integrated analysis of the *PATHWAY, GENES,* and *GENOME* databases. In contrast to efforts such as the *COG* database which attempts to classify all genes into clusters of orthologous genes, the concept of the ortholog group is applied here to sets of functionally correlated genes, such as orthologous operons, rather than to individual genes. The ortholog group table was first constructed for each conserved portion of the metabolic pathway that was identified as a correlated cluster (see below) of genes in the genome and gene products in the pathway; for example, a set of genes in an operon responsible for a biosynthetic pathway.

The collection of ortholog group tables was then expanded to other pathways and molecular complexes by examining correlated clusters of genes in multiple genomes also based on knowledge in the literature. The ortholog group table is an *HTML* table with an embedded manipulation program for row-wise and column-wise operations. Each row indicates whether genes are present or not for a given organism and also whether there are adjacent genes in the genome, possibly forming operons, by coloring. Each column contains a set of orthologous genes based not simply on sequence similarity but also on the positional correlation of genes and the completeness of the pathway. Thus, the compilation of ortholog group tables has been extremely useful in identifing unannotated or misannotated genes in the original databases. The table can also be viewed as a multiple alignment of organism-specific pathways, indicating a pathway motif or a functional unit of the cellular processes.

Expression Database

The EXPRESSION database is a new addition to the *KEGG* system. It is being developed for our ongoing project to analyze microarray gene expression profiles in *Saccharomyces cerevisiae, Synechocystis PCC6803 Bacillus subtilis,* and *Escherichia coli.* An entry in *EXPRESSION* corresponds to a piece of hybridization data, which can be viewed and analyzed in combination with the *PATHWAY* and *GENOME* information by the Java-based expression browser.

Brite Database

BRITE (Biomolecular Relations in Information Transmission and Expression) is a database of binary relations between proteins or other biological molecules. The concept of binary relations, which is equivalent to the concept of edges has also been used in the *DBGET*/LinkDB system to compute indirect (deduced) links between databases. *BRITE* is still at an early stage of development, but it aims at enhancing such deductive' database capabilities for biological relations of genes and gene products. In view of the developments in experimental technologies for protein-protein interactions, there will be a huge amount of biological binary relation data that will be part of the BRITE database.

Ligand Database

The role of the *LIGAND database* in the *KEGG* system has been to provide detailed molecular information about one type of the generalized protein-protein interaction, namely, the enzyme-enzyme relation. *LIGAND* is a composite database of *ENZYME* and *COMPOUND.* The *ENZYME* section stores the information about enzymatic reactions and enzyme molecules according to the up-to-date classification of the EC numbers, whereas the *COMPOUND* section is a collection of about six thousand chemical compounds, most of which are metabolites in the metabolic pathways. The *ENZYME* and *COMPOUND*, entries are linked from the *KEGG* reference pathways for metabolism, thus providing molecular details of network information.

The future role of *LIGAND* is to integrate the information about the environment of the network. We will organize data and knowledge of chemical compounds and chemical reactions that affect living cells and organisms, including drugs, environmental compounds, and their metabolisms, in the *COMPOUND* section and the third *REACTION* section of the *LIGAND* database.

Hierarchical Classifications

The data objects shown in table 3.2 are collected in the *KEGG* databases shown in table 3.3, except for the protein universe and the gene universe. These data objects have been derived from the sequence and 3D structure databases; *KEGG* just makes use of the existing compilations. For example, the hierarchy of protein folds and sequence similarities in the *SCOP* database can be used to analyze pathway information by the hierarchical text browser in *KEGG*.

In addition to hierarchically classified gene catalogs and protein catalogs, other types of classifications, such as diseases and cell types, are being integrated in *KEGG* in order to make links between genotypes and phenotypes.

NETWORK PREDICTION

Sequence comparison has been the most powerful method to identify molecular functions of proteins and nucleic acids. At the network level of interacting molecules, because all the

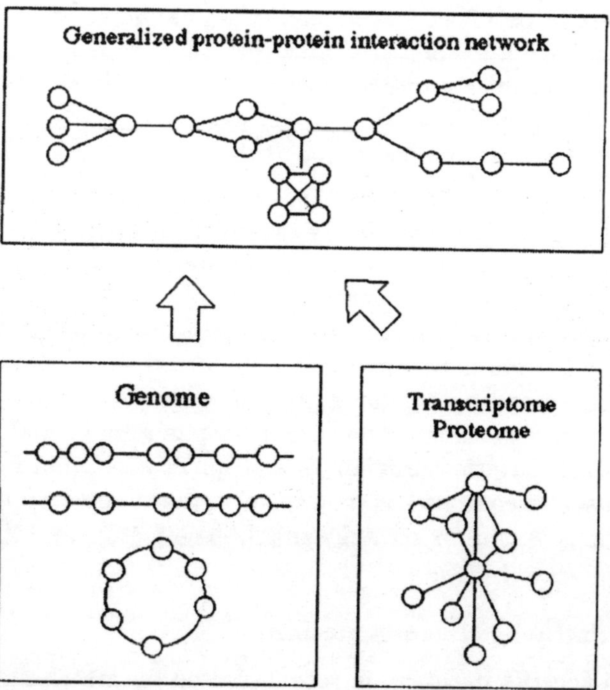

Fig. 3.4. The network prediction is formulated as a conversion of the genome graph with genes as nodes to the network graph with gene products as nodes. The prediction is based on the reference knowledge of similar networks as well as sets of binary relations in transcriptomes and proteomes.

data objects are represented by graphs the graph comparison is bound to become the most powerful method to understand higher order functions. We have developed a heuristic graph comparison algorithm to detect certain graph similarities called correlated clusters. In contrast to the standard notion of graph similarity, or graph isomorphism, this algorithm detects loose similarities that are biologically more relevant by allowing gaps and mismatches. A cluster is a set of nodes that are closely positioned in a graph.

A correlated cluster is a set of clusters in two or more graphs whose nodes are correlated by certain relations. For example, when comparing the genome graph and the pathway graph, the correspondence of nodes is given by the relation of genes to gene products. The resulting correlated cluster will be a set of genes that are adjacent in the genome and whose protein products are functioning at close positions in the pathway, such as a specific pathway coded by an operon. When comparing the genome graph of one organism to the genome graph of another organism, the correspondence may be given by the amino acid sequence similarity. The resulting cluster may then be a conserved operon consisting of similar genes. Note that the order of individual genes in each operon does not have to be conserved, which is the essence of loose similarity.

Genomic Information

The network prediction in *KEGG* is to compute the generalized protein-protein interaction network, or the network of gene products, from the catalog of genes in the genome. The prediction is based on the reference knowledge of real networks in the *PATHWAY* database and additional information of transcriptomes and proteomes in the *EXPRESSION* and *BRITE* databases. The problem can be viewed as a conversion of the genome graph to the network graph by integrating additional graphs of transcriptomes, proteomes, and similar networks. Thus, the graph comparison is an essential feature to integrate different information represented by different types of graphs. Although we do not yet have a fully automated method, the current *KEGG* databases and computational tools can be utilized for network prediction.

A general strategy is to first generate cores of known networks according to the knowledge in the *KEGG* reference pathways and then to extend the cores by searching for additional partners that are associated in the genome (*e.g.*, genes in the same operon), the transcriptome (*e.g.*, coexpressed genes), and the proteome (*e.g.*, binding partners). The first step is called pathway reconstruction, which is basically the matching of genes in the genome and gene products in the pathway. To enable this matching, the genes in the genome must be assigned the ortholog identifiers according to, for example, sequence similarities and positional correlations of genes in other genomes. The second step is a more ambitious step, which can be formulated as a path computation problem in a graph or a set of binary relations. The path computation has been used to compute alternative enzymatic reaction pathways from a set of substrate-product relations. A similar strategy should be effective and it is being implemented in the *BRITE database.*

Gene Annotation by Pathway Reconstruction

When an organism-specific pathway is reconstructed by matching genes in the genome against *KEGG* reference pathways, a few genes are often missing in an otherwise complete pathway. Most of the cases can be solved by reexamining gene annotations and assignments of ortholog identifiers. The information about pathways and complexes imposes an additional constraint of completeness, which is extremely useful for interpreting sequence similarity scores, especially when many paralogs are present, because in general there is no predefined level of sequence similarity that can safely be extended to functional identity. A case in point is the lysine biosynthesis pathway, in which an aminotransferase gene was missing.

Here each box is an enzyme (gene product) with the EC number inside and the shading indicates that the corresponding gene is present in the genome. This pathway was biochemically determined in *E. coli,* and the gene names were assigned by genetic studies, as indicated alongside the boxes. However, when the complete genome sequences were determined, the *dapC* gene for succinyldiaminopimelate aminotransferase (EC 2.6.1.17) could not be found in *E. coli* or any other genomes. Recently, it was reported that *N*-acetylornithine aminotransferase (EC 2.6.1.11) in *E. coli,* which is encoded in *argD* and which functions in the arginine biosynthesis pathway, had a dual role of catalyzing the reaction by *DapC* as well. Furthermore, *dapC* was found as part of the operon encoding *dapCDE* in *Bordetella pertussis*. Aminotransferases form a family of paralogous proteins.

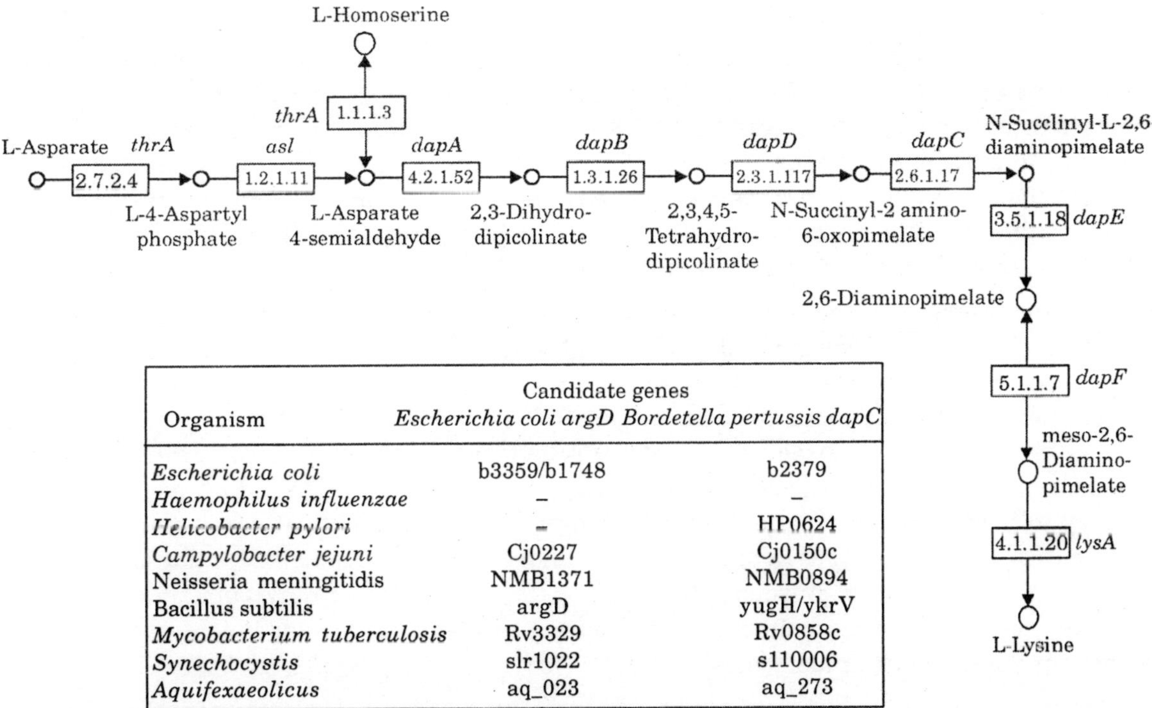

Organism	Candidate genes *Escherichia coli argD*	*Bordetella pertussis dapC*
Escherichia coli	b3359/b1748	b2379
Haemophilus influenzae	–	–
Helicobacter pylori	–	HP0624
Campylobacter jejuni	Cj0227	Cj0150c
Neisseria meningitidis	NMB1371	NMB0894
Bacillus subtilis	argD	yugH/ykrV
Mycobacterium tuberculosis	Rv3329	Rv0858c
Synechocystis	slr1022	s110006
Aquifexaeolicus	aq_023	aq_273

Fig. 3.5. The aminotransferase gene dapC in the lysine biosynthesis pathway had not been found in any of the completely sequenced genomes, but two recent reports found probable genes for this missing enzyme. The enclosed table summarizes homologs for these genes in different genomes.

It is impossible to predict substrate specificity from sequence similarity alone because the number of paralogs is different in different genomes and some aminotransferases must have dual roles. For example, aspartate aminotransferase (EC 2.6.1.1) and tyrosine aminotransferase (EC 2.6.1.5) are encoded by different genes with high sequence similarity in *E. coli,* but there is apparently no tyrosine aminotransferase gene in *Haemophilus influenzae,* and aspartate aminotransferase appears to function in the tyrosine pathway as well. In most genomes there are unassigned aminotransferases, especially those similar to aspartate aminotransferases, and *B. pertussis dapC* belongs to the aspartate aminotransferase subfamily. We have searched homologs of *E. coli argD* and *B. pertussis dapC* in the genomes of other organisms. The result shown in figure 12.5 has identified a homolog of *B. pertussis dapC* in *E. coli,* b1748, which is annotated as putative aminotransferase in the original database. It would be interesting to see if this gene product does have the *DapC* activity. Although the new findings did fill in the gaps in many genomes, some genomes are still unaccounted for, such as *H. influenzae.* It is still possible that aspartate aminotransferase or its paralog takes care of the lysine pathway as well in some genomes.

CONCLUDING REMARKS

For all the genomes that have been sequenced, there is a considerable number of genes whose functions are not yet understood. The fraction of unknown genes varies in the genomes and also depends on the definition of function. As we have seen, the assignment of a general molecular function like an aminotransferase, a kinase, or an ABC transporter, does not tell much about a specific role of the gene in a cellular function. When this is considered, perhaps one-half to two-thirds of the genes are still unknown in most genomes. There have been attempts to systematically uncover functions of unknown genes in functional genomics experiments. Although such experiments may be useful to obtain a rough draft of gene functions, they are unlikely to provide detailed pictures of molecular interactions and pathways that are responsible for specific cellular functions.

It is necessary to integrate with more accurate, traditional methods in biochemistry, molecular and cellular biology, and genetics. The *KEGG* resource should be useful for this integration. We have limited our discussions to the level of molecular interactions, but there are still more issues concerning the association of genes and higher level biological phenomena, such as brain functions, diseases, and human behaviors, where cellular interactions and organism interactions must play more dominant roles. Although *KEGC* does not attempt to move up to such higher level phenomena, it contains information about, for example, disease classification and cell lineages in order to better understand underlying molecular phenomena. The wiring information represented in *KEGG* pathway diagrams may appear to be static. However, there are two mechanisms to incorporate time- and space-dependent . behaviors of the network of interacting molecules.

One is the coloring mechanism used to generate organism-specific pathways. For example, the coloring of microarray hybridization data can be mapped onto the *KEGG* reference pathways and the time-course of gene expression changes can be followed by the changes in the coloring. The other mechanism is to simply draw additional pathway diagrams, each of which is considered to represent a snapshot of the dynamic change. *KEGG* is not suitable for simulating continuous behaviors of the cell because it does not contain any kinetic parameters. However, we still hope that *KEGG* will become useful to simulate perturbations to the cell, such as gene mutations and environmental changes, and their dynamic consequences.

4

Functional Genomics

Molecular biology has addressed functional questions by studying individual genes, either independently or a few at a time. Although it constituted a reductionistic approach, it was extremely successful in assigning functional properties and biological roles to genes and gene products. The recent possibility of obtaining information on thousands of genes or proteins in one sole experiment, thanks to high-throughput methodologies such as gene expression or proteomics has opened up new possibilities in querying living systems at the genome level that are beyond the old paradigm *'one gene-one postdoc'*. Relevant biological questions regarding gene or gene product interactions or biological processes played by networks of components, etc., can now for the first time be addressed realistically. Nevertheless, genomic technologies are at the same time generating new challenges for data analysis and demand a drastic change in the habits of data management.

Dealing with this overabundance of data must be approached cautiously because of the high occurrence of spurious associations if the proper methodologies are not used. Traditional molecular biology approaches tended to mix up the concepts of data and information. This was partially due to the fact that researchers had a great deal of information previously available about the typical data units they used (genes, proteins etc.).

Over the last few years the increasing availability of high-throughput methodologies has amplified by orders of magnitude the potential of data production. One direct consequence of this revolution in data production has been to clarify how fictitious the equivalence between data and information actually used to be. System biology approaches emerge then to convert the flood of data into information and knowledge. There are, however, several problems related to massive data management. One of them is the lack of accurate functional annotations for a considerable number of genes.

Another non-negligible difficulty stems from the fact that, even in the instance of availability of proper functional annotations, processing all the information corresponding to thousands of genes involved in a high-throughput experiment is beyond the human capabilities. Automatic processing of the information therefore becomes indispensable to draw out the biological significance behind the results in this type of experiment. As previously mentioned, the occurrence of false or spurious associations is common when dealing with thousands of elements. Unfortunately, these spurious associations are often considered as evidence of actual functional links, leading to misinterpretation of results. All these features of genomic data must be taken into account for any procedure aiming to properly identify functional roles in groups of genes with a particular experimental behaviour.

CURATED REPOSITORIES

Any approach using biological information for functional annotation purposes uses two main sources: free text or *curated repositories*. The use of techniques of automatic management of biological information to study the coherence of gene groups obtained from different methodologies has been addressed in recent years. Considerable effort has been focused on developing automatic procedures for extracting information from biomedical literature. Information extraction and text mining techniques in particular have been applied to the analysis of gene expression data. It has been claimed that free text processing, essentially using PubMed abstracts as a source of information, has the advantage of providing numerous gene-to-abstract correspondences. Nevertheless, text-mining methodologies still present many drawbacks, such as problems of interpreting terms due to the context of the sentence in which the gene is cited; the lack of a standardized nomenclature of genes in literature that makes it difficult to find all the citations for all the synonyms used for them; a profuse use of acronyms in literature that makes it hard to find accurate citations of genes (for example, for the term STC, corresponding to the gene *secretin,* 158 citations were found in the year 2001, 121 of them corresponded to stem cell transplantation and the rest to other concepts such as spiral computerized tomography, solid cystic tumour etc.; only one of them was a real reference to the gene *secretin*).

There are problems surrounding orthography and boundaries for identifying gene names and finally there are irrelevant terms related to non-functional features that appear, nonetheless, to be associated to gene names. On the other side of the spectra are the repositories with curated functional information, which contain fewer gene-to-term correspondences although these are reliable, consistent and standardized. There are diverse repositories such as pathway databases, among which the *KEGG* database is the paradigm, protein interaction databases protein motif databases. The most valuabe resource is most probably the Gene Ontology (GO) database of curated definitions and the annotations based on GO.

Diverse genome initiatives and databases are annotating genes according to GO terms constituting a priceless resource for information-mining implementations. Although some direct applications of free-text mining to data analysis have been proposed, the future of the practical application of these technologies probably resides in its use by information repository curators to help in the annotation process. Methods for predicting GO categories from the analysis of biomedical literature have therefore been proposed, and there are also similar methods based on the study of different biochemical and physical protein features.

GENE ONTOLOGY

Applied ontologies are centred around a specific domain of knowledge. These ontologies endeavour to represent a system of categories accounting for a particular vision of a given area, in order to establish rules that describe relationships between these categories, and to instantiate the objects in the categories. In practical terms, these ontologies provide an organizational framework of concepts about biological entities and processes in a hierarchical system in which, associative relations which provide reasoning behind biological knowledge, are included. One of the most powerful features of an ontology is the implementation of a controlled, unambiguous vocabulary. This is extremely useful in an inherently complex and heterogeneous discipline such as biology, where a great deal of sophisticated knowledge, in most cases of a hierarchical nature, needs to be integrated with molecular data. The most important ontologies in the domain of biology are included under the umbrella of the Open Biological Ontologies (OBO) initiative, which constitutes a *de facto* standard for them. Some known examples are the Unified Medical Language System which implements a hierarchy of medical terms used in the indexation of PubMed, or the Microarray Gene Expression Data Society (MGED) Ontology (http://mged.sourceforge.net/), recently popularized by the increasing production of gene expression data with microarrays, etc. Nevertheless, the most relevant ontology in the area of functional genomics is, undoubtedly.

The Gene Ontology which provides a controlled vocabulary for the description of molecular function, biological process and cellular component of gene products. Go terms are used as attributes of gene products by collaborating databases, facilitating uniform queries across them. Because of the existing homologies between proteins among different taxa, GO terms can be thoroughly used across species. The controlled vocabularies of terms are structured in a hierarchical manner that allows for both attribution and querying at different levels of granularity.

This hierarchical structure constitutes the representation of the ontology within which each term is a node of a directed acyclic graph (DAG), which is very similar to a tree - the only difference being that in a DAG it is possible for a node to have more than one parent. The deeper a node is in the hierarchy, the more detailed the description of the term. In GO, child to parent relationships can be of two types: 'is a', meaning the child is an instance of the parent (*e.g. chloroplast envelope GO:0009941* is a *membrane GO:0016020)* and 'part of, when the child is a component of the parent (*e.g. inner membrane GO:0019866* and *outer membrane GO:OO19867* are part of *membrane GO:0016020*). The success of an ontology relies largely upon the approval received from the scientific community.

The most important achievement of GO is perhaps that the GO consortium has been able to attract a large number of collaborating databases which are actively mapping gene products onto GO terms. These databases with controlled and curated annotations, which can easily be queried by computers, constitute an invaluable though not yet fully exploited resource, for the scientific community. Additional information on the quality of the annotation of gene products is provided by the collaborative databases through the evidence codes. The codes represent different types of evidence used in the annotation. The highest quality codes are for GO-gene correspondences supported by experimental functional assays (IDA, IMP codes) and the lowest quality corresponds to correspondences inferred from electronic annotations (IEA code). As previously mentioned, the representation of GO resides in its hierarchy. There are different tools available that are useful to browse this hierarchy. Such GO browsers allow the viewing of all gene products annotated with a given GO term, or searching for a gene product and view all its associations. In addition, by browsing the ontologies it is also possible to view relationships between terms.

FUNCTIONAL GENOMIC EXPERIMENTS

Functional genomics experiments allow the scaling of the classical functional experiments to a genomic level. Comparison of phenotypes (*e.g.* patients versus controls, studies of different clinical outcomes etc.) by means of techniques such as DNA microarrays or proteomics provides insight into their molecular basis. Nevertheless, the data obtained in these experiments are measurements of the gene or protein expression levels. To translate this data into information numerical analyses are firstly required to determine which genes (among the thousands analysed) can be considered as significantly related to the phenotypes. The second step is to interpret roles played by the targeted genes.

The availability of GO annotations for a considerable number of genes helps interpret these results from a biological point of view. The rationale commonly used is as follows: if some genes have been found to be differentially expressed when comparing two different phenotypes (or are correlated to a given continuous phenotypic trait, or to survival etc.) it is because the roles they play at molecular level account (to some extent) for the phenotypes analysed. The GO annotations available for the genes that present the same asymmetrical distribution or correlation serve as a more or less detailed description of these biological roles. For example, if 50 genes from an array of 6500 genes are differentially expressed and 40 of them (80 per cent - a high proportion) are annotated as response to 'external stimulus' (G0:0009605), it is intuitive to conclude that this process must be related to the phenotypes studied.

In addition, if the background distribution of this type of gene in the genome is, let us say, of four percent, one can conclude that most of the genes related to 'external stimulus' have been altered in their expression levels in the experiment. There are many tools listed on the GO consortium web page that extract lists of GO terms differentially represented when comparing two sets of genes and, in some cases, provide scores or even individual tests for comparisons between two sets of genes. For example, GoMiner, MAPPFinder GFINDer or eGON just to cite a few, generate tables that correlate groups of genes to biochemical or molecular functions or GO terms.

Some of them are specific for organisms, such as FunSpec which evaluates groups of yeast genes in terms of their annotations in diverse databases, or CLENCH, for *A. thaliana* Nevertheless, differences in the distribution of GO terms between groups must, in addition to being spectacular (which is quite a subjective concept), also be significant (which is an objective statistical concept related to the probability of drawing one's observations purely by chance).

STATISTICAL APPROACHES

As previously mentioned, much caution should be adopted when dealing with a large set of data because of the high occurrence of spurious associations Table 4.1 has been constructed using ten datasets obtained by the random sampling of 50 genes from the complete genome of *Saccharomyces cerevisiae*. For each random set, the proportions of all the GO terms (at GO level 4) have been compared between the two partitions (50 genes with respect to the remaining ones), and the GO term showing the most extreme differential distribution was displayed in each case (rows of the table).

The first column shows the percentage of genes annotated with the GO term in the random partition of 50 genes, the second column represents the corresponding percentage in the rest of the genome and the third column shows the *p*-value obtained upon the application of a Fisher exact test for 2 × 2 contingency tables.

Table 4.1. GO terms found to be differentially distributed when comparing 10 independent random partitions of 50 genes sampled from the complete genome of yeast.

% in random set	*% in genome*	*p-value*	*adjusted p-value*	*GO term*
8.33	1.86	0.0752	1	ion homeostasis (GO:0050801)
10.00	31.34	0.0096	0.6735	nucleobase, nucleoside, nucleotide and nucleic acid metabolism (GO:0006139)
3.33	0.24	0.075	1	one-carbon compound metabolism (G0:0006730)
4.04	8.00	0.0177	0.6599	energy pathways (G0:0006091)
3.45	0.22	0.0669	1	metabolic compound salvage (GO:0043094)
5.88	0.67	0.024	1	vesicle fusion (GO:0006906)
6.45	1.60	0.09	1	negative regulation of gene expression, epigenetic (GO:0045814)
13.79	3.97	0.028	1	response to external stimulus (G0:0009605)
16.13	4.23	0.0097	1	response to endogenous stimulus (G0:0009719)
2.70	0.13	0.054	1	host-pathogen interaction (G0:0030383)

For many people it still seems staggering that most of the random partitions present asymmetrical distributions of GO terms with significant individual p-values.' This apparent paradox stems from the fact that we are not conducting a single test in each partition, but as many tests as GO terms are being checked (several hundreds). Nevertheless, in this situation the researcher tends to forget about the many hypotheses rejected and only focus on the term for which an apparent asymmetrical distribution was found. In some cases this situation is caused by the way in which some of the above mentioned programs work.

To some extent the fact that many tests are really being conducted is hidden to the user and the result is presented as if it were the case of a unique test. If we conduct several hundreds of tests simultaneously, the probability of finding an apparently asymmetrical distribution for a given GO term increases enormously.

A very simple example can be used here to illustrate this concept: let us imagine you flip a coin 10 times and you get 10 heads. You would certainly suspect that something was wrong with the coin. If the same operation were repeated with 10000 different coins one or even several occurrences of 10 heads would not be considered surprising. We intuitively accept this because of the probability of having an unexpected result just by chance is high. If we were interested in checking whether an observation is significantly different from what we could expect simply by chance in a multiple testing situation then the proper correction must be applied. The fourth column of Table 4. 1 shows an adjusted *p*-value using one of the most popular multiple-testing corrections, the false discovery rate and it is obvious that none of the situations depicted in columns 1 and 2 can be attributed to anything other than random occurrence. Table 4.1 shows how random partitions, for which no functional enrichment should be expected, yield apparent enrichments in GO terms because the most asymmetrically distributed GO terms among several hundreds are chosen *a posteriori*. These values occur simply by chance and cannot be considered as either biologically authentic or statistically significant. This clearly shows, beyond any doubt, that multiple testing adjustment must be used if several hypotheses are simultaneously tested. Multiple testing has been addressed in different ways depending on particular cases and the number of simultaneous hypotheses tested.

Thus, corrections such as Bonferroni or Sidak are of very simple application but are too conservative if the number of simultaneous tests is high. Another family of methods that allow less conservative adjustments is the family wise error rate (*FWER*) that controls the probability that one or more of the rejected hypotheses (GOterms whose differences cannot be attributed to chance) is true (that is, a false positive). The minP step-down method a permutation-based algorithm, provides a strong control (*i.e.* under any mix of false and true null hypothesis) of the *FWER*. Approaches that control the *FWER* can be used in this context although they are dependent on the number of hypotheses tested and tend to be too conservative for a high number of simultaneous tests. Aside from a few cases in which FWER control could be necessary, the multiplicity problem in prospective functional assignation does not require protection against even a single false positive. In this case, the drastic loss of power involved in such protection is not justified.

It would be more appropriate to control the proportion of errors among the identified GO terms whose differences among groups of genes cannot be attributed to chance instead. The expectation of this proportion is the false discovery rate (FDR). Different procedures offer strong control of the FDR under independence and some specific types of positive dependence of the test statistics or under arbitrary dependency of test statistics. We have shown how important multiplicity issues are in finding functional associations to clusters of genes. Any procedure that does not take this into account is as a consequence considering a high number of spurious relationships as reliable.

CLUSTERS OF GENES

The *FatiGO* (Fast Assignment and Transference of Information using GO tool was the first application for finding significant differences in the distribution of GO terms between groups of genes taking the multiple testing nature of the contrast into account FatiGO takes two lists of genes (ideally a group of interest and the rest of the genome, although any two groups formed in any way can be tested against each other) and convert them into two lists of GO terms using the corresponding gene-GO association table. Since distinct genes are annotated with more or less detail at the different levels of the hierarchy, it is meaningless to test for different terms that are really descriptions in different detail of the same functional property (*e.g.*, why test apoptosis versus regulation of apoptosis?). To deal with this, FatiGO implements the 'inclusive analysis', in which a level in the DAG hierarchy is chosen for the analysis. Genes annotated with terms that are descendant of the parent term corresponding to the level chosen therefore take the annotation from the parent.

Figure 4.1 illustrates this procedure. If apoptosis node level is chosen for the analysis, eight genes, annotated in descendant nodes, will be assigned to the term apoptosis. If inclusive analysis is not used, then four terms: apoptosis (with two genes), regulation of apoptosis (three), negative regulation of apoptosis (one) and induction of apoptosis (two) are taken into account, with the obvious decrease in the power of the test. A Fisher exact test tor 2 × 2 contingency tables is used. For each GO term the data are represented as a 2 × 2 contingency table with the rows being presence/absence of the GO term, and each column representing each of the two clusters (so that the numbers in each cell would be the number of genes of the first cluster where the GO term is present, the number of genes in the first cluster where the GO term is absent, and so on). In addition to the unadjusted p-values (which are given just because they are obtained as part of the process, but should not be considered as evidence of significant differential distribution of GO terms between clusters), FatiGO returns adjusted *p*-values based on three different ways of accounting for multiple testing: FDR under independence or under arbitrary dependency of test statistics as well as FWER control by the minP step-down method. Results are arranged by p-value to facilitate the identification of GO terms with a significant asymmetrical distribution between the groups of genes studied.

GO:0003673 : Gene Ontology (130309)
GO:0008150 : biological process (78842)
GO:0009987 : cellular process (29557)
GO:0008219 : cell death (1398)
GO:0012501 : programmed cell death (1256)
GO:0006915 : apoptosis (1189)
GO:0008632 : apoptotic program (148)
GO:0045476 : nurse cell apoptosis (16)
GO:0006924 : programmed cell death, activated T-cells (2)
GO:0001783 : programmed cell death, B-cells (0)
GO:0006925 : programmed cell death, inflammatory cells (0)
GO:0006927 : programmed cell death, transformed cells (3)
GO:0006926 : programmed cell death, virus-infected cells (1)
GO:0042981 : regulation of apoptosis (688)
GO:0043066 : negative regulation of apoptosis (323)
GO:0006916 : anti-apoptosis (322)
GO:0001719 : inhibition of caspase activation (17)
GO:0045767 : regulation of anti-apoptosis (30)
GO:0045884 : regulation of survival gene product activity (9)
GO:0019050 : viral inhibition of apoptosis (0)
GO:0043065 : positive regulation of apoptosis (368)
GO:0006917 : induction of apoptosis (368)
GO:0008624 : induction of apoptosis by extracellular signals (64)
GO:0008620 : induction of apoptosis by intracellular signals (40)
GO:0019051 : induction of apoptosis by virus (1)
Level chosen for inclusive analysis
BCL2-like
plasmtnogen
Caspase 6
Caspase 7
Caspase 8
Interleukin 9
Caspase 12

Fig. 4.1. Representation of the inclusive analysis concept.

OTHER TOOLS

Recently, other tools have included some multiple-testing possibilities. For example, the latest versions of Onto-Express include Bonferroni and Sidak corrections as well as a permutation test, or GeneMerge which implements Bonferroni correction. New tools such as FunAssociate include unspecified permutation tests, although others include more established multiple testing controls such as FDR, which is the case of GO Surfer or GOStat which has exactly the same functionalities as FatiGO.

Examples

As previously mentioned, a research scientist is continuously interested in understanding the molecular roles played by potentially relevant genes in a given experiment. One of the most popular hypotheses in microarray data analysis is that coexpression of genes across a given series of experiments is most probably explained through some common functional role. Actually, this causal relationship has been used to predict gene function from patterns of coexpression. Here is an example using the data from DeRisi, lyer and Brown (1997), in which they analyse the complete genome of *Saccharomyces cerevisiae* to carry out a comprehensive study of the temporal programme of gene expression accompanying the metabolic shift from fermentation to respiration.

With the aim of finding groups of genes that coexpress across the seven time points measured, gene expression patterns were clustered using the SOTA algorithm as implemented in the GEPAS suite of web tools. The clusters of genes obtained. The parameters used were coefficient of correlation as distance measure and the growth was stopped at 95 percent of variability. The cluster with 21 genes that are initially active and suffer a late repression was analysed with FatiGO. Seventy-five percent of these genes were annotated as biosynthesis, and the differences in *j* proportion with respect to the background (30 per cent) were clearly significant.

It can be claimed that genes with the described temporal behaviour are involved in the biosynthesis biological process. In the event of not performing p-value adjustment, another three processes (sexual reproduction, conjugation and aromatic compound metabolism) would have been considered as important despite the differences in the proportions between the cluster and the rest of genes that can occur simply by chance. Genes showing significant differential expression when comparing two or more phenotypes, or genes significantly correlated to a trait (*e.g.* the level of a metabolite) or to survival, can be analysed in the same way. Comparison of distributions of GO terms helps to understand what makes these genes different from the rest.

CONCLUDING REMARKS

The importance of using biological information as an instrument to understand the biological roles played by genes targeted in functional genomics experiments has been highlighted in this chapter. There are situations in which the existence of noise and/or the weakness of the signal hamper the detection of real inductions or repressions of genes. Improvements in methodologies of data analysis, dealing exclusively with expression values, can to some extent help. Recently, the idea of using biological knowledge as part of the analysis process is gaining in support and popularity. The rationale is similar to the justification of using biological information to understand the biological roles of differentially expressed genes. What differs here is that genes are no longer the units of interest, but groups of genes with a common function. Let us consider a list of genes arranged according their degree of differential expression between two conditions (*e.g.* patients versus controls). If a given biological process is accounting for the observed phenotypic differences we should then expect to find most genes involved in this process overexpressed in one of the conditions against the other.

In contrast if the process has nothing to do with the phenotypes, the genes will be randomly distributed amongst both classes (for example if genes account for physiological functions unrelated to the disease studied, they will be active or inactive in both patients and controls). Diaz-Uriarte, Al-Shahrour and Dopazo (2003) proposed the use of a sliding window across the list of genes to compare the distribution of GO terms corresponding to genes within the window against genes outside the window. If terms (but not necessarily individual genes) were found differentially represented in the extremes of the list, one could conclude that these biological processes are significantly related to the phenotypes. Al-Shahrour *et al.* (2003) generalized this approach to other types of arrangement based on other types of experiment. Recently, Mootha *et. al.* (2003) proposed a different statistic with the same goal. This is part of a more general question, which would be the study of differences on prespecified groups of genes. Different creative uses of information in the gene selection process as well as the availability of more detailed annotations will enhance our capability of translating experimental results into biological knowledge.

5

Microarrays

Technologies for generating high-density arrays of cDNAs and oligonucleotides are developing rapidly, changing the landscape of biological and biomedical research. They enable, for the first time, a global, simultaneous view of the transcription levels of many thousands of genes, when the cell undergoes specific conditions or processes. For several organisms that have had their genomes completely sequenced, the full set of genes can already be monitored this way today. The potential of such technologies is tremendous .'The information obtained by monitoring gene expression levels in different developmental stages, tissue types, clinical conditions, and different organisms can help the understanding of gene function and gene networks, and assist in the diagnostic of disease conditions and effects of medical treatments. Undoubtedly, other applications will emerge in coming years. A key step in the analysis of gene expression data is the identification of groups of genes that manifest similar expression patterns.

The translates to the algorithmic problem of clustering gene expression data. A clustering problem consists of elements and a characteristic vector for each element. A measure of similarity is defined between pairs of such vectors. (In gene expression, elements are usually genes, the vector of each gene contains its expression levels under each of the monitored conditions, and similarity can be measured, for example, by the correlation coefficient between vectors). The goal is to partition the elements into subsets, which are called *clusters,* so that two criteria are satisfied: *homogeneity*—elements in the same cluster are highly similar to each other; and *separation*—elements from different clusters have low similarity to each other.

BIOLOGICAL RESUME

In this section we outline three technologies that generate large-scale gene expression data. All three are based on performing of a large number of hybridization experiments in parallel on high-density arrays (*a.k.a.* "DNA chips") between probes and targets. They differ in the nature of the probes and the targets and in other technological aspects, which raise different computational issues in analyzing the data. For more on the technologies and their applications see, for example, Marshall and Hodgson (1998), Ramsay (1998), Eisen and Brown (1999), Chipping (1999), Lockhart and Winzeler (2000).

cDNA Microarrays

cDNA microarrays are microscopic arrays that contain large sets of cDNA sequences immobilized on a solid substrate. In an array experiment, many gene-specific cDNAs are spotted on a single matrix. The matrix is then simultaneously probed with fluo-rescently tagged cDNA representations of total RNA pools from test and reference cells, allowing one to determine the relative amount of transcript present in the pool by the type of fluorescent signal generated. Current technology can generate arrays with over ten thousand cDNAs per square centimeter.

cDNA microarrays are produced by spotting PCR products of approximately 0.6-2.4 KB, representing specific genes onto a matrix. The spotted cDNAs are usually chosen from appropriate databases, such as GenBank (Benson *et. al.*, 1999) and UniGene.

Additionally, cDNAs from any library of interest (whose sequences may be known or unknown) can be used. Each array element is generated by the deposition of a few nanoliters of purified PCR product. Printing is carried out by a robot that spots a sample of each gene product onto a number of matrices in a serial operation. To maximize the reliability and precision with which quantitative differences in the abundance of each RNA species are detected, one directly compares two samples by labeling them with spectrally distinct fluorescent dyes and mixing the two probes for simultaneous hybridization to one array. The relative representation of a gene in the two samples is assayed by measuring the ratio of the (normalized) fluorescent intensities of the two dyes at the target element. Cy3-dUTP and Cy5-dUTP are frequently used as the fluorescent labels. For the comparison of multiple samples, such as in time-course experiments, one often uses the same reference sample with each of the test samples.

Oligonucleotide Microarrays

In oligonucleotide microarrays each spot on the array contains a short synthetic oligonucleotide (oligo), typically 20-30 bases long. The oligos are designed based on the knowledge of the DNA (or EST) target sequences, to ensure high affinity and specificity of each oligo to a particular target gene.

Moreover, they should not be near-complementary to other RNAs that may be highly abundant in the sample (*e.g.*, rRNAs, tRNAs, alu-like sequences etc. One of the leading approaches to the construction of high-density DNA probe arrays employs photolithography and solid-phase DNA synthesis. First, synthetic linkers, modified with a photochemically removable protecting groups are attached to a glass substrate. At each phase, light is directed through a photolithographic mask to specific areas on the surface to produce localized deprotection. Specific Hydroxyl protected deoxynucleosides are incubated with the surface, and chemical coupling occurs at those sites that have been illuminated.

Current technology allows for approximately 300,000 oligos to be synthesized on a 1.28 × 1.28 cm array. Key to this approach is the use of multiple distinct oligonucleotides designed to hybridize to different regions of the same RNA. This use of multiple detectors greatly improves signal-to-noise ratio and accuracy of RNA quantitation, and reduces the rate of false-positives and miscalls. An additional level of redundancy comes from the use of mismatch control probes that are identical to their perfect match partners except for a single base difference in a central position. These probes act as specificity controls. They allow the direct subtraction of both background and cross-hybridization signals, and allow discrimination between "real" signals and those due to non-specific or semi-specific hybridizations.

Oligonucleotide Fingerprinting

Historically, the Oligonucleotide Fingerprinting (ONF) method preceded the other two. It was initially proposed in the context of sequencing by hybridization, as an alternative to DNA sequencing. Although that approach to sequencing is currently not competitive, ONF has found other good applications, including gene expression. It can be used to extract gene expression information about a cDNA library from a specific tissue under analysis, without prior knowledge of the genes involved. Conceptually, it takes the "reverse" approach to that of the oligo microarrays: The target is on the array, and the oligos are "in the air". The ONF method is based on spotting the cDNAs on high density nylon membranes (about 31,000 different cDNA can be spotted currently in duplicates on one filter.

A large quantity of a short synthetic oligo, typically 7-12 bases long, radioactively labeled, is put in touch with the membrane in proper conditions. The oligos hybridize to those cDNAs that contain a DNA sequence complementary to that of the oligo. By inspecting the filter one can detect to which of the cDNAs the oligo hybridized. Hence, ideally, the result of such an experiment is one 1/0 bit for each of the cDNAs. The experiment is repeated with p different oligos, giving rise to a p-long vector for each cDNA spot, indicating which of the (complements of) oligo sequences are contained in each cDNA. This *fingerprint* vector, similar to a barcode, identifies the cDNA. Thus, distinct spots of cDNAs originating from the same gene should have similar fingerprints. By clustering these fingerprints, one can identify cDNAs originating from the same gene. The larger that number, the higher the expression level of the corresponding gene.

Gene identification can subsequently be obtained by sample sequencing or by comparison of average cluster fingerprints to a sequence database. Because of the short oligos used, the hybridization information is rather noisy, but this can be compensated by using a longer fingerprint. The method is probably less efficient than the other two methods, which measure abundance directly in a single spot. However, it has the advantage of applicability to species with unknown genomes, which oligo microarrays cannot handle, and it requires relatively lower mRNA quantities than cDNA microarrays.

MATHEMATICAL FORMULATIONS AND BACKGROUND

Let $N = \{e_1, ..., e_n\}$ be a set of n elements, and let $\wp = (C_1, ..., C_l)$ be a *partition* of N into subsets. That is, the subsets are disjoint and their union is N. Each subset is called a *cluster*, and $\wp$ is called a *clustering solution*, or simply a *clustering*. Two elements, e_i and e_j, are called *mates with respect to* $\wp$ if they are members of the same cluster in $\wp$. In the gene expression context, the elements are the genes and we often assume that there exists some correct partition of the genes into "true" clusters. When $\wp$ is the true clustering of N, elements that belong to the same true cluster are simply called mates. The input data for a clustering problem is typically given in one of two forms: (1) *Fingerprint data* — each element is associated with a real-valued vector, called its *fingerprint*, or *pattern*, which contains p measurements on the element, such as expression levels of an mRNA at different conditions, or hybridization intensities of a cDNA with different oligos. (2) *Similarity data*—pairwise similarity values between elements.

These values can be computed from fingerprint data, such as by correlation between vectors. Alternatively, the data can represent pairwise dissimilarity, for example, by computing distances. Fingerprints contain more information than similarity, but the latter is completely generic and can be used to represent the input to clustering in any application. (Note that there is also a practical consideration regarding the presentation: the fingerprint matrix is of order $n \times p$, whereas the similarity matrix is of order $n \times n$, and in gene expression applications often $n \gg p$). The goal in a clustering problem is to partition the set of elements N into homogeneous and well-separated clusters.

That is, we require that elements from the same cluster will be highly similar to each other, whereas elements from different clusters will have a low similarity to each other. Note that this formulation does not define a single optimization problem: homogeneity and separation can be defined in various ways, leading to a variety of optimization problems. Note also that even when the homogeneity and separation are precisely defined, those are two objectives that are typically conflicting: the higher the homogeneity, the lower the separation, and vice versa.

The lack of a single objective agreed upon by the community is inherent in the clustering problem, a point we will return to in the sequel. Clustering problems and "algorithms are often represented in graph-theoretic terms. We therefore include some basic definitions on graphs. We refer the readers to Golumbic (1980), and Even (1979) for more background and terminology on graphs.

Let $G = (V, E)$ be a weighted graph. We denote the vertex set of G also by $V(G)$. For a subset $R \subseteq V$, *the subgraph induced by R*, denoted G_R, is obtained from G by deleting all vertices not in R and the edges incident on them. That is, $G_R = (R, E_R)$ where $E_R = \{(i, j) \in E \mid i, j \in R\}$. For a vertex $v \in V$ define the *weight* of v to be the sum of weights of the edges incident on v. A *cut* C in G is a subset of its edges, whose removal disconnects G. The *weight* of C is the sum of the weights of its edges. A *minimum weight cut* is a cut in G with minimum weight. In case of non-negative edge weights, a minimum weight cut C partitions the vertices of G into two disjoint nonempty subsets $A, B \subset V, A \cup B = V$, such that $E \cap \{(u, v) : u \in A, v \in B\} = C$. For a pair of vertices $u, v \in V$, the *distance* between u and v is the length of the shortest path that connects them. The *diameter* of G is the maximum distance between a pair of vertices in G. For an example of these definitions. For a set of elements $K \subseteq N$, we define the *fingerprint* or *centroid* of K to be the mean vector of the fingerprints of the members of K. For two fingerprints x and y, we denote their similarity by $S(x,y)$ and their dissimilarity by $d(x, y)$.

A *similarity graph* is a weighted graph in which vertices correspond to elements and edge weights are derived from the similarity values between the corresponding elements. Hence, the similarity graph is just another representation of the similarity matrix.

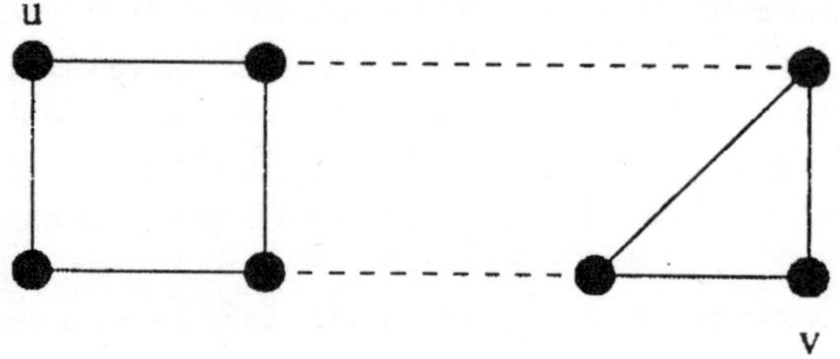

Fig. 5.1. A graph and a corresponding minimum weight cut, assuming that all edge weights are 1. Minimum cut edges are denoted by broken lines. The length of the shortest path between u and v is 3, which is also the diameter of the graph.

An alternative formulation of the clustering problem is hierarchical: rather than asking for a single partition of the elements, one seeks an iterated partition: A *den-dogram* is a rooted weighted tree, with leaves corresponding to elements. Each edge 11 defines the cluster of elements contained in the subtree below that edge. The edge's weight (or length) reflects the dissimilarity between that cluster and the remaining elements.

In this formulation the clustering solution is the dendogram, and each non-singleton cluster, corresponding to a rooted subtree, is split into subclusters. The determination of disjoint clusters is left to the judgment of the user. Typically, one tends to consider as genuine clusters elements of a subtree just below a connecting edge of high weight. Irrespective of the representation of the clustering problem input, judicious preprocessing of the raw data is key to meaningful clustering. This preprocessing is application dependent and must be chosen in view of the expression technology used and the biological questions asked. The goal of preprocessing is to normalize the data and calculate the pairwise element (dis)similarity, if applicable. Common procedures for normalizing fingerprint data include transforming each fingerprint to have mean of 0 and variance of 1, a fixed norm, or a fixed maximum entry. Statistically based methods for data normalization have also been developed recently.

ALGORITHMS

Several algorithmic techniques were previously used in clustering gene expression data, including hierarchical clustering, self-organizing maps, and graph theoretic approaches. We describe these approaches in the sequel. For other approaches to clustering expression patterns, Alon *et. al.*, (1999); Getz et al. (2000b); and Heyer *et. al.*, (1999). Much more information and background on clustering is available. Several algorithms for clustering were developed by first designing a "clean" algorithm that has proven properties, either in terms of time complexity, or in terms of (deterministic or probabilistic) solution quality. Then a more efficient yet heuristic algorithm is developed based on the same idea. We shall describe here the heuristics used in practice, but refer also briefly to the properties of the theoretical algorithm that motivated them.

1. Find a minimal entry d_{i*j*} in D, and merge clusters $i*$ and $j*$.
2. Modify D by deleting rows and columns i, j and adding a new row and column $i* \cup j*$, with their dissimilarities defined by:

$$d_{k, i* \cup j*} = d_{i* \cup j*, k} = a_{i*} d_{kj*} + \gamma |d_{ki*} - d_{kj*}|$$

3. If there is more than one cluster, then go to Step 1.

Fig. 5.2. The agglomerative hierarchical clustering scheme.

Hierarchical Clustering

Hierarchical clustering solutions are typically represented by a dendogram. Algorithms for generating such solutions often work either in a top-down manner, by repeatedly partitioning the set of elements, or in a bottom-up fashion. We shall describe here the latter. Such *agglomerative* hierarchical clustering algorithms are among the oldest and most popular clustering methods. They proceed from an initial partition into singleton clusters by successive merging of clusters until all elements belong to the same cluster. Each merging step corresponds to joining two clusters. The general scheme due to Lance and Williams (1967) is presented in figure 5.2. It is assumed that $D = (d_{ij})$ is the input dissimilarity matrix.

Common variants of this scheme are the following:

- *Single linkage:* $d_{k, i* \cup j*} = \min\{d_{ki*}, d_{kj*}\}$. Here $\alpha_{i*} = \alpha_{j*} = 1/2$ and $\gamma = -1/2$.
- *Complete linkage:* $d_{k, i* \cup j*} = \max\{d_{ki*}, d_{kj*}\}$. Here $\alpha_{i*} = \alpha_{j*} = 1/2$ and $\gamma = 1/2$.

- *Average linkage:* $d_{k,\, i* \cup j*} = n_{i*} d_{ki*}/(n_{i*} + n_{j*}), + n_{j*} d_{kj*}/(n_{i*} + n_{j*})$, where n_i denotes the number of elements in cluster i. Here $\alpha_{i*} = n_{i*}/(n_{i*} + n_{j*})$, $\alpha_{j*} = n_{j*}/(n_{i*} + n_{j*})$, and $\gamma = 0$.

1. Start with an arbitrary partition P of N into k clusters.
2. For each element i and cluster $j \neq P(i)$ let E_P^{ij} be the cost of a solution in which i is moved to cluster j. If $E_P^{i*j*} = min_{ij} E_P^{ij} < E_P$ then move $i*$ to cluster $j*$ and repeat Step 2. Otherwise halt.

Fig. 5.3. The K-means algorithm.

Average linkage: $d_{k,i*\cup j*} = n_{i*}d_{ki*}/(n_{i*}d_{kj*}/(n_{i*}+n_{j*})$, where n_i denotes the number of elements in cluster i. Here $\alpha_{i*} = n_{i*}/(n_{i*} + n_{j*})$, $\alpha_{j*} = n_{j*}\ (n_{i*} + n_{j*})$, and $\gamma = 0$.

Eisen *et. al.*, (1998) developed a clustering software package based on the average-linkage hierarchical clustering algorithm. The software package is called Cluster, and the accompanying visualization program is called Tree View. Both programs are available at http://rana.Stanford.EDU/software/. The gene similarity metric used is a form of correlation coefficient. The algorithm iteratively merges elements whose similarity value is the highest, as explained above. The output of the algorithm is a dendogram and an ordered fingerprint matrix. The rows in the matrix are permuted based on the dendogram, so that groups of genes with similar expression patterns are adjacent. The ordered matrix is represented graphically by coloring each cell according to its content. Cells with log ratios of 0 are colored black, increasingly positive log ratios with reds of increasing intensity, and increasingly negative log ratios with greens of increasing intensity.

K-Means

K-means is another classical clustering algorithm. It assumes that the number of clusters k is known, and aims to minimize the distances between elements and the centroids of their assigned clusters. Let M be the $n \times m$ fingerprint matrix. For a partition P of the elements in $\{1, ..., n\}$, denote by $P(i)$ the cluster assigned to i, and by $c(j)$ the centroid of cluster j. Let $d(v_1, v_2)$ denote the Euclidean distance between the fingerprint vectors v_1 and v_2. K-means tries to find a partition P for which the error-function $E_P = \Sigma_{i=1}^{n} d(i, c(P(i)))$ is minimum.

Each iteration of K-means modifies the current partition by checking all possible modifications of the solution in which one element is moved to another cluster, and making a switch that reduces the error function. Figure 5.3 describes the most basic scheme. A more efficient variant moves in one iteration all elements that would benefit from a move: for each i simultaneously, *if* $min_j\ E_P^{ij} < E_P$, move i to the cluster j minimizing E_P^{ij}. This algorithm is very easy to implement and is used in many applications. Another heuristic inspired by K-means was developed by Herwig *et. al.*, (1999) to cluster cDNA oligo-fingerprints.

1. Start with a set of sufficiently different elements as clusters.
2. For each remaining element i do:
   - For each cluster C s.t. $S(i, C) \geq \rho$ do:
     - — add i to G.
     - — While there exists a cluster C' s.t. $S(C, C') > \gamma$, merge C' into C.
   - If i was not added to any duster then form a new cluster $\{i\}$.
3. Assign each element to the cluster to which it is most similar.

Fig. 5.4. The algorithm of Herwig et. al.,

Unlike the regular K-means algorithm, this algorithm does not require a prespecified number of clusters. Instead, it uses two parameters: γ is the maximal admissible similarity of two distinct clusters, and ρ is the maximal admissible similarity between an element and a cluster different from its own cluster. (Similarity to a cluster is similarity to its centroid.) Elements are handled one at a time, added to sufficiently close clusters, or otherwise, form a new cluster. Whenever centroids become too close, their clusters are merged. Unlike the K-means algorithm, an element may be tentatively assigned to more than one cluster, and thus influence the location of several centroids to which it is sufficiently close. The algorithm is shown in figure 5.4. Here $S(i, C)$ is the similarity between element i and cluster C.

HCS and Click

The HCS and CLICK algorithms use a similar graph theoretic approach to clustering. The input data is represented as a similarity graph. The algorithm recursively partitions the current set of elements into two subsets. Before a partition, the algorithm considers the subgraph induced by the current subset of elements. If the subgraph satisfies a stopping criterion, then it is declared a *kernel*. Otherwise, a minimum weight cut is computed in that subgraph, and the set is split into the two subsets separated by that cut. The output is a list of kernels that serve as a basis for the eventual clusters.

```
Form-Kcrnels(G):
If V(G) = {v} then move v to the singleton set.
Else if G kernel then output V(G)
Else
        (H, H) ←MinWeightCut(G).
        Form-Kernels(H).
        Form-Kernels(H).
```

Fig. 5.5. The basic scheme of HCS and CLICK. Procedure MinWeightCut (G) computes a minimum weight cut of G and returns a partition of G into two subgraphs H and H according to this cut.

HCS and CLICK differ in the similarity graph they construct, their stopping criteria, and the postprocessing of the kernels. We describe each of the algorithms below.

HCS The HCS algorithm (Hartuv et al. 2000; Hartuv and Shamir 1999) builds from the input data an *unweighted* similarity graph G (each edge has weight 1 and each non-edge has weight 0) in which there is an edge between two vertices if and only if the similarity between their corresponding elements exceeds a predefined threshold. The following notion is key to the algorithm: A *highly connected subgraph* (HCS) is an induced subgraph H of G, whose minimum cut value exceeds $|V(H)|2$. That is, H remains connected if any $[|V(H)|/2]$ of its edges are removed. The algorithm identifies highly connected subgraphs as kernels. The HCS algorithm possesses several good properties for clustering. The diameter of each cluster it produces is at most two, and each cluster is at least half as dense as a clique. Both properties indicate strong cluster homogeneity. Inter-cluster separation is harder to prove, but it is argued that if errors are random, any nontrivial set split by the algorithm is unlikely to have diameter two unless the involved sets are small.

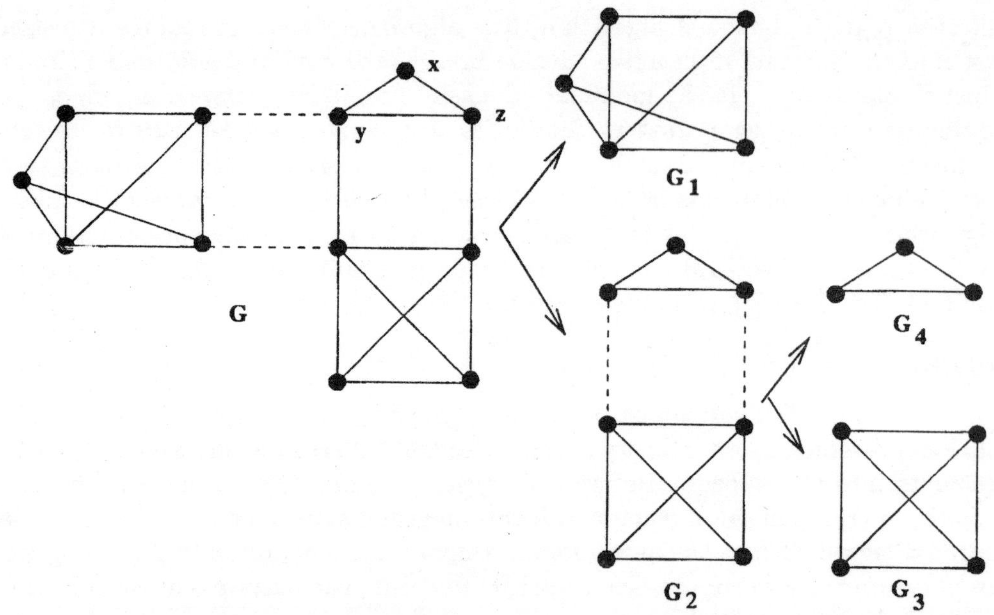

Fig. 5.6. An example of applying the HCS algorithm to a graph. Minimum cut edges are denoted by broken lines.

To improve separation in practice, several heuristics are used to expand the kernels and speed up the algorithm:

Iterated-HCS

When the minimum cut value is obtained by several distinct cuts, the HCS algorithm chooses one arbitrarily. This process may break small clusters into singletons. To overcome this, several (1-5) HCS iterations are carried out until no new cluster is found.

Singletons Adoption

Singletons can be "adopted" by clusters. For each singleton element x we compute the number of neighbors it has in each cluster and in the singletons set y. If the maximum number of neighbors is sufficiently large, and is obtained by one of the clusters (rather than by y) then x is added to that cluster. The process is repeated several times.

Removing Low-degree Vertices

When the similarity graph contains vertices with low degrees, one iteration of the minimum cut algorithm may simply separate a low degree vertex from the rest of the graph. This is computationally very expensive, not informative in terms of the clustering, and may happen many times if the graph is large. Removing low-degree vertices from G eliminates such iterations, and significantly reduces the running time. The process is repeated with several thresholds on the degree. This simple procedure is very powerful for large problems.

Click

The *CLICK* algorithm (CLuster Identification via Connectivity Kernels) (Sharan and Snamir 2000b), available at http://www.math.tau.ac.il/~rshamir/click/ click.html, builds on a statistical model. The model gives probabilistic meaning to edge weights in the similarity graph and to the stopping criterion. The key probabilistic assumption of *CLICK*,is that pairwise similarity values between elements are normally distributed: similarity values between mates are Normally distributed with mean μ_T and variance σ_F^2, and similarity values between nonmates are normally distributed with mean μ_F and variance σ_F^2, where $\mu_T > \mu_F$. This situation often holds on real data, and can be asymptotically justified. The algoritnm uses the values of μ_T, μ_F, σ_T, and σ_F, as well as the probability p_{mates}, that two randomly chosen elements are mates.

These parameters can be com-puted directly from a known solution on a subset of the elements (which is often available in ONF experiments or estimated using the EM algorithm, assuming the above probabilistic model for similarity values.

Let $S = (S_{ij})$ be the input similarity matrix. Form a weighted similarity graph $G = (V, E)$, in which the weight w_{ij} of the edge (i, j) reflects the probability that i and j are mates, and is derived from the normal density function $f(x) = \frac{1}{\sqrt{2\pi\sigma}} e^{-((x-\mu)^2/2\sigma^2)}$ and Bayes theorem:

$$w_{ij} = \text{In} \frac{Prob(i, j \text{ are mates} \mid S_{ij})}{Prob(i, j \text{ are non-mates} \mid S_{ij})}$$

$$= \frac{p_{mates}\sigma_F}{(1 - p_{mates})\sigma_T} + \frac{(S_{ij} - \mu_F)^2}{2\sigma_F^2} - \frac{(S_{ij} - \mu_T)^2}{2\sigma_T^2}$$

CLICK uses the same basic scheme as HCS to form kernels. The current subgraph is determined to be a kernel if the value of a minimum cut in it is positive. This is the case if and only if for every cut C in the current subgraph, the probability that it contains only edges between mates exceeds the probability that C contains only edges between nonmates. The actual implementation omits from the graph all edges with values below some predefined non-negative threshold, computes the minimum cut in that simplified graph, and corrects the solution value for the missing edges.

$R \leftarrow N$.

While some change occurs do:

Form-Kernels(G_R).

Adoption(L, R).

Merge(L),

Adoption(L, R).

Fig. 5.7. The CLICK algorithm. N is the complete set of elements (all the vertices in the similarity graph). Throughout the algorithm, $\mathcal{L}$ is the current list of kernels and R is the set of singletons. G_R is the subgraph of G induced by the vertex set R. Adoption ($\mathcal{L}$, R) performs the iterative singletons adoption procedure. Merge ($\mathcal{L}$) is the iterative merging procedure.

CLICK first produces kernels that form the basis of the eventual clusters. Subsequent processing includes singleton adoption, recursive clustering process on the set of remaining singletons, and an iterative *merging step*. The singletons adoption step is based on computing similarities between singletons' and clusters' fingerprints. The merging step iteratively merges two kernels whose fingerprint similarity is the highest, provided that this similarity exceeds a predefined threshold. The use of the fingerprints (rather than average similarity values) here is very powerful.

Similar ideas were employed in Milosavljevic et al. (1995) and Hartuv et al. (2000). Finally, a last singleton adoption step is performed. In order to reduce the running time of CLICK on very big instances, a screening heuristic is applied, similar to the low-degree heuristic of the HCS algorithm. Low-weight vertices are screened from large components in the following manner: First, the average vertex weight W of the component is computed, and is multiplied by a factor that is proportional to the logarithm of the size of the component. Denote the resulting threshold by $W*$. Then vertices whose weight is below $W*$ are removed repeatedly, each time updating the weight of the remaining vertices, until the updated weight of every (remaining) vertex is greater than $W*$. The removed vertices are marked as singletons and handled at a later stage.

Cast

Ben-Dor et al. (1999) developed a polynomial algorithm for finding true clustering with high probability, under the following stochastic model of the data. The underlying correct cluster structure is represented by a graph that is a disjoint union of cliques, and errors are subsequently introduced in the graph by independently removing and adding edges between pairs of vertices with probability α. If all clusters are of size at least cn, for some constant $c > 0$, the algorithm solves the problem to a desired accuracy with high probability.

While there are unclustered elements do:

Pick an unclustered element to start a new cluster G.

Repeat ADD and REMOVE until no changes occur:

    ADD: add an unclustered element v with maximum affinity to C if $a(v) > t|C|$.

    REMOVE: remove an element u from C with minimum affinity if $a(u) \leq t|C|$.

Add C to the list of final cluster.

Fig. 5.8. The CAST algorithm.

CAST uses as input the similarity matrix S. The *affinity* of an element v to a putative cluster C is $a(v) = \Sigma_{i \in C} S(i, v)$. The polynomial algorithm motivated the use of affinity to develop a faster heuristic called CAST (Clustering Affinity Search Technique), which is implemented in the package BIOCLUST. The algorithm uses a single parameter t. Clusters are generated one by one. The next cluster is started with a single element, and elements are added or removed from the cluster if their relative affinity is larger or lower than t, respectively, until the process stabilizes. An additional heuristic is employed at the end of the algorithm. A series of moving steps aims at a clustering in which the affinity of every element is higher to its assigned cluster than to any other cluster.

Self-organizing Maps

The self-organizing maps were developed by Kohonen (1997) as a method for fitting a number of ordered discrete reference vectors to the distribution of vectorial input samples. A self-organizing map (SOM) assumes that the number of clusters is known. Those clusters are organized as a set of nodes in a hypothetical "elastic network," with a simple neighborhood structure on the nodes, for example, a two-dimensional $k \times l$ grid. Each of these nodes is associated with a reference vector in $\mathcal{R}^n$. In the process of running the algorithm, the input vectors direct the movement of the reference vectors, so that an organization of the input vectors over the network emerges.

Arbitrarily set the reference vectors $f_1(v) \in R^n$ for each node v.

For $i = 1$ until no node location is changed by more than ε do;

Randomly pick a data point p.

Compute the node n_p with reference vector $f(n_p)$ closest to p.

Update all reference vectors: $f_{i+1}(n) = f_i(n) + T(n, n_p, i)\,[p - f_i(n)]$.

Assign each data point to the cluster with the closest reference vector.

Fig. 5.9. The Self Organizing Map algorithm. The learning function $\tau(\cdot)$ *monotonically decreases with* $d(n, n_p)$ *and with the iteration number i.*

In the following we describe the SOM algorithm in the Euclidean space, and use $d(x, y)$ to denote the distance between points x and y.

The SOM process is iterative. Denote by $f_i(n)$ the position of node n at the i-th iteration. The initial positioning f_1 is random. The algorithm iteratively selects a random data point p, identifies the nearest reference node n_p, and updates the reference nodes according to a learning function $\tau(\cdot)$, where nodes closer to n_p are updated more. The updates also decrease with the iteration number. The function $\tau(\cdot)$ represents the "stiffness" of the network. The intuition for this learning process is that the nodes that are close enough to p will "activate" each other to learn something from p.

Two popular choices for the learning function are:

- Neighborhood function: For each node n we denote by $N_i(n)$ the set of nodes within some distance from n. (These distances are with respect to the neighborhood structure in the network.) We then define $\tau(n, n_p, i) = 0$ if $n = \notin N(n_p)$ and $\tau(n, n_p, i) = \alpha(i)$ otherwise, $\alpha(i)$ is called the learning rate, and it decreases with i.
- Gaussian function: $\tau(n, n_p, i) = \alpha(i) \cdot \exp\left(-\frac{d(n, n_p)^2}{2\sigma^2(i)}\right)$, where $\alpha(i)$ and $\sigma(i)$ decrease with i.

Tamayo *et. al.*, (1999) devised a gene expression clustering software, GeneCluster, which uses the SOM algorithm. The software is available at http://waldo.wi.mit.edu/ MPR/. In their implementation, they incorporated a neighborhood learning function, for which $\alpha(i) = 0.02T/(T + 100i)$, where T is the maximum number of iterations; and $N_i(n_p)$ contains all nodes whose distance to n_p is at most $\rho(i)$, where $\rho(i)$ decreases linearly with i, $\rho(0) = 3$.

GeneCluster accepts an input file of expression levels together with a two-dimensional grid geometry for the nodes. The number of grid points is the prescribed number of clusters. The resulting clusters are visualized by presenting for each cluster its average expression pattern with error-bars. Clusters are presented in their grid order, as clusters of close nodes tend to be similar.

ASSESSMENT OF SOLUTIONS

A key question in the design and analysis of clustering techniques is how to evaluate solutions. We present in this section figures of merit for measuring the quality of a clustering solution. Different measures are applicable in different situations, depending on whether a partial true solution is known or not, and whether the input is fingerprint or similarity data. We describe below some of the applicable measures in each case. For other possible figures of merit, we refer the reader to Everitt (1993), Hansen and Jaumard (1997), and Yeung et al. (2000).

Assessment Given the True Solution

Suppose at first that the true solution is known, and we wish to compare it to a suggested solution. Any clustering solution can be represented by a binary $n \times n$ matrix C, in which $C_{ij} = 1$ if and only if i and j belong to the same cluster in that solution. Let T and C be the matrices for the true solution and the suggested solution, respectively. Let n_{kl}, $k, l = 0, 1$, denote the number of pairs (i, j) $(i \neq j)$ for which $T_{ij} = k$ and $C_{ij} = l$. Thus, n_{11} is the number of true mates that are also mates in the suggested solution, n_{00} is the number of nonmates correctly identified as such, whereas n_{01} and n_{10} count the disagreements between the true solution and the suggested one.

The *Minkowski measure* (see, *e.g.*, Sokal 1977) is defined as $\sqrt{\frac{n_{01} + n_{10}}{n_{11} + n_{10}}}$.

Hence, it measures the proportion of disagreements to the total number of mates in the true solution. A perfect solution has score of 0, and the lower the score, the better the solution.

The *Jaccard coefficient* is the ratio $\frac{n_{11}}{n_{11} + n_{10} + n_{01}}$.

It is the proportion of correctly identified mates to the sum of the correctly identified mates plus the total number of disagreements. Hence, a perfect solution has score of 1, and the higher the score, the better the solution. This measure is a lower bound for both sensitivity $\left(\frac{n_{11}}{n_{11} + n_{10}}\right)$ and specificity $\left(\frac{n_{11}}{n_{11} + n_{01}}\right)$ of the solution.

Note that both measures do not (directly) involve the term n_{00}, as solution matrices tend to be sparse and this term would dominate the other three in good and bad solutions alike. When the true solution is known only for a subset $N* \subset N$, the Minkowski and Jaccard measures can be computed on the submatrices corresponding to $N*$. In some cases, such as for cDNA oligo-fingerprint data, we have the additional information that no element of $N*$ has a mate in $N/N*$. In such a case, the Minkowski and Jaccard measures are evaluated using all the pairs $\{(i, j): i \in N*, j \in N \cup N*, i \neq j\}$.

Assessment When the True Solution is Unknown

When the true solution is not known, we evaluate the quality of a suggested solution by computing two figures of merit that measure its homogeneity and separation. For fingerprint data, homogeneity is evaluated by the average similarity between the fingerprint of an element and that of its cluster. Precisely, if $cl(u)$ is the cluster of u, $F(X)$ and $F(u)$ are the fingerprints of a cluster X and an element u, respectively, then

$$H_{Ave} = \frac{1}{|N|} \sum_{u \in N} S(F(u), F(cl(u)))$$

Separation is evaluated by the weighted average similarity between cluster fingerprints. That is, if the clusters are X_1, ... X_t, then

$$S_{Ave} = \frac{1}{\Sigma_{i \neq j} |X_i||X_j|} \sum_{i \neq j} |X_i||X_j| S(F(X_i), F(X_j))$$

Related measures that take a worst case instead of average case approach are minimum homogeneity: $H_{\min} = \min_{u \in N} S(F(u), F(cl(u)))$; and minimum separation: $S_{\max} = \max_{i \neq j} S(F(X_i), F(X_j))$. Hence, a solution improves if H_{Ave} or $H_{\min}$ increase, and if S_{ave} or $S_{\max}$ decrease. In computing all the above measures, singletons are considered as additional one-member clusters.

A CASE STUDY

In order to highlight the characteristics of each of the methods described above, we applied them to a yeast cell-cycle dataset containing the gene expression levels of yeast ORFs over 79 conditions. The original dataset contains samples from yeast cultures synchronized by four independent methods: a factor arrest (samples taken every seven minutes for 119 minutes), arrest of a cdc!5 temperature sensitive mutant (samples taken every 10 minutes for 290 minutes), arrest of a cdc28 temperature sensitive mutant (this part of the data is from Cho et al. 1998; samples taken every 10 minutes for 160 minutes), and elutriation. It also contains separate experiments in which Gl cyclin Cln3p or B-type cyclin Clb2p were induced. Spellman *et. al.*, identified in this data eight hundred genes that are cell-cycle regulated (Spellman *et. al.*, 1998).

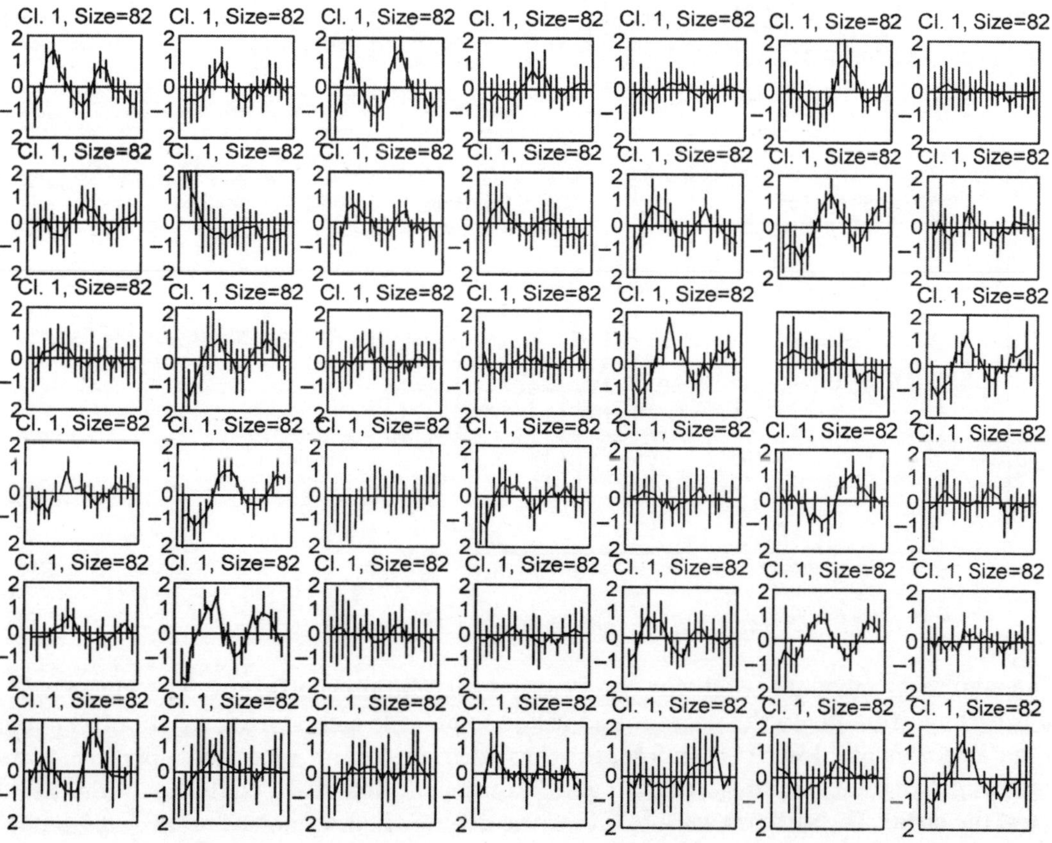

Fig. 5.10. The clustering produced by the K-means algorithm of Herwing et.al., x axis: time points 1-18 for the a factor experiment. y axis: normalized expression levels. The solid line in each subfigure plots the average pattern for that cluster. Error bars display the measured standard deviation. The cluster size is printed above each plot.

The dataset that we used contains the expression levels of 698 out of those eight hundred genes, which have no missing entries, over the 72 conditions that cover the α factor, cdc28, cdc15, and elutriation experiments. Each row of the 698 × 72 matrix was normalized to have mean 0 and variance 1. Based on the analysis conducted by Spellman *et. al.*, we expect to find in the data five main clusters: G1-peaking genes, S-peaking genes, G2-peaking genes, M-peaking genes, and M/G1-peaking genes. Each of these was shown to contain biologically meaningful subclusters. The 698 × 72 dataset was clustered using five of the methods described above: K-means, SOM, CAST, hierarchical, and CLICK.

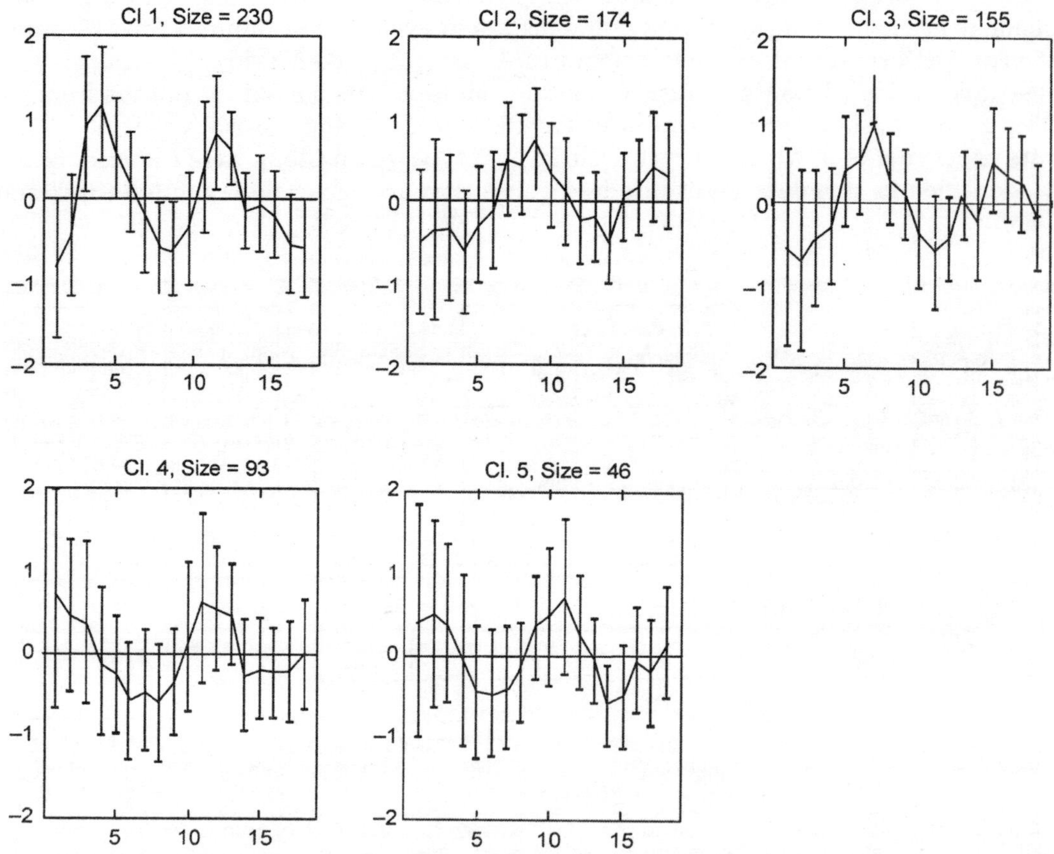

Fig. 5.11. The clustering produced by the CAST alogrithm of Ben-Dor et.al.

The similarity measure used was the Pearson correlation coefficient. The authors of each of the programs were given the dataset and asked to provide a clustering solution. The identity of the dataset was not described and genes were permuted in an attempt to perform a "blind" test (although anyone familiar with the gene expression literature could have identified the nature of the data). The authors were told that the average homogeneity and average separation would be used to evaluate the quality of the solutions. To allow the reader an impression of the results, we added for each of the clusterings (except the hierarchical one, which does not produce a hard partition of the elements) a reference figure prepared using MATLAB. This figure depicts the average pattern of the clusters along with error-bars for the first 18 datapoints, which correspond to the α factor experiment. We have chosen not to show full expression patterns over all the 72 conditions, as these are much harder to interpret visually. Over the first 18 datapoints, one expects to view periodic behavior, with a distinct, typical pattern in each cluster. We also omitted from these figures small clusters with fewer than four members.

As most programs output a variation of this figure, we have chosen to include the characteristic graphical output only for the programs Cluster and CAST. The following table summarizes the solutions produced by each program (except for Cluster), and their homogeneity and separation parameters. The so-called "True" clustering, reported by Spellman *et. al.*, (1998) is that obtained manually by inspecting the expression patterns and comparing to the literature. The solution produced by CLICK contains 67 unclustered singletons. The reference figure for each of the solutions-are given in figures 5.10 to 5.14. The output of CAST is shown in figure 5.15. It depicts the similarity matrix before and after ordering its rows and columns based on the clustering. It includes a dendogram and a graphical representation of the ordered fingerprint matrix.

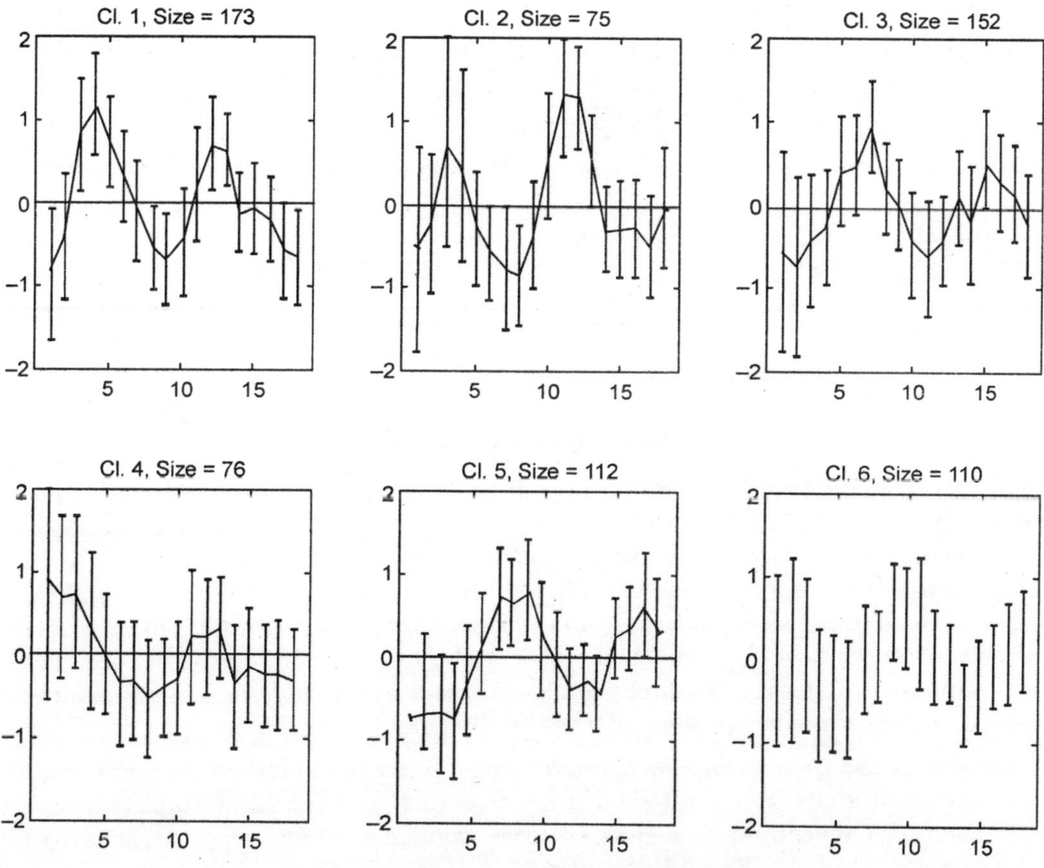

Fig. 5.12. The clustering produced by the GeneCluster algorithm of Tamayo *et.al.*

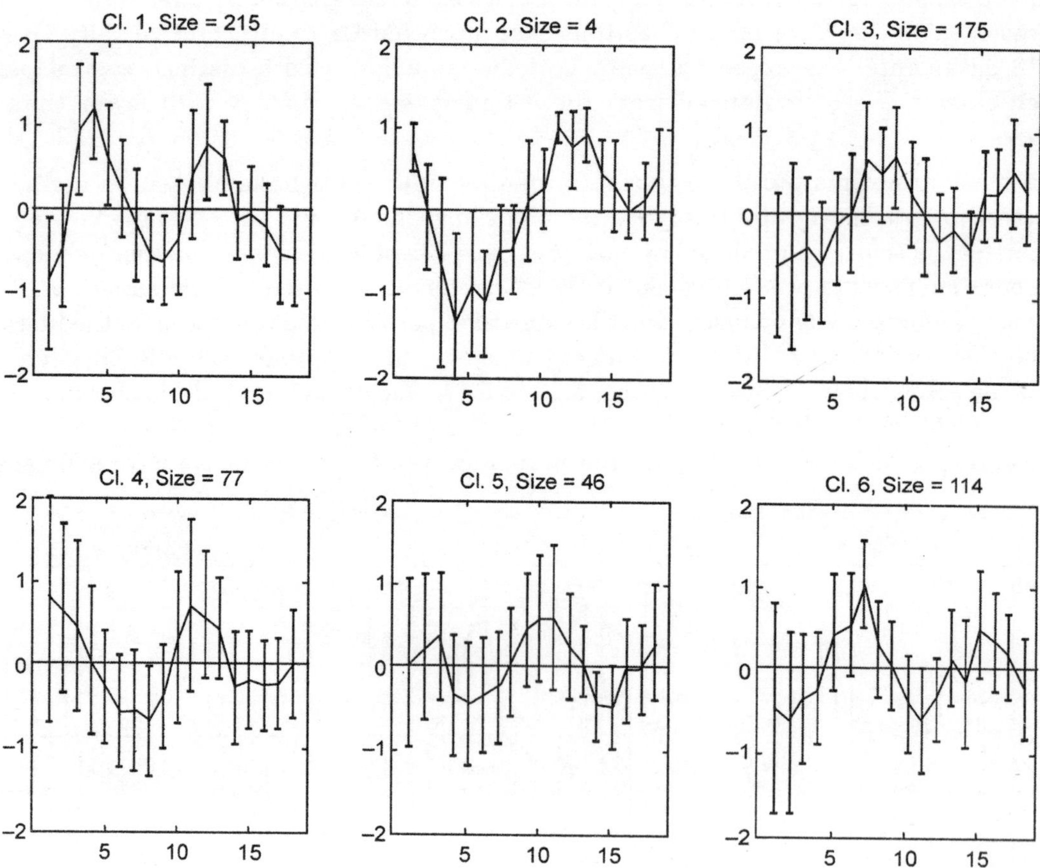

Fig. 5.13. The clustering produced by the CLICK algorithm.

CONCLUDING REMARKS

Clustering remains, to certain extent, an art. There are no universal, agreed-upon criteria for evaluating solutions, and there is no ultimate algorithm. The clustering problem is so general, covering diverse disciplines and applications, that it is impossible to choose a single, *"best"* algorithm for solving the problem. This holds true even for the specific application of gene expression that we have addressed here. The eventual decision on what solution and what algorithm works best depends on the user and on the specific questions the clustering process is supposed to answer. Each of the algorithms that we have described has its strong points and its disadvantages. We shall address briefly below several key issues.

(*i*) ***Choosing the Clustering Approach :*** The hierarchical method is exceptional in our review, as it gives an overall view of the structure without an attempt to force a hard clustering. This can be viewed as an advantage or a disadvantage, depending on the experimental goals. The other methods aim to split the universe of elements into clusters, either by geometric approaches that move cluster centers (SOM, K-Means) or by using a graph theoretic approach. The latter may take a global view (CLICK) or single out one affinity-stable cluster at a time (CAST). As noted above, many other approaches were developed in other applications.

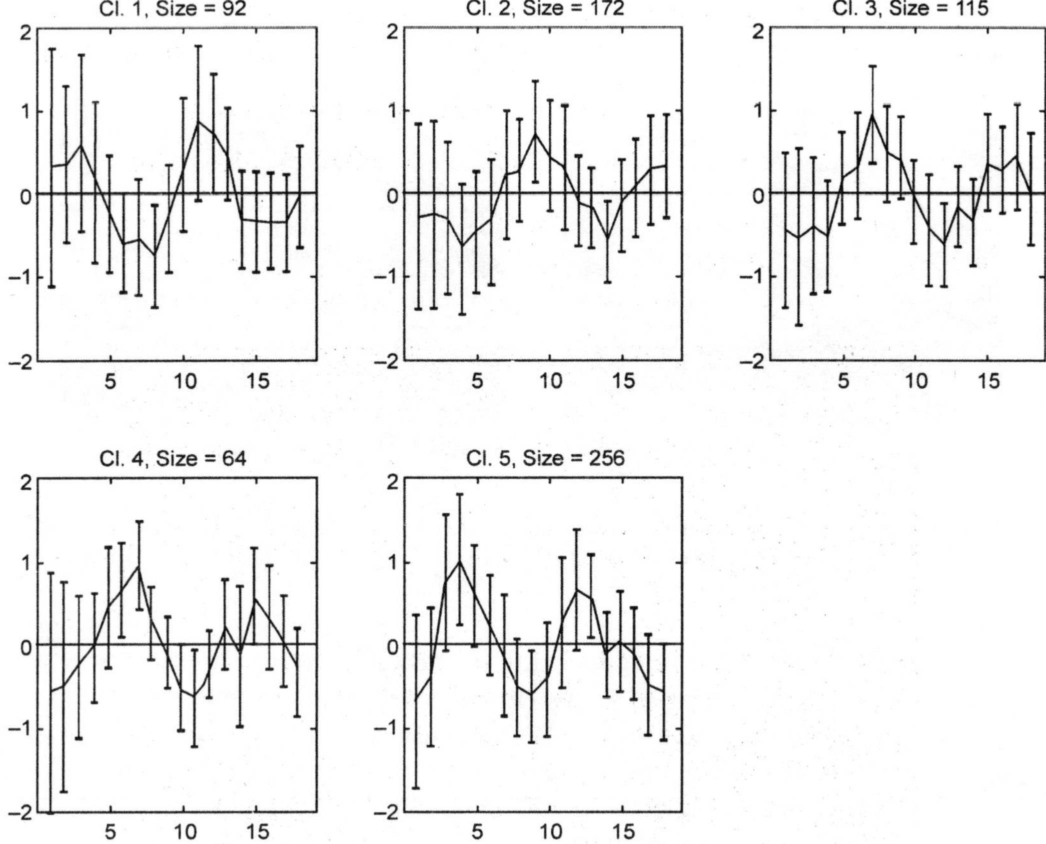

Fig. 5.14. The "True" clustering of Spellman et.al.

(ii) *How should we Evaluate Solution Quality :* We have described above several measures that evaluate solutions, both in the presence of a *"correct"* solution and in its absence. The obvious advantage of having an objective function is the ability to compare solutions and measure progress in algorithm development. The caveat is that the measures may not reflect exactly the intuition that the biologist may have. Even if one accepts the need of a numerical measure, the clustering literature is not in agreement on which measure to use, so we have presented two measures instead: an intra-cluster measure (homogeneity) and an inter-cluster measure (separation).

The two are inherently conflicting, as an improvement in one will correspond to worsening of the other. One idea of overcoming this is by presenting a curve of homogeneity versus separation. Such a curve can naturally be obtained in CAST (by varying the single threshold parameter used) and can also be obtained by multiple runs of other algorithms. This curve can tell that one algorithm dominates another if it provides better homogeneity for all separation values, but typically each algorithm will dominate in a particular range. One way of getting around the "two objectives" problem is to fix the number of clusters. This is done by SOM and the classical K-means.

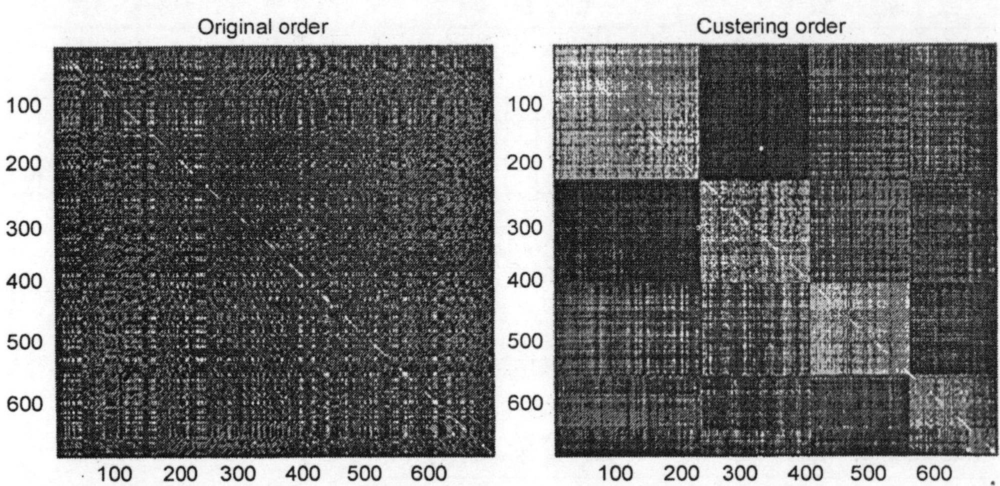

Fig. 5.15. Representation of the solution produced by CAST. Left: the original similarity matrix. Right: the same matrix reordered according to the clustering. Grey level is inversely proportional to similarity. Genes belonging to the same cluster appear contiguously.

When the number of clusters is known this is of course the way to go. When it is not known, what users often do is run such algorithms several times with several numbers of clusters (or grid topologies, in the case of SOM). However, this brings back the problem of evaluating and comparing solutions, so algorithms that seek a globally optimal solution seem preferable. Alternative methods of determining the number of clusters are given, for example, by Hartigan (1975) and Tibshirani et al. (2000).

Fig. 5.16. The output of Cluster. Actual output is in color, where red denotes increase and the reference level, and green denotes decrease. The gene clustering dendogram is on the top, and the experiment clustering dendogram is on the right.

(*iii*) *Should we Cluster all Elements* : The SOM, K-means, and hierarchical algorithms require that the solution will constitute a partition of all the elements. Other algorithms, such as CLICK, allow some singletons to be left unclustered. By allowing singletons to be discarded, intra-cluster deviations can be reduced, perhaps at the expense of weaker separation. In gene expression applications, one often does not seek an identification of *all* the genes involved, particularly as many genes have already been discarded in preprocessing steps, because of insignificant fingerprint variations. It is thus desirable to allow some room for discarding elements from a solution. It is not hard to add such flexibility into virtually all clustering algorithms that we have discussed.

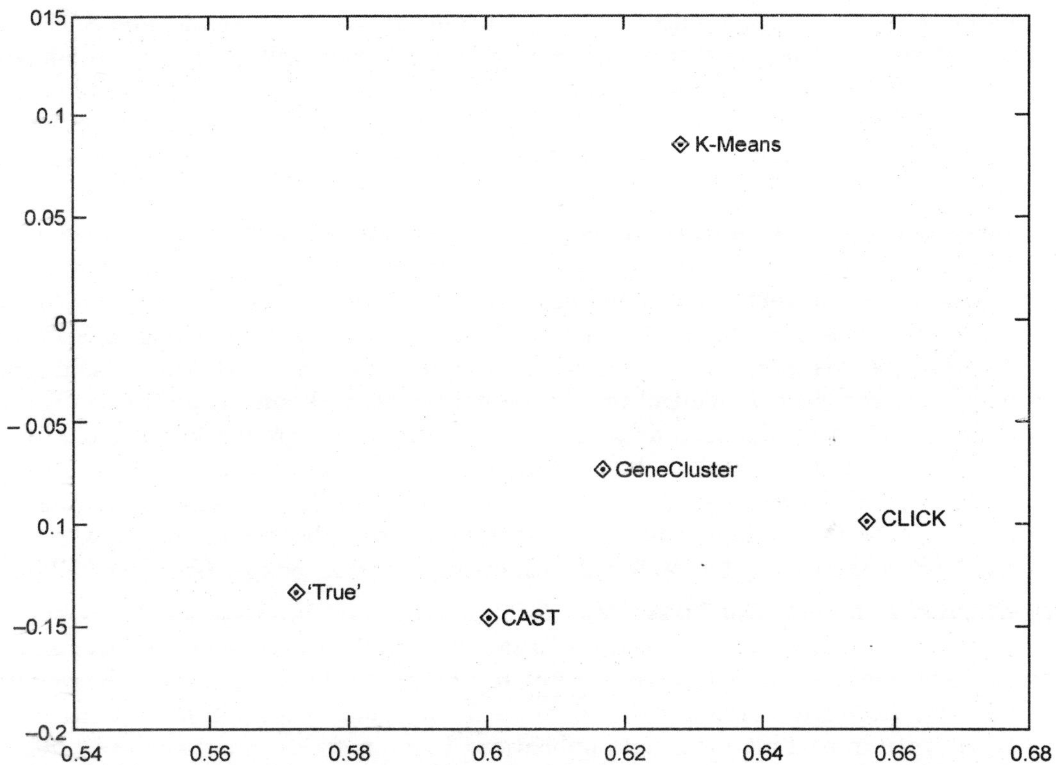

Fig. 5.17. A comparison of homogeneity (x-axis) and separation (y-axis) values for all solutions. Recall that a solution improves if homogeneity increases or separation decreases.

Table 5.1. A summary of the clustering solutions and their figures of merit.

Program	*No. of Clusters*	*Homogeneity*		*Separation*	
		H_{Ave}	H_{Min}	S_{Ave}	S_{Max}
K Means	40	0.620	0.330	0.086	0.011
CAST	5	0.6	0.037	–0.146	0.322
GeneCluster	6	0.617	0.067	–0.073	0.584
CLICK	6	0.656	0.097	–0.098	0.546
"True"	5	0.572	–0.322	–0.133	0.73

(*iv*) ***Fingerprints versus Similarity*** **:** Some algorithms use only similarity values between elements, whereas others use the fingerprints themselves. Obviously, one loses some information by using the fingerprints to compute pairwise similarities only. One of the advantages of *CLICK* over *HCS* for example, is by explicit use of the fingerprints for merging and adoption. Geometric algorithms like K-means and *SOM* use only fingerprints. Other algorithms like *CAST* may benefit from using such information more.

(*v*) ***Visualization is Crucial*** **:** As the datasets and the solutions are very large, it is imperative to have tools to visualize summaries of the data and its solution from various viewpoints. The average patterns figures are useful to show trends, and *SOM* goes a step further by putting similar patterns in neighboring cells in a grid, generating a convenient *"executive summary."* The dendograms of Eisen et al. (1998) viewed together with the color-coded expression patterns of the genes are also very useful. Yet, devising additional novel, sophisticated (and ideally interactive) visualizations is an important challenge.

(vi) ***We need more Testing data :*** In order to improve the algorithms, we need more data. The best kind is actual gene expression data, along with a known clustering solution, so that it can be compared to the algorithmic solution. This is quite hard to obtain (except perhaps for oligofingerprint data) in the current status of biological knowledge. A second best is generating synthetic (simulated) datasets with known solutions, in which one can directly control individual parameters (cluster structure, errors, etc.). Some initial work has been done in this direction, but more work is needed in order to understand how to make the simulations realistic. Generating a publicly accessible benchmark of datasets—both synthetic and real—with known solutions, would be of great benefit to developing better algorithms. In the absence of such resources, the available real data can be combined with evaluation methods as demonstrated here.

(vii) ***Clustering is only the First Step :*** In analyzing gene expression data, clustering is an essential initial step, but there is a lot more that can be done with the data. For example, one can use supervised learning techniques to cluster or classify the conditions. Such methods were recently shown to yield very good results in determining cancer types, with important potential applications to diagnostics. Another useful idea is to cluster both the genes and the conditions, and to pinpoint subsets of the genes and the conditions ("biclustering"). Given the clusters, a variety of biological inference steps are possible. For example, identification of common control regions of upstream regions of genes from the same cluster.

6

Integrating Genomic Analysis

Protein-protein interactions are fundamental to cellular functions, and comprehensively identifying them is important towards systematically defining the biological role of proteins. New experimental and computational methods have produced a vast number of known or putative interactions catalogued in databases, such as MIPS, DIP and BIND. Unfortunately, interaction datasets are often incomplete and contradictory. In the context of genome-wide analyses these inaccuracies are greatly magnified because the protein pairs that do not interact (negatives) far outnumber those that do (positives). For instance, in yeast the ~6000 proteins allow for ~18 million potential interactions, but the estimated number of actual interactions is below 100000. Thus, even reliable techniques can generate many false positives when applied on a genomic scale.

An analogy to this situation would be a diagnostic with a one per cent false-positive rate for a rare disease occurring in 0.1 per cent of the population, which would roughly produce one true positive for every 10 false ones. Consequently, when evaluating protein-protein interactions, one needs to integrate evidence from many different sources to reduce errors. In the era of post-genomic biology, it becomes particularly useful to think of cells as a complex network of interacting proteins. Biology is increasingly moving from the study of the individual parts of the system separately to the study of the emergent properties of the entire system.

Most biological functions are the result of the interactions of many different molecules. The challenge of systems biology is to develop models of biological functions that incorporate and elucidate this complexity. This chapter has thus been written with the aim of introducing our recent development of a nai've Bayes approach in the protein complex membership prediction, and recent progress in the protein interaction network analysis. The first half of the chapter will provide a detailed description of the nai've Bayesian approach developed by Jansen *et al.* (2003) that probabilistically combines multiple information sources to predict protein interactions in yeast.

GENOMIC FEATURES

Jansen *et. al.* (2003) recently showed how protein complexes can be predicted *de novo* with high confidence when multiple datasets are integrated, and demonstrated the application to yeast. These multiple datasets can be either noisy interaction datasets or circumstantial genomic evidence. The genomic data sources used in above study are the correlation of mRNA amounts in two expression datasets, two sets of information on biological function, and information about whether proteins are essential for survival. Although none of these information sources are interaction *per se,* they contain information weakly associated with interaction: two subunits of the same protein complex often have co-regulated mRNA expression and similar biological functions and are more likely to be both essential or non-essential.

mRNA expression

Two sets of publicly available expression data –*a* time course of expression fluctuations during the yeast cell cycle and the Rosetta compendium, consisting of the expression profiles of 300 deletion mutants and cells under chemical treatments have been used in the above study. These data are useful for the prediction of protein-protein interaction because proteins in the same complex are often co-expressed. The Pearson correlation for each protein pair for both the Rosetta and cell cycle datasets indicates that these two datasets are strongly correlated. This problem can be circumvented by computing the first principal component of the vector of the two correlations. This first principal component is a stronger predictor of protein-protein interactions than either of the two expression correlation datasets by themselves.

In order to perform Bayesian networks analysis, this first principal component of expression correlations is divided into 19 bins, and the overlap of each bin with the gold standard datasets (see below) is assessed. The first column of Table 6.1 bears the name of the genomic feature and the number of bins we divide this feature into. The second column gives the number of protein pairs that this feature covers in the yeast interactome (~18 million pairs of proteins). The third column, which contains five subcolumns, shows the overlap between the genomic feature and the gold-standard (positive and negative) sets. The subcolumns positive (+) and negative (–) show how many protein pairs in the present bin of the genomic feature are among the protein pairs in the gold-standard positive set and negative set, respectively. The subcolumns sum(+) and sum(–) show the cumulative number of overlaps of the present and above bins. The subcolumn sum(+)/sum(–) is the ratio of sum(+) and sum(–). The next two columns are the conditional probabilities of the feature, and the last column is the likelihood ratio *L,* which is the ratio of the conditional probabilities in the two preceding columns.

Biological Functions

Interacting proteins often function in the same biological process. This means that two proteins acting in the same biological process are more likely to interact than two proteins involved in different processes. In addition, proteins functioning in small, specific biological processes are more likely to interact than those functioning in large, general processes. Two catalogues of functional information about proteins are collected from the MIPS functional catalogue – which is separate from the MIPS complexes catalogue -and the data on biological processes from Gene Ontology (GO). Most classification systems have the structure of a tree (*e.g.* MIPS) or a directed acyclic graph (DAG) (*e.g.* GO). Obviously, a pair of proteins should be very similar if there are only a few descendants of a given ancestor, whereas the similarity will be less significant if many proteins descend from it.

Given two proteins that share a specific set of lowest common ancestor nodes in the classification structure, one can count the total number of protein pairs n that also have the exact same set of lowest common ancestors. This number is expected to be low for proteins that share a very detailed functional description, but very high for proteins that have no

Table 6.1. Combining genomic features to predict protein-protein interactions in yeast.

mRNA expression correlation		*Number of protein paris*	*Gold-standard overlap*							*Likelihood ratio(L)*
			positive(+)	*negative(–)*	*sum(+)*	*sum(–)*	*sum(+)/sum(–)*	*P(Exp/+)*	*P(Exp/–)*	
Bins	0.9	678	16	45	16	45	0.36	2.10×10^{-3}	1.68×10^{-5}	124.9
	0.8	4827	137	563	153	608	0.25	1.80×10^{-2}	2.10×10^{-4}	85.5
	0.7	17626	530	2117	683	2725	0.25	6.96×10^{-2}	7.91×10^{-4}	88.0
	0.6	42815	1073	5597	1756	8322	0.21	1.41×10^{-1}	2.09×10^{-3}	67.4
	0.5	96650	1089	14459	2845	22781	0.12	1.43×10^{-1}	5.40×10^{-3}	26.5
	0.4	225712	993	35350	3838	58131	0.07	1.30×10^{-1}	1.32×10^{-2}	9.9
	0.3	529 268	1028	83483	4866	141614	0.03	1.35×10^{-1}	3.12×10^{-2}	4.3
	0.2	1200331	870	183356	5736	324 970	0.02	$1.14 - 10^{-1}$	6.85×10^{-2}	1.7
	0.1	2575103	739	368 469	6475	693439	0.01	9.71×10^{-2}	1.38×10^{-1}	0.7
	0	9 363 627	894	1 244 477	7369	1937916	0.00	1.17×10^{-1}	4.65×10^{-1}	0.3
	–0.1	2753735	164	408 562	7533	2 346 478	0.00	2.15×10^{-2}	1.53×10^{-1}	0.1
	–0.2	1 241 907	63	203 663	7596	2550141	0.00	8.27×10^{-3}	7.61×10^{-2}	0.1
	–0.3	484 524	13	84957	7609	2 635 098	0.00	1.71×10^{-3}	3.18×10^{-2}	0.1
	–0.4	160 234	3	28870	7612	663 968	0.00	3.94×10^{-4}	1.08×10^{-2}	0.0
	–0.5	48852	2	8091	7614	2672 059	0.00	2.63×10^{-4}	3.02×10^{-3}	0.1
	–0.6	17423	N/A	2134	7614	2674193	0.00	0.00	7.98×10^{-4}	0.0
	–0.7	7602	N/A	807	7614	2 675 000	0.00	0.00	3.02×10^{-4}	0.0
	–0.8	2147	N/A	261	7614	2 675 261	0.00	0.00	9.76×10^{-5}	0.0
	–0.9	67	N/A	12	7614	2675273	0.00	0.00	4.49×10^{-6}	0.0
Total number		18773128	7614	2675273	N/A	N/A	N/A	1.00	1.00	1.0

Go biological process similarity		*Number of protein paris*	*Gold-standard overlap* positive(+)	*negative(–)*	*sum(+)*	*sum(–)*	*sum(+)/sum(–)*	*P(GO/+)*	*P(GO/–)*	*Likelihood ratio(L)*
Bins	1–9	4789	88	819	88	819	0.11	1.17×10^{-2}	1.27×10^{-3}	9.2
	10–99	20467	555	3315	643	4134	0.16	7.38×10^{-2}	5.14×10^{-3}	14.4
	100–999	58738	523	10232	1166	14366	0.08	6.95×10^{-2}	1.59×10^{-2}	4.4
	1000–9999	152850	1003	28225	2169	42591	0.05	1.33×10^{-1}	4.38×10^{-2}	3.0
	10000–∞	2909442	5351	602434	7520	645025	0.01	7.12×10^{-1}	9.34×10^{-1}	0.8
Total number		3146286	7520	645025	N/A	N/A	N/A	1.00	1.00	1.0

MIPS functional similarity		*Number of protein paris*	*Gold-standard overlap* positive(+)	*negative(–)*	*sum(+)*	*sum(–)*	*sum(+)/sum(–)*	*P(MIPS/+)*	*P(MIPS/–)*	*Likelihood ratio(L)*
Bins	1–9	6.584	171	1094	171	1094	0.16	2.12×10^{-2}	8.33×10^{-4}	25.5
	10–99	25823	584	4229	755	5323	0.14	7.25×10^{-2}	3.22×10^{-3}	22.5
	100–999	88548	688	13011	1443	18334	0.08	8.55×10^{-2}	9.91×10^{-3}	8.6
	1000–9999	255096	6146	47126	7589	65460	0.12	7.63×10^{-1}	3.59×10^{-2}	21.3
	10000–∞	5785754	462	1248119	8051	1313579	0.01	5.74×10^{-2}	9.50×10^{-1}	0.1
Total number		6161805	8051	1313579	N/A	N/A	N/A	1.00	1.00	1.0

Co-essentiality		*Number of protein*	*Gold-standard overlap* positive(+)	*negative(–)*	*sum(+)*	*sum(–)*	*paris* sum(+)/sum(–)	*P(Ess/+)*	*P(Ess/–)*	*Likelihood ratio(L)*
Bins	EE	384126	1114	81924	1114	81924	0.014	5.18×10^{-1}	1.43×10^{-1}	3.6
	NE	2767812	624	285487	1738	367411	0.005	2.90×10^{-1}	4.98×10^{-1}	0.6
	NN	4978590	412	206313	2150	573724	0.004	1.92×10^{-1}	3.60×10^{-1}	0.5
Total number		8130528	2150	573724	N/A	N/A	N/A	1.00	1.00	1.0

function in common. For instance, if a functional class contains only two proteins, then the count would yield $n = 1$. On the other hand, if the root node is the lowest common ancestor of two proteins, n is on the order of the number of protein pairs contained in the classification. The functional similarity between two proteins is thus quantified by the following procedure. First, two proteins of interest are assigned to a set of functional classes two proteins share, given one of the functional classification systems. Then the number of the ~18 million protein pairs in yeast that share the exact same functional classes as the interested protein pairs is counted (yielding a count between 1 and ~18 million).

In general, the smaller this count, the more similar and specific is the functional description of the two proteins, while large counts indicate a very non-specific functional relationship between the proteins. Low counts (*i.e.* high functional similarity) are found to correlate with a higher chance of two proteins being in the same complex.

Essentiality

Protein essentiality is also considered in the study. It should be more likely that both of two proteins in a complex are essential or non-essential, but not a mixture of these two attributes. This is because a deletion mutant of either one protein should by and large produce the same phenotype: they both impair the function of the same complex. Indeed, such a relationship is supported by the data. Finally, protein-protein interaction datasets generated by high-throughput experiments can also be seen as a special type of genomic feature.

Gold-Standard Datasets

The basic idea of how to integrate different sources of information is to assess each source of evidence for interactions by comparing it against samples of known positives and negatives, yielding a statistical reliability. Then, extrapolating genome-wide, the chance of possible interactions for every protein pair can be predicted by combining each independent evidence source according to its reliability. Thus, reliable reference datasets that serve as gold standards of positives (proteins that are in the same complex) and negatives (proteins that do not interact) are essential. An ideal gold-standard dataset should satisfy the three following criteria: (1) independent from the data sources serving as evidence, (2) sufficiently large for reliable statistics and (3) free of systematic bias.

It is important to note that different experimental methods carry with them different systematic errors – errors that cannot be corrected by repetition. Therefore, the gold-standard dataset should not be generated from a single experimental technique. Positive gold standards are extracted from the MIPS (Munich Information Center for Protein Sequences) complexes catalogue. It consists of a list of known protein complexes based on the data collected from the biomedical literature (most of these are derived from small-scale studies, in contrast to the high-throughput experimental interaction data). Only classes that are on the second level of MIPS complex code are considered. F or instance, the MIPS class 'translation complexes' (500) contains the subclasses 'mitochondrial ribosome' (500.60), the 'cytoplasmic ribosome' (500.40) and a number of other subclasses related to translation-related complexes; we only considered pairs among proteins in those subclasses (500.*) as positives.

Overall, this yielded a filtered set of 8250 protein pairs that are within the same complex. A negative gold standard is harder to define, but essential for successful training. There is no direct information about which proteins do not interact. However, protein localization data provide indirect information if we assume that proteins in different compartments do not interact. A list of ~2.7 million protein pairs in different compartments are compiled from the current yeast localization data in which proteins are attributed to one of five compartments as has been done previously. These compartments are the nucleus (*N*), mitochondria (*M*), cytoplasm (*C*), membrane (*T* for transmembrane), and secretory pathway (*E* for endoplasmic reticulum or extracellular) .

PROTEIN-PROTEIN INTERACTIONS

A wide spectrum of supervised methods can be applied to integrate genomic features in order to predict protein-protein interactions. Among them, machine-learning approaches, including simple unions and intersections of datasets, neural networks, decision trees, support-vector machines and Bayesian networks have been successfully applied to this goal.

Below we try to elaborate basic concepts in machine learning, and provide a basic tutorial on how to employ decision trees and Bayesian networks in protein-protein interaction analysis. According to Merriam-Webster's *Collegiate Dictionary,* learning is a process in which people 'gain knowledge or understanding of or skill in by study, instruction, or experience'. The key idea of 'learning' is to perform better based on past experience. Ever since computers were invented, people have tried to program them to learn (*i.e.* machine learning). Precisely, 'a computer program is said to *learn* from experience *E* with respect to some class of tasks T and performance measure *P,* if its performance at tasks in *T,* as measured by *P,* improves with experience *E*'. For example, researchers have used computer programs to recognize tumours based on biopsy results:

- task *T,* to determine whether an examinee has cancer or not
- performance measure *P,* percentage of correct predictions
- training experience *E,* biopsy results from cancer patients and normal people.

Let *X* and *Y* denote the sets of possible inputs and outputs. The learning algorithm needs to find the (approximate) target function *V* that takes each input $x_i \in X$ and gives the corresponding prediction $y_i, \in Y$, *i.e.*, output. If *Y is* a subset of the real numbers, we have a regression problem. Otherwise, we have a classification problem (binary or multiclass). In this case, the algorithm will determine whether a patient has cancer based on his/her biopsy result, which is a binary classification problem.

There are, of course, other learning problems (*e.g. reinforcement learning*). Here, we are mainly interested in classification problems. Why do we need machine learning? First, for many complicated problems, there is no known method to compute the accurate output from a set of inputs. Second, for other problems, computation according to known exact methods may be too expensive. In both cases, good approximate methods with reasonable amounts of computational demand are desired. Machine learning is a field in which computer algorithms are developed to learn approximate methods for solving different problems. Obviously, machine learning is a multidisciplinary field, including computer science, mathematics, statistics and so on.

Supervised versus Unsupervised Learning

Learning algorithms are usually divided into two categories: supervised and unsupervised. In *supervised learning*, a set of input/output example pairs is given, which is called the training set. The algorithms learn the approximate target function based on the training set. Once a new case comes in, the algorithms will calculate the output value based on the target function learned from the training set. By contrast, in *unsupervised learning*, a set of input values are provided, without the corresponding output. The learning task is to gain understanding of the process that generates input distribution. In this section, we will focus our discussion on supervised learning algorithms.

Decision Trees

Decision tree learning is one of the most widely used algorithms to search for the best discrete-valued hypothesis (h) within H. Figure 6.1 illustrates a decision tree for the protein-protein interaction classification. Only the yeast protein pairs without missing values in genomic features and in gold-standard sets are considered. The decision tree tries to predict protein-protein interactions based on three genomic features using the ID3. S is a set of examples, in this case a collection of the protein pairs. E stands for entropy and G stands for information gain calculated according to the formula. Each diamond node is one attribute or genomic feature.

The learned decision tree classifies a new instance by going down the tree from the root to a certain leaf node. Each leaf node provides a classification for all instances within it. A test of a specific attribute is performed at each node, and each branch descending from that node corresponds to one of the possible values for this attribut. The basic idea behind decision tree learning is to determine which attribute is the best classifier at a certain node to split the training examples. Many algorithms have been developed to solve this problem, such as ID3, C4.5, ASSITANT and CART. Here, we focus on the ID3 algorithm.

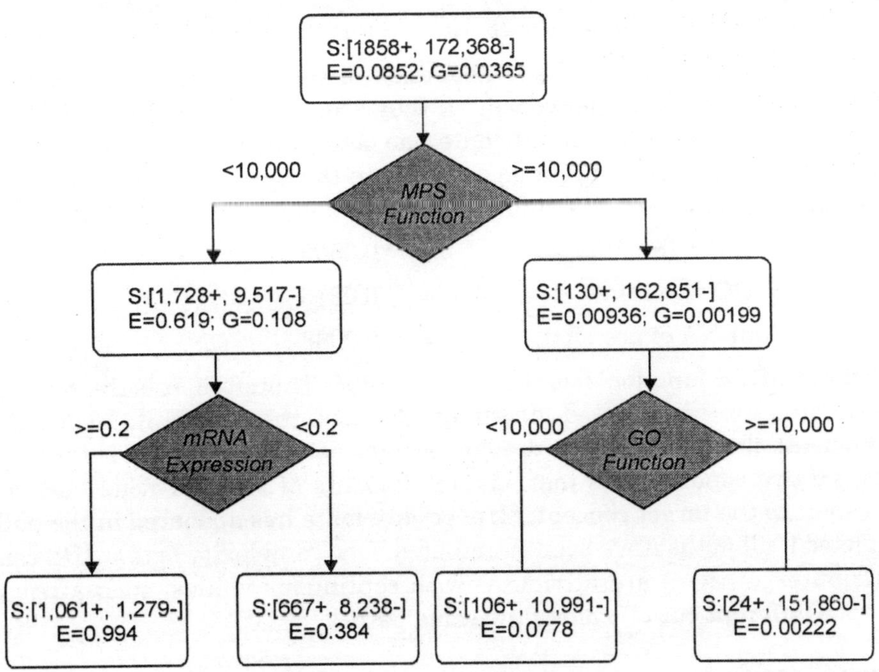

Fig. 6.1. A typical decision tree.

In order to determine the best classifier, ID3 uses a statistical property, named *information gain,* to measure how well a given attribute separates the training examples with respect to the target classification. To define information gain, we need to introduce the concept of *entropy* (*E*) in information theory. Entropy is used to measure the impurity of a set of examples, which is calculated by the formula

$$E(S) = {}^{-}p + \log_2 p_+ {}^{-}p_- \operatorname{los}_2 p_- \qquad ...(6.1)$$

where S is a set of examples (positives and negatives) regarding some target concept, p_+ is the proportion of positives in S and p_- is the proportion of negatives in S. There are 1858 positives and 172368 negatives in the example shown in Figure 6.1. Therefore, the entropy is

$$E(1858_+, 172368\_]) = -\frac{1858}{174226}\log_2\left(\frac{1858}{174226}\right) \cdot - \frac{172368}{174226}\log_2\left(\frac{172368}{174226}\right) = 0.0852.$$

More generally, if the target concept can take on n different values (*i.e.*, n different classes), the entropy of S relative to this n-wise classification is defined as

$$E(S) \equiv \sum_{i=1}^{n} {}^{-}p_i \log_2 p_i \qquad ...(6.2)$$

where p_i is the proportion of S belonging to class i.

Having defined entropy, we can now define information gain (G):

$$G(S, A) \equiv E(S) - \sum_{v \in \text{values}(A)} \frac{|S_v|}{|S|} E(s_v) \qquad ...(6.3)$$

where A is an attribute associated with each instance. Here value (A) is the set of all possible values *that* A could take on, v is an element of values (A), and S_v *is* the set of instances whose value of A is v. Clearly, S_v is a subset of S.

The information gain measures the deduction of the impurity (entropy) of the training examples with respect to the target concept if they are split by a certain attribute. Therefore, the higher the information gain of an attribute, the better it classifies the examples. As a result, ID3 uses the value of the information gain to choose the best classifier at each node. For the training examples in Figure 6.1, the information gain values for all three attributes are

$G(S$, MIPS function$)$ = 0.0365

$G(S$, GO function$)$ = 0.0216

G(S,mRNA expression) = 0.0088

The attribute 'MIPS function' has the highest value. Therefore, it is the root node in Figure 6.1. The same procedure is iterated for the child nodes, then the child nodes of these nodes, and so on. Each attribute can only be used once along each path. A path is terminated if either of the following two conditions is met: (1) all elements of the leaf node belong to the same class with respect to the target concept; (2) every attribute has appeared in the path. The whole tree is completed if all paths have been terminated. One complexity is that IDS can only handle nominal attributes. If there are attributes with continuous values, such attributes could be used twice with different cut-offs along the same path.

Naive Bayes Classifier

Besides decision tree learning, another commonly used method is naive Bayes learning, often called the naive Bayes classifier. It is also applied to the kind of data in which each instance is associated with a set of nominal attributes. The naive Bayes classifier assigns the most probable target value to the instance with the attribute values $\langle f_1, f_2, ..., f_n \rangle$:

$$h = \arg\max_{h_j \in H} P(h_j \mid f_1, f_2, ..., f_n) \qquad ...(6.4.)$$

Using the Bayes theorem, the formula can be rewritten as

$$h = \arg\max_{h_j \in H} \frac{P(f_1, f_2, ..., f_n \mid h_j) P(h_j)}{P(f_1, f_2, ..., f_n)} = \arg\max_{h_j \in H} P(f_1, f_2, ..., f_n \mid h_j) P(h_j) \qquad ...(6.5)$$

The most important assumption in naive Bayes learning is that all attributes are conditionally independent of each other with respect to every hypothesis h_j. Therefore, the joint probability of all attributes is the product of the individual probabilities:

$$h = \arg\max_{h_j \in H} P(h_j) \prod_{i=1}^{n} P(f_1 / h_j) \qquad ...(6.6.)$$

The Bayesian approach has been widely used in biological problems. Jansen *et. al.*, (2003) described an approach using Bayesian networks to predict protein-protein interactions. A pair of proteins that interact is defined as *'positive'*. Given some positives among the total number of protein pairs, the 'prior' odds of finding one are

$$O_{\text{prior}} = \frac{P(\text{pos})}{P(\text{neg})} = \frac{P(\text{pos})}{1 - P(\text{pos})} \qquad ...(6.7)$$

In contrast, 'posterior' odds are the chance of finding a positive after considering N features with values $f_1 ... f_n$:

$$O_{\text{post}} = \frac{P(\text{pos}) \mid f_1 \cdots f_n)}{P(\text{neg}) \mid f_1 \cdots f_n)} \qquad ...(6.8)$$

(The terms 'prior' and 'postenrior' refer to the situation before and after knowing the information in the N features). The likelihood ratio L is defined as

$$L(f_1 ... f_N) = \frac{P(f_1 \cdots f_n \mid \text{pos})}{P(f_1 \cdots f_n \mid \text{neg})} \qquad ...(6.9)$$

It relates prior and posterior odds according to Bayes' rule, $O_{\text{post}} = L(f_1 ... f_n)\, O_{\text{prior}}$. In the special case where the N features are conditionally independent (*i.e.*, they provide uncorrelated evidence), the Bayesian network is a so-called 'naive' network, and L can be simplified to

$$L(f_1 ... f_n) = \prod_{i=1}^{N} L(f_1) = \prod_{i=1}^{N} \frac{P(f_i / \text{Pos})}{P(f_i / neg)} \qquad ...(6.10)$$

L can be computed from contingency tables relating positive and negative examples with the N features (by binning the feature values $f_1 ... f_n$ into discrete intervals). Simply put, consider a genomic feature f expressed in binary terms (*i.e.*, *'present'* or *'absent'*). The likelihood ratio $L(f)$ is then defined as the fraction of gold-standard positives having feature f divided by the fraction of negatives having f. For two features f_1 and f_2 with uncorrelated evidence, the likelihood ratio of the combined evidence is simply the product $L(f_1, f_2) = L(f_1)\, L(f_2)$. A protein pair is predicted as positive if its combined likelihood ratio exceeds a particular cut-off ($L > L_{\text{cutoff}}$)

(negative otherwise). The likelihood ratios are computed for all possible protein pairs in the yeast genome. Based on previous estimates, we think that 30,000 positives is a conservative lower bound for the number of positives (*i.e.* pairs of proteins that are in the same complex). Given that there are approximately 18 million protein pairs in total, the prior odds would then be about 1 in 600. With $L > L_{cutoff} = 600$ we would thus achieve $O_{post} > 1$.

Cross-validation with the reference datasets shows that naive Bayesian integration of multiple genomic sources leads to an increase in sensitivity over the high-throughput data sets it combined for comparable true positive (TP)/false positive (FP) ratios. ('Sensitivity' measures coverage and is defined as TP over the number of gold-standard positives, *P.)* This means that the Bayesian approach can predict, at comparable error levels, more complex interactions *de novo* than are present in the high-throughput experimental interaction datasets. The predicted dataset (PIP) was also compared with a voting procedure where each of the four genomic features contributes an additive vote towards positive classification. The results showed that the Bayesian network achieved greater sensitivity for comparable TP/FP ratios.

MISSING VALUE

The naive Bayes procedure presented above lends itself naturally to the addition of more features, possibly further improving results. As more sparse data are incorporated, however, the *missing value* problem becomes severe: the number of protein pairs with complete feature data decreases, and the number of possible missing feature patterns increases exponentially with the number of features. Some classification methods, such as decision trees, can handle missing values in an automated fashion.

Most other classification methods, however, require a full data matrix as Input. It is therefore necessary to first fill in missing data with plausible values, a process called imputation. It is important for the imputed values to be consistent with the observed data and preserve the overall statistical characteristics of the feature table, for example the correlations between features. Below we will discuss different mechanisms of missing values, followed by a brief description of several representative methods for missing data imputation.

Mechanisms of Missing Values

There are two broad categories of missing value mechanisms. The first category is called Missing at Random (MAR). Here, the probability of a feature being missing can be determined entirely by the observed data. A special case is Missing Completely at Random (MCAR), where the patterns of missing data are completely random. Since most missing value analysis methods assume MAR, it is important to assess whether or not this assumption holds for a given set of missing values.

In general, missing values are approximately MAR for pair protein features. In one example, synthetic lethal features for some protein pairs are missing because the experiments were not performed due to limited resources. In another example, structure-based features, such as multimeric threading scores, can only be computed for proteins with a solved structural homologue, and will become missing otherwise. These missing values can all be approximated as MAR. In certain cases, however, the probability of a feature being missing is directly related to the missing feature itself and cannot be determined entirely by the observed data. In this case, the missing data are not MAR. For example, consider a situation where a protein pair feature is computed for all protein pairs, and only the best scores (indicative of protein interaction) are kept and the rest of the scores are thrown away and thus become missing.

The missing data are no longer MAR, and they cannot be treated in the same way as missing due to incomplete coverage. On the other hand, simply recording all the scores, no matter good or bad, will solve this problem. Most methods for missing value analysis assume MAR; we briefly summarize a few representative methods below. Let us suppose that instance x has a missing attribute A.

Mean Substitution and Regression Imputation

In *mean substitution* the missing attribute A in instance x is replaced with the most common value of attribute A in the whole data matrix, or in the subset of instances that x belongs to. A major disadvantage of this simple method is that the correlations between features are not preserved.

In *k nearest neighbours* (*KNN*), the distance between instance x and all instances with complete attributes are calculated, based on the observed attributes in x. The k nearest neighbours are identified, and the missing attribute A in instance x is replaced with the weighted average of attribute A in these k nearest neighbours.

In *regression imputation* attribute A is regressed against all other attributes based on all instances with complete attributes. Afterwards, the missing attribute A in instance x is replaced with its predicted value based on the regression model.

Bayesian Multiple

In SVD imputation, all missing values in the data matrix are filled with initial guesses. Using singular value decomposition (SVD), all instances are then projected to a low-dimensional feature space spanned by the first k principal components. To update the missing attribute A in instance x, instance x is regressed against the first k principal components using observed attributes in x, and the resulting regression model is used to predict the missing attribute A. After all missing values in the data matrix are updated, another round of SVD is performed and the whole process is iterated until convergence. In the case of imputing missing values in DNA microarrays, SVD imputation is found to perform better than mean substitution, but worse than KNN.

The expectation maximization (EM) algorithm is a popular method for finding maximum-likelihood estimates for parametric models with missing data. Here, all instances are assumed to be independently and identically distributed based on a parametric model (for example, a normal distribution) with unknown parameters θ. The EM algorithm makes point estimates for missing data and parameters θ in the following way. First, parameters θ are initialized. In the E-step, the missing attribute A in instance x is replaced by its expected value calculated from the estimates for θ and observed attributes in x. In the M-step, parameters θ are estimated that maximize the complete-data likelihood. This process is iterated till convergence.

The EM algorithm only provides point estimates for missing data and parameters θ. In Bayesian multiple imputation, the posterior probability distributions for the missing values and parameters θ can be directly simulated using Markov chain Monte Carlo methods (MCMC), thereby taking into account the uncertainties associated with the missing values and parameters θ. Details on the EM algorithm and Bayesian multiple imputation can be found in the work of Schafer (1997).

Network Analysis

The recent explosion of genome-scale protein interaction screens has made it possible to construct a map of the interactions within a cell. These interactions form an intricate network. A crucial challenge as these data continue to flood in is how to reduce this complex tangle of interactions into the key components and interconnections that control a biological process. For example, in developing a drug to attack a disease, a molecular target that is a central player in the disease is required. The target must be as specific as possible to reduce unanticipated side-effects.

Topological Analysis

In addition to protein networks, complex networks are also found in the structure of a number of wide-ranging systems, such as the internet, power grids, the ecological food web and scientific collaborations. Despite the disparate nature of the various systems, it has already been demonstrated that all these networks share common features in terms of topology. In this sense, networks and topological analysis can provide the framework for describing biological systems in a format that is more transferable and accessible to a broader scientific audience. As mentioned previously, the topological analysis of networks is a means of gaining quantitative insight into their organization at the most basic level.

Of the many methods in topological statistics, four are particularly pertinent to the analysis of networks. They are average degree (K), clustering coefficient (C), characteristic path length (L) and diameter (D). Chapter 8 will give formal definitions to these method. Earlier analyses of complex networks were based on the theory of classical random networks. The idea was introduced by Erdos and Renyi (1959). The theory assumes that any two given nodes in a network are connected at random, with probability p, and the degrees of the nodes follow a Poisson distribution. This means that there is a strong peak at the average degree, K. Most random networks are of a highly homogenous nature, that is, most nodes have the same number of links, $k(i) = K$, where $k(i)$ is the ith node. The chance of encountering nodes with k links decreases exponentially for large values of k, *i.e.*, $P(k) = e^{-k}$. This shows that it is highly unlikely to encounter nodes of a degree that is significantly higher than the average. Recently, theories other than the classical random network theory were proposed. One such attempt is the 'scale-free' model by Barabasi and Albert (1999) to explain the heterogeneous nature of some complex networks.

In the 'scale-free' model, the degree distribution of networks is assumed to follow a power-law relationship ($P(k) = k^{-r}$), rather than the Poisson distribution assumed under earlier classical random network theory. One advantage of having such an assumption is that most of the nodes within such networks are highly connected via hubs, with very few links between them. This attribute makes the model particularly applicable to complex biological networks such as those involving protein-protein interactions. Many aspect of genomic biology have such a scale-free structure. In a concurrent effort by Watts and Strogatz (1998), it is found that many networks can be attributed with a 'small- world' property, which means that they are both highly clustered in nature and contain small characteristic path lengths (*i.e.*, large values of C, and small values of L). Finally, the analysis of complex networks can be further divided into two broad categories: that is, undirected versus directed.

In the former, there is a commutative property: the statement 'node A is linked to node B' is the exact equivalent to the statement 'node B is linked to node A'. In contrast, in a directed network, the edges have a defined direction, and thus the clustering coefficient is not applicable for directed networks. There are many complex networks in biology that can be analysed using graph-topological tools. Recent advances in large-scale experiments have generated a great variety of genome-wide interaction networks, especially for *S. cerevisiae*.

Moreover, there exist a number of databases (*e.g.*, MIPS, BIND, DIP) that provide manually curated interactions for the yeast organism. Beyond experimentally derived protein-protein interactions, there are also predicted interactions literature-derived interactions and regulatory interactions. All of these networks are amenable to topological analysis. In order to facilitate the topological analysis of interaction networks, we constructed a web tool, TopNet, to perform automatic comparisons. TopNet takes an arbitrary undirected network and a group of node classes as an input to create sub-networks. Then it computes all four topological statistics mentioned above and draws a power-law degree distribution for each sub-network.

The results of these calculations are plotted in the same format for each statistic to facilitate direct comparison. TopNet also enables the user to explore complex networks by sections. For example, all neighbours of a certain node can be shown on a simple graph. Alternatively, the user can select two nodes and request that all paths not exceeding some specified length be displayed as an independent graph. Clearly, the great variety and complexity of biological networks present a wealth of interesting problems and challenges for the application of topological analysis, which would lead to better understanding of many aspects of modern biology.

Modelling Networks

Cellular interaction networks are composed of modules. Biological modules are conserved groups of proteins and other molecules that are responsible for a common structure and function. Experimental study of the signalling network of the budding yeast, *S. cerevisiae,* sparked the conception of *modular signalling networks*. These well studied pathways are an ideal proving ground for computational study of modularity in biological networks. Yeast signalling pathways are composed of five distinct mitogen-activated protein kinase (*MAP kinase*) pathways.

The yeast MAP kinase pathways are composed of two modules, an upstream sensing component that is responsible for detecting signals in the environment and a downstream kinase module that amplifies and propagates the signal while maintaining its specificity. MAP kinase pathways and their modules are highly conserved in eukaryotes. These modules are key determinants of the specificity of signalling. The filamentation pathway is one of the least understood of the pathways. Under certain conditions yeast cells undergo a morphological transition from a round form to a filamentous invasive form. This is accompanied by an altered cell cycle and bipolar budding pattern.

Little is known about how the signal that mediates the altered cell cycle is transmitted from the MAP kinase module. Modelling a biological system as a network of modules reduces the complexity of the network and pinpoints the crucial interconnections between modules. These connections can be reprogrammed in evolution, or the laboratory to create new biological responses. For example a drug that targeted the connection between a growth factor detection module and a signal amplification module could stop the progression of cancer by ablating the link responsible for transmitting the aberrant growth signal.

Rives and Galitski (2003) show an example of a network clustering method that has been successfully applied to biological networks to model their modular structure. The goal of the clustering method is to define a similarity metric between each possible pair of proteins in the network. This similarity metric is a function that can range from –1 to 1, where a higher score represents a greater prediction that the two proteins are in the same module. Proteins can then be clustered based on their similarity scores. This network clustering method was applied to the yeast filamentation response network to model its modular structure. Modelling a biological system as a network of modules identifies key proteins and interconnections. Two Two important features are hubs and intermodule connections. Modules tend to have one or a proteins that are highly connected within the module. These hubs are essential to the functions of their modules. While there are many interactions within modules, there is a relative paucity between modules.

The biological functions of proteins that appear as connections between modules suggest they are crucial points of information flow constriction and cross-talk. Protein interaction networks reflect the modular structure of biological systems. They have a high clustering coefficient and a low frequency of direct connections between high-connectivity nodes. Members of biological modules have a high frequency of interactions with other members of their module and a paucity of interactions with members of other modules. Modules form clusters in the interaction network that can be identified using network-clustering methods. Biological modules can be identified in complex protein interaction networks. They can be used to reduce the complexity of a network by moving up in the biological hierarchy. An induced graph is a graph in which some nodes are collapsed together into a single node that shares all of their interactions. It allows a complex interaction network to be reduced to the interactions between emergent modular components. This preserves important information while reducing the complexity.

Modular modelling opens many avenues for the investigation of biological systems. As genome-scale interaction data continue to be produced, methods are required which can identify the important interactions and proteins. Computational network clustering approaches can be used to identify these essential proteins and crucial points of cross talk. Furthermore they can be used to generate testable biological hypotheses.

DISCUSSION

Among the machine-learning approaches that could be applied to predicting interactions described above, Bayesian networks have clear advantages: (1) they allow for combining highly dissimilar types of data (*i.e.* numerical and categorical), converting them to a common probabilistic framework, without unnecessary simplification; (2) they readily accommodate missing data and (3) they naturally weight each information source according to its reliability.

In contrast to 'black-box' predictors, Bayesian networks are readily interpretable, as they represent conditional probability relationships among information sources. In a naive Bayesian network, the assumption is that the different sources of evidence (*i.e.* our datasets with information about protein interactions) are conditionally independent. Conditional independence means that the information in the N datasets is independent given that a protein pair is either positive or negative. From a computational standpoint, the naive Bayesian network is easier to compute than the fully connected network. As we add more features, we will find more sources of evidence that are strongly correlated. This issue can be addressed in two ways:

(1) we will use a fully connected network or subnetwork to handle the correlated features; (2) we will use principal component analysis (PCA), in which the first principal component of the vector of the two correlations will be used as one independent source of evidence for the protein interaction prediction.

For example, in analysing expression correlations, we found that two of the main datasets were strongly correlated; however, using the first component of the PCA removed this issue. Another challenge in extending our naive Bayesian integration to incorporate additional genomic features is the missing value problem. Determination on protein interactions is the initial step and cornerstone towards mapping molecular interaction networks. Three challenges in the network analysis remain: (1) 3D view of interaction networks in a cell; (2) dynamics and context-dependent nature of interaction networks; (3) quantitative measures of networks. Molecular interaction networks lay the foundation for analysis of the cell in systems biology. With combined experimental, computational and theoretical efforts, a complete mapping of interaction networks, and ultimately a rational understanding of cellular behaviour, will become reality.

7

Data Base Mapping

A few years ago, only a handful of ready-made maps of the human genome existed, and these were low-resolution maps of small areas. Biomedical researchers wishing to localize and clone a disease gene were forced, by and large, to map their region of interest, a time-consuming and painstaking process. This situation has changed dramatically in recent years, and there are now high-quality genome-wide maps of several different types containing tens of thousands of DNA markers. With the pending availability of a finished human sequence, most efforts to construct genomic maps will come to a halt; however, integrated maps, genome catalogues, and comprehensive databases linking positional and functional genomic data will become even more valuable.

Genome projects in other organisms are at various stages, ranging from having only a handful of available maps to having a complete sequence. By taking advantage of the available maps and DNA sequence, a researcher can, in many cases, focus in on a candidate region by searching public mapping databases in a matter of hours rather than by performing laboratory experiments over a course of months. Subsequently, the researcher's burden has now shifted from mapping the genome to navigating a vast *terra incognita* of Web sites, FTP servers, and databases. There are large databases such as the National Center for Biotechnology Information (NCBI) Entrez Genomes Division, Genome Database (GDB), and *Mouse Genome Database* (MGD), smaller databases serving the primary maps published by genome centers, sites sponsored by individual chromosome committees, and sites used by smaller laboratories to publish highly detailed maps of specific regions. Each type of resource contains information that is valuable in its own right, even when it overlaps with the information found at others. Finding one's way around this information space is not easy.

A recent search for the word *"genome"* using the AltaVista Web search engine turned up 400,000 potentially relevant documents. This chapter is intended as a *"map of the maps,"* a way to guide readers through the maze of publicly available genomic mapping resources. The different types of markers and methods used for genomic mapping will be reviewed and the inherent complexities in the construction and utilization of genome maps will be discussed. Several large community databases and method-specific mapping projects will be presented in detail. Finally, practical examples of how these tools and resources can be used to aid in specific types of mapping studies such as localizing a new gene or refining a region of interest will be provided. A complete description of the mapping resources available for all species would require an entire book. Therefore, this chapter focuses primarily on humans, with some references to resources for other organisms.

MAPPING AND SEQUENCING

The recent advent of whole-genome sequencing projects for humans and select model organisms is dramatically impacting the use and utility of genomic map-based information and methodologies. *Genomic maps* and *DNA sequence* are often treated as separate entities, but large, uninterrupted DNA sequence tracts can be thought of and used as an ultra-high-resolution mapping technique. Traditional genomic maps that rely on genomic markers and either clone-based or statistical approaches for ordering are precursory to finished and completely annotated DNA sequences of whole chromosomes or genomes. However, such completed genome sequences are predicted to be publicly available only in 2002 for humans, 2005 for the mouse, and even later for other mammalian species, although complete sequences are now available for some individual human chromosomes and selected lower eukaryotes.

Until these completed sequences are available, mapping and sequencing approaches to genomic analysis serve as complementary approaches for chromosome analysis. Before determination of an entire chromosome's sequence, the types of sequences available can be roughly grouped into marker/gene-based tags [*e.g.*, expressed sequence tags (ESTs) and sequence-tagged sites (STSs)], single gene sequences, prefinished DNA clone sequences, and completed, continuous genomic sequence tracts. The first two categories provide rich sources of the genomic markers used for mapping, but only the last two categories can reliably order genomic elements.

The human genome draft sequence is an example of a prefinished sequence, in which >90% of the entire sequence is available, but most continuous sequence tracts are relatively short (usually <100 kb and often <10 kb), thus providing high local resolution but little long-range ordering information. Genomic maps can help provide a context for this sequence information. Thus, two or more sequences containing unique genomic markers can be oriented if these markers are ordered on a map. In this way, existing maps serve as a scaffold for orienting, directing, and troubleshooting sequencing projects. Similarly, users can first define a chromosomal region of interest using a traditional map approach and then can identify relevant DNA sequences to analyze by finding long sequences containing markers mapping within the defined region. NCBI tools such as BLAST and electronic PCR (e-PCR) are valuable for finding marker/sequence identities, and several of the resources discussed below provide marker/sequence integration. As large sequence tracts emerge from the human and model organism projects, sequence-based ordering of genomic landmarks will eventually supplant map-based ordering methods.

The evolution from a mapped chromosome to the determination of the chromosome's complete sequence is marked by increasing incorporation of partial genomic sequence tracts into the underlying map. Once complete, finished sequences can be used to confirm map-determined marker orders. Given the error rates inherent in both map and sequence-assembly methodology, it is good practice to use both map and sequence information simultaneously for independent verification of regional order.

GENOMIC MAP ELEMENTS

DNA Markers

A DNA *marker* is simply a uniquely identifiable segment of DNA. There are several different types of markers, usually ranging in size from one to 300-400 nucleotide bases in size. Markers can be thought of as landmarks, and a set of markers whose relative positions (or order) within a genome are known comprises a *map*. Markers can be categorized in several ways. Some markers are polymorphic, and others are not (monomorphic). Detection of markers may be either PCR based or hybridization based.

Some markers lie in a sequence of DNA that is expressed; some do not, or their expression status may be unknown. PCR-based markers are commonly referred to as sequence-tagged sites (STSs). An STS is defined as a segment of genomic DNA that can be uniquely PCR amplified by its primer sequences. STSs are commonly used in the construction of physical maps. STS markers may be developed from any genomic sequence of interest, such as from characterized and sequenced genes, or from expressed sequence tags. Alternatively, STSs may be randomly identified from total genomic DNA. The EST database (dbEST) at NCBI stores information on most STS markers.

Polymorphic Markers

Polymorphic markers are those that show sequence variation among individuals. Polymorphic markers are used to construct genetic linkage maps. The number of alleles observed in a population for a given polymorphism, which can vary from two to >30, determines the degree of polymorphism. For many studies, highly polymorphic markers (>5 alleles) are most useful.

Polymorphisms may arise from several types of sequence variations. One of the earlier types of polymorphic markers used for genomic mapping is a *restriction fragment length polymorphism* (RFLP). An RFLP arises from changes in the sequence of a restriction enzyme recognition site, which alters the digestion patterns observed during hybridization-based analysis. Another type of hybridization-based marker arises from a *variable number of tandem repeat units* (VNTR).

A VNTR locus usually has several alleles, each containing a different number of copies of a common motif of at least 16 nucleotides tandemly oriented along a chromosome. A third type of polymorphism is due to *tandem repeats* of short sequences that can be detected by PCR-based analysis. These are known variously as *microsatellites, short tandem repeats* (STRs), *STR polymorphisms* (STRPs), or *short sequence length polymorphisms* (SSLPs). These repeat sequences usually consist of two, three, or four nucleotides and are plentiful in most organisms. All PCR-converted STR markers (those for which a pair of oligonucleotides flanking the polymorphic site suitable for PCR amplification of the locus has been designed) are considered to be STSs.

The advent of PCR-based analysis quickly made microsatellites the markers of choice for mapping. Anotner polymorphic type of PCR-based marker is a single nucleotide polymorphism (SNP), which results from a base variation at a single nucleotide position. Most SNPs have only two alleles (*biallelic*). Because of their low heterozygosity, maps of SNPs require a much higher marker density than maps of microsatellites. SNPs occur frequently in most genomes, with one SNP occurring on average approximately once in every 100-300 bases in humans. SNPs lend themselves to highly automated fluidic or DNA chip-based analyses and have quickly become the focus of several large-scale development and mapping projects in humans and other organisms. Further details about all of these types of markers can be found elsewhere.

DNA Clones

The possibility of physically mapping eukaryotic genomes was largely realized with the advent of cloning vehicles that could efficiently and reproducibly propagate large DNA fragments. The first generation of large-insert cloning was made possible with yeast artificial chromosome (YAC) libraries. Because YACs can contain fragments up to 2 Mb, they are suitable for quickly making low-resolution maps of large chromosomal regions, and the first whole-genome physical maps of several eukaryotes were constructed with YACs.

Although YAC libraries work well for ordering STSs and for joining small physical maps, the high rate of chimerism and instability of these clones makes them unsuitable for DNA sequencing. The second and current generation of large-insert clones consists of bacterial artificial chromosomes (BACs) and PI-artificial chromosomes, both of which act as episomes in bacterial cells rather than as eukaryotic artificial chromosomes. Bacterial propagation has several advantages, including higher DNA yields, ease-of-use for sequencing, and high integrity of the insert during propagation.

Despite the relatively limited insert sizes (usually 100-300 kb), BACs and PACs have largely replaced YACs as the clones of choice for large-genome mapping and sequencing projects. DNA fingerprinting has been applied to BACs and PACs to determine insert overlaps and to construct clone contigs. In this technique, clones are digested with a restriction enzyme, and the resulting fragment patterns are compared between clones to identify those sharing subsets of identically sized fragments. In addition, the ends of BAC and PAC inserts can be directly sequenced; clones whose insert-end sequences have been determined are referred to as sequence-tagged clones (STCs). Both DNA fingerprinting and STC generation now play instrumental roles in physical mapping strategies.

TYPES OF MAPS

Cytogenetic Maps

Cytogenetic maps are those in which the markers are localized to chromosomes in a manner that can be directly imaged. Traditional cytogenetic mapping hybridizes a radioactively or fluorescently labeled DNA probe to a chromosome preparation, usually in parallel with a chromosomal stain such as Giemsa, which produces a banded karyotype of each chromosome. This allows assignment of the probe to a specific chromosomal band or region. Assignment of cytogenetic positions in this manner is dependent on some subjective criteria (variability in technology, methodology, interpretation, reproducibility, and definition of band boundaries). Thus, inferred cytogenetic positions are often fairly large and occasionally overinterpreted, and some independent verification of cytogenetic position determinations is warranted for crucial genes, markers, or regions.

Probes used for cytogenetic mapping are usually large-insert clones containing a gene or polymorphic marker of interest. Despite the subjective aspects of cytogenetic methodology, karyotype analysis is an important and relatively simple clinical genetic tool; thus, cytogenetic positioning remains an important parameter for defining genes, disease loci, and chromosomal rearrangements. Newer cytogenetic techniques such as interphase fluorescence in situ hybridization (FISH) and fiber FISH instead examine chromosomal preparations in which the DNA is either naturally or mechanically extended.

Studies of such extended chromatin have demonstrated a directly proportional relationship between the distances measured on the image and the actual physical distance for short stretches, so that a physical distance between two closely linked probes can be determined with some precision. However, these techniques have a limited ordering range ($\leq 1 - 2$ Mb) and are not well-suited for high-throughput mapping.

Genetic Linkage Maps

Genetic linkage (GL) maps (also called meiotic maps) rely on the naturally occurring process of recombination for determination of the relative order of, and map distances between, polymorphic markers. Crossover and recombination events take place during meiosis and allow rearrangement of genetic material between homologous chromosomes. The likelihood of recombination between markers is evaluated using genotypes observed in multigenerational families. Markers between which only a few recombination occur are said to be linked and such markers are usually located close to each other on the same chromosome.

Markers between which many recombinations take place are unlinked and usually lie far apart, either at opposite ends of the same chromosome or on different chromosomes. Because the recombination events cannot be easily quantified, a statistical method of maximum likelihood is usually applied in which the likelihood of two markers being linked is compared with the likelihood of being unlinked. This likelihood ratio is called a *"lod"* score (for "*log of the odds*"), and a lod score greater than 3 (corresponding to odds of 1,000:1 or greater) is usually taken as evidence that markers are linked. The lod score is computed at a range of recombination fraction values between markers (from 0 to 0.5), and the recombination fraction at which the lod score is maximized provides an estimate of the distance between markers. A map function is then used to convert the recombination fraction into an additive unit of distance measured in centiMorgans (cM), with 1 cM representing a 1% probability that a recombination has occurred between two markers on a single chromosome.

Because recombination events are not randomly distributed, map distances on linkage maps are not directly proportional to physical distances. The majority of linkage maps are constructed using multipoint linkage analysis, although multiple pairwise linkage analysis and minimization of recombination are also valid approaches. Commonly used and publicly available computer programs for building linkage maps include LINKAGE, CRI-MAP MultiMap MAPMAKER and MAP. The MAP-O-MAT Web server is available for estimation of map distances and for evaluation of statistical support for order. Because linkage mapping is a based on statistical methods, linkage maps are not guaranteed to show the correct order of markers. Therefore, it is important to be critical of the various available maps and to be aware of the statistical criteria that were used in map construction.

Typically, only a subset of markers (framework or index markers) is mapped with high statistical support. The remainder are either placed into well-supported intervals or bins or placed into unique map positions but with low statistical support for order. To facilitate global coordination of human linkage mapping, DNAs from a set of reference pedigrees collected for map construction were prepared and distributed by the Centre d'Etude du Polymorphism Humain (CEPH; Dausset *et. al.*, 1990). Nearly all human linkage maps are based on genotypes from the CEPH reference pedigrees, and genotypes for markers scored in the CEPH pedigrees are deposited in a public database maintained at CEPH. Most recent maps are composed almost entirely of highly polymorphic STR markers. These linkage maps have already exceeded the maximum map resolution possible given the subset of CEPH pedigrees that are commonly used for map construction, and no further large-scale efforts to place STR markers on human linkage maps are planned.

Thousands of SNPs are currently being identified and characterized, and a subset are being placed on linkage maps. Linkage mapping is also an important tool in experimental animals, with many maps already produced at high resolution and others still under development.

Radiation Hybrid Maps

Radiation hybrid (RH) mapping is very similar to linkage mapping. Both methods rely on the identification of chromosome breakage and reassortment. The primary difference is the mechanism of chromosome breakage. In the construction of radiation hybrids, breaks are induced by the application of lethal doses of radiation to a donor cell line, which is then rescued by fusion with a recipient cell line (typically mouse or hamster) and grown in a selective medium such that only fused cells survive. An RH panel is a library of fusion cells, each of which has a separate collection of donor fragments.

The complete donor genome is represented multiple times across most RH panels. Each fusion cell, or radiation hybrid, is then scored by PCR to determine the presence or absence of each marker of interest. Markers that physically lie near each other will show similar patterns of retention or loss across a panel of RH cells and behave as if they are linked, whereas markers that physically lie far apart will show completely dissimilar patterns and behave as if they are unlinked. Because the breaks are largely randomly distributed, the break frequencies are roughly directly proportional to physical distances. The resulting data set is a series of positive and negative PCR scores for each marker across the hybrid panel.

These data can be used to statistically infer the position of chromosomal breaks, and, from that point on, the procedures for map construction are similar to those used in linkage mapping.

A map function is used to convert estimates of breakage frequency to additive units of distance measured in centirays (cR), with 1 cR representing a 1% probability that a chromosomal break has occurred between two markers in a single hybrid. The resolution of a radiation hybrid map depends on the size of the chromosomal fragments contained in the hybrids, which in turn is pro-proportional the amount of irradiationc to which the human cell line was exposed. Most RH maps are built using multipoint linkage analysis, although multiple-pairwise linkage analysis and minimization of recombination are also valid approaches.

Three genome-wide RH panels exist for humans and are commercially available, and RH panels are available for many other species as well. Widely used computer programs for RH mapping are RHMAP, RHMAPPER and MultiMap and on-line servers that allow researchers to place their RH mapped markers on existing RH maps are available. The Radiation Hybrid Database (RHdb) is the central repository for RH data on panels available in all species. The Radiation Hybrid Information Web site also contains multi-species information about available RH panels, maps, ongoing projects, and available computer programs.

Transcript Maps

Of particular interest to researchers chasing disease genes are maps of transcribed sequences. Although the transcript sequences are mapped using one of the methods described in this section, and thus do not require a separate mapping technology, they are often set apart as a separate type of map. These maps consist of expressed sequences and sequences derived from known genes that have been converted into STSs and usually placed on conventional physical maps. Recent projects for creating large numbers of ESTs have made tens of thousands of unique expressed sequences available to the mapping laboratories. Transcribed sequence maps can significantly speed the search for candidate genes once a disease locus has been identified. The largest human transcript map to date is the GeneMap '99, described below.

Physical Maps

Physical maps include maps that either are capable of directly measuring distances between genomic elements or that use cloned DNA fragments to directly order elements. Many techniques have been created to develop physical maps. The most widely adopted methodology, due largely to its relative simplicity, is STS content mapping. This technique can resolve regions much larger than 1 Mb and has the advantage of using convenient PCR-based positional markers. In STS content maps, STS markers are assayed by PCR against a library of large-insert clones. If two or more STSs are found to be contained in the same clone, chances are high that those markers are located close together. (The fact that they are not close 100% of the time is a reflection of various artifacts in the mapping procedure, such as the presence of chimeric clones.) The STS content mapping technique builds a series of contigs (*i.e.*, overlapping clusters of clones joined together by shared STSs).

The resolution and coverage of such a map are determined by a number of factors, including the density of STSs, the size of the clones, and the depth of the clone library. Maps that use cloning vectors with smaller insert sizes have a higher theoretical resolution but require more STSs to achieve coverage of the same area of the genome. Although it is generally possible to deduce the relative order of markers on STS content maps, the distances between adjacent markers cannot be measured with accuracy without further experimentation, such as by restriction mapping. However, STS content maps have the advantage of being associated with a clone resource that can be used for further studies, including subcloning, DNA sequencing, or transfection. Several other techniques in addition to STS content and radiation hybrid mapping have also been used to produce physical maps.

Clone maps rely on techniques other than STS content to determine the adjacency of clones. For example, the CEPH YAC map used a combination of fingerprinting, inter-Alu product hybridization, and STS content to create a map of overlapping YAC clones. Fingerprinting is commonly used by sequencing centers to assemble and/or verify BAC and PAC contigs before clones are chosen for sequencing, to select new clones for sequencing that can extend existing contigs, and to help order genomic sequence tracts generated in whole-genome sequencing projects. Sequencing of large-insert clone ends (STC generation), when applied to a whole-genome clone library of adequate coverage, is very effective for whole-genome mapping when used in combination with fingerprinting of the same library. Deletion and somatic cell hybrid maps relying on large genomic reorganizations (induced deliberately or naturally occurring) to place markers into bins defined by chromosomal breakpoints have been generated for some human chromosomes.

Optical mapping visualizes and measures the length of single DNA molecules extended and digested with restriction enzymes by high-resolution microscopy. This technique, although still in its infancy, has been successfully used to assemble whole chromosome maps of bacteria and lower eukaryotes and is now being applied to complex genomes.

Comparative Maps

Comparative mapping is the process of identifying conserved chromosome segments across different species. Because of the relatively small number of chromosomal breaks that have occurred during mammalian radiation, the order of genes usually is preserved over large chromosomal segments between related species. Orthologous genes (copies of the same genes from different species) can be identified through DNA sequence homology, and sets of orthologous genes sharing an identical linear order within a chromosomal region in two or more species are used to identify conserved segments and ancient chromosomal breakpoints. Knowledge about which chromosomal segments are shared and how they have become rearranged over time greatly increases our understanding of the evolution of different plant and animal lineages.

One of the most valuable applications of comparative maps is to use an established gene map of one species to predict positions of orthologous genes in another species. Many animal models exist for diseases observed in humans. In some cases, it is easier to identify the responsible genes in an animal model than in humans, and the availability of a good comparative map can simplify the process of identifying the responsible genes in humans. In other cases, more might be known about the gene(*s*) responsible in humans, and the same comparative map could be used to help identify the gene(*s*) responsible in the model species. There are several successful examples of comparative candidate gene mapping As mapping and sequencing efforts progress in many species, it is becoming possible to identify smaller homologous chromosome segments, and detailed comparative maps are being developed between many different species.

Fairly dense gene-based comparative maps now exist between the human, mouse, and rat genomes and also between several agriculturally important mammalian species. Sequence- and protein-based comparative maps are also under development for several lower organisms for which complete sequence is available. A comparative map is typically presented either graphically or in tabular format, with one species designated as the index species and one or more others as comparison species. Homologous regions are presented graphically with nonconsecutive segments from the comparison species shown aligned with their corresponding segments along the map of the index species.

Integrated Maps

Map integration provides interconnectivity between mapping data generated from two or more different experimental techniques. However, achieving accurate and useful integration is a difficult task. Most of the genomic maps and associated Web sites discussed in this section provide some measure of integration, ranging from the approximate cytogenetic coordinates provided in the Genethon GL map to the inter-associated GL, RH, and physical data provided by the Whitehead Institute (WICGR) Web site. Several integration projects have created truly integrated maps by placing genomic elements mapped by differing techniques relative to a single map scale. The most advanced sources of genomic information provide some level of genomic cataloguing, where considerable effort is made to collect, organize, and map all available positional information for a given genome.

COMPLEXITIES AND PITFALLS OF MAPPING

It is important to realize that the genomic mapping information currently available is a collection of a large number of individual data sets, each of which has unique characteristics. The experimental techniques, methods of data collection, annotation, presentation, and quality of the data differ considerably among these data sets. Although most mapping projects include procedures to detect and eliminate and/or correct errors, there are invariably some errors that occur, which often result in the incorrect ordering or labeling of individual markers.

Although the error rate is usually very low (5% or less), a marker misplacement can obviously have a great impact on a study. A few mapping Web sites are beginning to flag and correct (or at least warn) users of potential errors, but most errors cannot be easily detected. Successful strategies for minimizing the effects of data error include (1) simultaneously assessing as many different maps as possible to maximize redundancy (note that ideally "different" maps use independently-derived data sets or different techniques); (2) increased emphasis on utilizing integrated maps and genomic catalogues that provide access to all available genomic information for the region of interest (while closely monitoring the map resolution and marker placement confidence of the integrated map); and (3) if possible, experimentally verifying the most critical marker positions or placements. In addition to data errors, several other, more subtle complexities are notable.

Foremost is the issue of nomenclature, or the naming of genomic markers and elements. Many markers have multiple names, and keeping track of all the names is a major bioinformatics challenge. For example, the polymorphic marker D1S243 has several assigned names: AFM214yg7, which is actually the name of the DNA clone from which this polymorphism was identified; SHGC-428 and stSG729, two examples of genome centers renaming a marker to fit their own nomenclature schemes; and both GDB:201358 and GDB:133491, which are database identifier numbers used to track the polymorphism and STS associated with this marker, respectively, in the Genome Database (GDB). Genomic mapping groups working with a particular marker often assign an additional name to simplify their own data management, but, too often, these alternate identifiers are subsequently used as a primary name. Furthermore, many genomic maps display only one or a few names, making comparisons of maps problematic.

Mapping groups and Web sites are beginning to address these inherent problems, but the difficulty of precisely defining "markers," "genes," and *"genomic elements."* adds to the confusion. It is important to distinguish between groups of names defining different elements. A gene can have several names, and it can also be associated with one or more EST clusters, polymorphisms, and STSs. Genes spanning a large genomic stretch can even be represented by several markers that individually map to different positions. Web sites providing genomic cataloguing, such as LocusLink, UniGene, GDB, GeneCards, and eGenome, list most names associated with a given genomic element. Nevertheless, collecting, cross-referencing, and frequently updating one's own sets of names for markers of interest is also a good practice as even the genomic cataloguing sites do not always provide complete nomenclature collections. Each mapping technique yields its own resolution limits. Cytogenetic banding potentially orders markers separated by $\geq 1-2$ Mb, and genetic linkage (GL) and RH analyses yields long-range resolutions of $\geq 0.5-1$ Mb, although localized ordering can achieve higher resolutions.

The confidence level with which markers are ordered on statistically based maps is often overlooked, but this is crucial for assessing map quality. For genomes with abundant mapping data such as human or mouse, the number of markers used for mapping often far exceeds the ability of the technique to order all markers with high confidence (often, confidence levels of 1,000:1 or lod 3 are used as a cutoff, which usually means that a marker is ≥ 1,000: 1 times more likely to be in the given position than in any other). Mappers have taken two approaches to address this issue. The first is to order all markers in the best possible linear order, regardless of the confidence for map position of each marker.

Alternatively, the high confidence linear order of a subset of markers is determined, and the remaining markers are then placed in high confidence "intervals," or regional positions.

The advantage of the first approach is that resolution is maximized, but it is important to pay attention to the odds for placement of individual markers, as alternative local orders are often almost equally likely. Thus, beyond the effective resolving power of a mapping technique, increased resolution often yields decreased accuracy, and researchers are cautioned to strike a healthy balance between the two. Each mapping technique also yields very different measures of distance. Cyto-genetic approaches, with the exception of high-resolution fiber FISH, provide only rough distance estimates, GL and STS content mapping provide marker orientation but only relative distances, and RH mapping yields distances roughly proportional to true physical distance. For GL analysis, unit measurements are in centMiorgans, with 1 cM equivalent to a 1% chance of recombination between two linked markers.

The conversion factor of 1 cM ≈1 Mb is often cited for the human genome but is overstated, as this is just the *average* ratio genome-wide, and many chromosomal regions have recombination hotspots and coldspots in which the cM-to-Mb ratio varies as much as 10-fold. In general, cytogenetic maps provide subband marker regionalization but limited localized ordering, GL and STS content maps provide excellent ordering and limited-to-moderate distance information, and RH maps provide the best combination of localized orderning and distance estimates. Finally, there are various levels at which genomic information can be presented. *Single-resource maps* such as the Genethon GL maps use a single experimental technique and analyze a homogeneous set of markers. Strictly *comparative maps* make comparisons between two or more different single-dimension maps either within or between species but without combining data sets for integration. GDB's Mapview program can display multiple maps in this fashion. *Integrated maps* recalculate or completely integrate multiple data sets to display the map position of all genomic elements relative to a single scale; GDB's Comprehensive Maps are an example of such integration.

Lastely, *genome cataloguing* is a relatively new way to display genomic information, in which many data sets and/or Web sites are integrated to provide a comprehensive listing and/ or display of all identified genomic elements for a given chromosome or genome. Completely sequenced genomes such as *C. elegans* and *S. cerevisiae* have advanced cataloguing efforts but catalogues for complex genome organisms are in the early stages. Examples include the interconnected NCBI databases, MGD, and eGenome Catalogues provide a "one-stop shopping" solution to collecting and analyzing genomic data and are recommended as a maximum-impact means to begin a regional analysis. However, the individual data sets provide the highest quality positional information and are ultimately the most useful for region definition and refinement.

DATA REPOSITORIES

There are several valuable and well-developed data repositories that have greatly facilitated the dissemination of genome mapping resources for humans and other species. This section covers three of the most comprehensive resources for mapping in humans: the *Genome Database* (GDB), the *National Center for Biotechnology* Information (NCBI), and the *Mouse Genome Database* (MGD). More focused resources are mentioned in the *Mapping Projects and Associated Resources* section of this chapter.

GDB

The *Genome Database* (GDB) is the official central repository for genomic mapping data created by the Human Genome Project. GDB's central node is located at the Hospital for Sick

Children (Toronto, Ontario, Canada). Members of the scientific community as well as GDB staff curate data submitted to the GDB. Currently, GDB comprises descriptions of three types of objects from humans: Genomic Segments (genes, clones, amplimers, breakpoints, cytogenetic markers, fragile sites, ESTs, syndromic regions, contigs, and repeats), Maps (including cytogenetic, GL, RH, STS-content, and integrated), and Variations (primarily relating to polymorphisms). In addition, contributing investigator contact information and citations are also provided.

The GDB holds a vast quantity of data submitted by hundreds of investigators. Therefore, like other large public databases, the data quality is variable. A more detailed description of the GDB can be found in Talbot and Cuticchia (1994). GDB provides a full-featured query interface to its database with extensive online help. Several focused query interfaces and predefined reports, such as the Maps within a Region search and Lists of Genes by Chromosome report, present a more intuitive entry into GDB. In particular, GDB's Mapview program provides a graphical interface to the genetic and physical maps available at GDB. A Simple Search is available on the home page of the GDB Web site. This query is used when searching for information on a specific genomic segment, such as a gene or STS (amplimer, in GDB terminology) and can be implemented by entering the segment name or GDB accession number.

Depending on the type of segment queried and the available data, many different types of segment-specific information may be returned, such as alternate names (aliases), primer sequences, positions in various maps, related segments, polymorphism details, contributor contact information, citations, and relevant external links. At the bottom of the GDB home page is a link to Other Search Options. From the Other Search Options page there are links to three customized search forms (Markers and Genes within a Region, Maps within a Region, and Genes by Name or Symbol), sequence-based searches, specific search forms for subclasses of GDB elements, and precompiled lists of data. A particularly useful query is the Maps within a Region search. This search allows retrieval of all maps stored in GDB that span a defined chromosomal region. In a two-step process, the set of maps to be retrieved is first determined, and, from these, the specific set to be displayed is then selected. Select the Maps within a Region link to display the search form.

To view an entire chromosome, simply select it from the pop-up menu. However, entire cnromosomes may take considerable time to download and display; therefore, it is usually best to choose a subchromosomal region. To view a chromosomal region, type the names of two cytogenetic bands or flanking genetic markers into the text fields labeled From and To. If the flanking markers used in the query are stored in GDB as more than one type of object, the next form will request selection of the specific type of element for each marker. The resulting form lists all maps stored in GDB that overlap the selected region. Given the flanking markers specified above, there are a total of 21 maps.

The user selects which maps to display by marking the respective checkboxes. Note that GDB's Comprehensive Map is automatically selected. If a graphical display is requested, the size of the region and the number of maps to be displayed can significantly affect the time to fetch and display them. The resulting display will appear in a separate window showing the selected maps in side-by-side fashion. While the Mapview display is loading, a new page is shown in the browser window.

If your system is not configured to handle Java properly, a helpful message will be displayed in the browser window. *(Important:* Do not close the browser window behind Mapview. Because of an idiosyncrasy of Java's security specification, the applet cannot interact properly with GDB unless the browser window remains open. Mapview has many useful options, which are well described in the online help. Some maps have more than one *tier,* each displaying different types of markers, such as markers positioned with varying confidence thresholds on a linkage or radiation hybrid map. It is possible to zoom in and out, highlight markers across maps, color code different tiers, display markers using different aliases, change the relative position of the displayed maps, and search for specific markers.

To retrieve additional information on a marker from any of the maps, double-click on its name to perform a *Simple Search* (as described above). A separate browser window will then display the GDB entry for the selected marker. Two recently added GDB tools are CDB BLAST and e-PCR. These are available from the Other Search Options page and enable users to employ GDB's many data resources in their analysis of the emerging human genome sequence. GDB BLAST returns GDB objects associated with BLAST hits against the public human sequence. GDB's e-PCR finds which of its many amplimers are contained within queried DNA sequences and is thereby a quick means to determine or refine gene or marker localization. In addition, the GDB has many useful genome resource Web links on its Resources page.

NCBI

The NCBI has developed many useful resources and tools, several of which are described throughout this book. Of particular relevance to genome mapping is the Genomes Division of Entrez. Entrez provides integrated access to several different types of data for over 600 organisms, including nucleotide sequences, protein structures and sequences, PubMed/ MEDLINE, and genomic mapping information. The NCBI Human Genome Map Viewer is a new tool that presents a graphical view of the available human genome sequence data as well as cytogenetic, genetic, physical, and radiation hybrid maps. Because the Map Viewer provides displays of the human genome sequence for the finished contigs, the BAG tiling path of finished and draft sequence, and the location of genes, STSs, and SNPs on finished and draft sequences, it is an especially useful tool for integrating maps and sequence. The only other organisms for which the Map Viewer is currently available is *M. musculus* and *D. Melanogaster.* The NCBI Map Viewer can simultaneously display up to seven maps that are selected from a set of 19, including cytogenetic, linkage, RH, physical, and sequence-based maps.

Some of the maps have been previously published, and others are being computed at NCBI. An extensive set of help pages is available. There are many different paths to the Map Vieweron the NCBI Web site, as described in the help pages. The Viewer supports genome-wide or chromosome-specific searches. A good starting point is the *Homo sapiens* Genome View page. This is reached from the NCBI home page by-connecting to Human Genome Resources (listed on the right side), followed by the link to the Map Viewer (listed on the left side). From the Genome View page, a genome-wide search may be initiated using the search box at the top left, or a chromosome-specific search may be performed by entering a chromosome number(s) in the top right search box or by clicking on a chromosome idiogram.

The searchable terms include gene symbol or name and marker name or alias. The search results include a list of hits for the search term on the available maps. Clicking on any of the resulting items will bring up a graphical view of the region surrounding the item on the specific map that was selected. For example, a genome-wide search for the term CMT* returns 33 hits, representing the loci for forms of Charcot-Marie-Tooth neuropathy on eight different chromosomes. Selecting the Genes_seq link for the PMP22 gene (the gene symbol for CMT1A, on chromosome 17) returns the view of the sequence map for the region surrounding this gene. The Display Settings window can then be used to select simultaneous display of additional maps. The second search box at the top right may be used to limit a genome-wide search to a single chromosome or range of chromosomes.

Alternatively, to browse an entire chromosome, click on the link below each idiogram. Doing so will return a graphical representation of the chromosome using the default display settings. Currently, the default display settings select the STS map (shows placement of STSs using electronic PCR), the GenBank map (shows the BAC tiling path used for sequencing), and the contig map (shows the contig map assembled at NCBI from finished high-throughput genomic sequence) as additional maps to be displayed. To select a smaller region of interest from the view of the whole chromosome, either define the range (using base pairs, cytogenetic bands, gene symbols or marker names) in the main Map Viewer window or in the display settings or click on a region of interest from the thumbnail view graphic in the sidebar or the map view itself. As with the GDB map views, until all sequence is complete, alignment of multiple maps and inference of position from one map to another must be judged cautiously and should not be overinterpreted. There are many other tools and databases at NCBI that are useful for gene mapping projects, including *e*-PCR, BLAST the GeneMap '99 and the LocusLink, OMIM dbSTS, dbSNP, dbEST and UniGene databases, *e*-PCR and BLAST can be used to search DNA sequences for the presence of markers and to confirm and refine map localizations.

In addition to EST alignment information and DNA sequence, UniGene reports include cytogenetic and RH map locations. The GeneMap '99 is a good starting point for finding approximate map positions for EST markers, although additional fine-mapping should be performed to confirm order in critical regions. LocusLink, OMIM, and UniGene are good starting points for genome catalog information about genes and gene-based markers. LocusLink presents information on official nomenclature, aliases, sequence accessions, phenotypes, EC numbers, MIM numbers, UniGene clusters, homology, map locations, and related Web sites.

The dbSTS and dbEST databases themselves play a lesser role in human and mouse gene mapping endeavors as their relevant information has already been captured by other more detailed resources but are currently the primary source of genomic information for other organisms. The dbSNP database stores population-specific information on variation in humans, primarily for single nucleotide repeats but also for other types of polymorphisms. In addition, the NCBIs Genomic Biology page provides genomic resource home pages for many other organisms, including mouse, rat, *Drosophila,* and zebrafish.

MGI/MGD

The *Mouse Genome Initiative Database*, (MGI) is the primary public mouse genomic catalogue resource. Located at The Jackson Laboratory, the MGI currently encompasses three cross-linked topic-specific databases: the Mouse Genome Database (MGD), the mouse *Gene Expression Database* (GXD), and the *Mouse Genome Sequence project* (MGS). The MGD has evolved from a mapping and genetics resource to include sequence and genome information and details on the functions and roles of genes and alleles (Blake et al., 2000). MGD includes information on mouse genetic markers and nomenclature, molecular segments (probes, primers, YACs and MIT primers), phenotypes, comparative mapping data, graphical displays of linkage, cytogenetic, and physical maps; experimental mapping data, and strain distribution] patterns for recombinant inbred strains (RIs) and cross haplotypes. As of November 2000, there were over 29,500 genetic markers and 11,600 genes in MGD, with 85% and 70% of these placed onto the mouse genetic map, respectively. Over 4,800 genes have been matched with their human ortholog and over 1,800 matched with their rat ortholog. Genes are easily searched through the Quick Gene Search box on the MGD home page.

Markers and other map elements may also be accessed through several other search forms. The resulting pages contain summary information such as element type, official symbol, name, chromosome, map positions, MGI accession ID, references, and history. Additional element-specific information may also be displayed, including links to outside resources. A thumbnail linkage map of the region is shown to the right, which can be clicked on for an expanded view. The MGD contains many different types of maps and mapping data, including linkage data from 13 different experimental cross panels and the WICGR mouse physical maps, and cytogenetic band positions are available for some markers. The MGD also computes a linkage map that integrates markers mapped on the various panels. A very useful feature is the ability to build customized maps of specific regions using subsets of available data, incorporating private data, and showing homology information where available (see *Comparative Resources* section below). The MGD is storing radiation hybrid scores for mouse markers, but to date, no RH maps have been deposited at MGD.

MAPPING PROJECTS

In addition to the large-scale mapping data repositories outlined in the previous section, many invaluable and more focused resources also exist. Some of these are either not appropriate for storage at one of the larger-scale repositories or have never been deposited in them. These are often linked to specific mapping projects that primarily use only one or a few different types of markers or mapping approaches. For most studies requiring the use of genome maps, it remains necessary to obtain maps or raw data from one or more of these additional resources. By visiting the resource-specific sites outlined in this section, it is usually possible to view maps in the form preferred by the originating laboratory, download the raw data, and review the laboratory protocols used for map construction.

Cytogenetic Resources

Cytogenetic-based methodologies are instrumental in defining inherited and acquired chromosome abnormalities, and (especially gene-based) chromosomal mapping data is often expressed in cytogenetic terms. However, because cytogenetic markers are not sequence based and the technique is less straightforward and usually more subjective than GL, RH, or physical mapping, there is only a modicum of integration between chromosomal band assignments and map coordinates derived from other techniques in humans and very little or none in other species. Thus, it is often difficult to determine the precise cytogenetic location of a gene or region.

Useful human resources can be divided into displays of primary cytogenetic mapping data, efficient methods of integrating cytogenetic and other mapping data, and resources pertaining to specific chromosomal aberrations. The central repository for human cytogenetic information is GDB, which offers several ways to query for marker and map information using cytogenetic coordinates GDB is a useful resource for cross-referencing cytogenetic positions with genes or regions of interest. NCBI's LocusLink and UniGene catalogues, as well as their other integrated mapping resources, are also valuable repositories of cytogenetic positions. LocusLink and NCBI's Online Mendelian Inheritance in Man (OMIM) list cytogenetic positions for all characterized genes and genetic abnormalities, respectively.

The National Cancer Institute (NCI)-sponsored project to identify GL-tagged BAC clones at 1 Mb density through-out the genome is nearing completion. This important resource, which is commercially available both as clone sets and as individual clones, provides the first complete integration of cytogenetic band information with other genome maps. At this site, BACs can be searched for individually by clone name, band position, or contained STS name, and chromosome sets are also listed. Each clone contains one or more microsatellite markers and has GL and/or RH mapping coordinates along with a FISH-determined cytogenetic band assignment. This information can be used to quickly determine the cytogenetic position of a gene or localized region and to map a cytogenetic observation such as a tumor-specific chromosomal rearrangement using the referenced GL and physical mapping reagents. Three earlier genome-wide efforts to cytogenetically map large numbers of probes are complementary to the NCI site. The Lawrence Berkeley National Laboratory-University of California, San Francisco, Resource for Molecular Cytogenetics has mapped large-insert clones containing polymorphic and expressed markers using FISH to specific bands and also with fractional length (flpter) coordinates, in which the position of a marker is measured as a percentage of the length of the chromosome's karyotype.

Similarly, the Genetics Institute at the University of Bari, Italy, and the Max Planck Institute for Molecular Genetics have independently localized large numbers of clones, mostly YACs containing GL-mapped microsatellite markers, onto chromosome bands by FISH. All three resources have also integrated the mapped probes relative to existing GL and/or RH maps. Many data repositories and groups creating integrated genome maps list cytogenetic localizations for mapped genomic elements. These include GDB, NCBI, the Unified Database (UDB), the Genetic Location Database (LDB), and eGenome, all of which infer approximate band assignments to many or all markers in their databases. These assignments rely on determination of the approximate boundaries of each band using subsets of their marker sets for which accurate cytogenetic mapping data are available. The NCI's cancer Chromosome Aberration Project (CCAP; Wheeler *et. al.*, 2000), Infobiogen the Southeastern Regional Genetics Group (SERGG), and the Coriell Cell Repositories all have Web sites that display cytogenetic maps or descriptions of characterized chromosomal rearrangements.

These sites are useful resources for determining whether a specific genomic region is frequently disrupted in a particular disease or malignancy and for finding chromosomal cell lines and reagents for regional mapping. However, most of these rearrangements have only been mapped at the cytogenetic level. Nonhuman resources are primarily limited to displays or simple integrations of chromosome idiograms. ArkDB is an advanced resource for displaying chromosomes of many amniotes; MGD incorporates mouse chromosome band assignments into queries of its database; and the Animal Genome Database has clickable chromosome idiograms for several mammalian genomes. A recent work linking the mouse genetic and cytogenetic maps consists of 157 BAC clones distributed genome-wide and an associated Web site is available for this resource at the Cedars-Sinai Medical Center.

Linkage Map

Even with the *"sequence era"* approaching rapidly, linkage maps remain one of the most valuable and widely used genome mapping resources. Linkage maps are the starting point for many disease-gene mapping projects and have served as the backbone of many physical mapping efforts. Nearly all human linkage maps are based on genotypes from the standard CEPH reference pedigrees. There are three recent sets of genome-wide GL maps currently in use, all of which provide high-resolution, largely accurate, and convenient mapping information. These maps contain primarily the conveniently genotyped PCR-based microsatellite markers, use genotypes for only 8-15 of the 65 available CEPH pedigrees, and contain few, if any, gene-based or cytogenetically mapped markers. Many chromosome-specific linkage maps have also been constructed, many of which use a larger set of CEPH pedigrees and include hybridization- and gene-based markers.

Over 11,000 markers have been genotyped in the CEPH pedigrees, and these genotypes have been deposited into the CEPH genotype database and are publicly available. The first of the three genome-wide maps was produced by the Cooperative Human Linkage Center (CHLC; Murray *et. al.*, 1994). Last updated in 1997, the CHLC has identified, genotyped, and/or mapped over 3,300 microsatellite repeat markers. The CHLC Web site currently holds many linkage maps, including maps comprised solely of CHLC-derived markers and maps combining CHLC markers with those from other sources, including most markers in CEPHdb. CHLC markers can be recognized by unique identifiers that contain the nucleotide code for the tri- or tetranucleotide repeat units. For example, CHLC.GATA49A06 (D1S1608) contains a repeat unit of GATA, whereas CHLC.ATA28C07 (D1S1630) contains an ATA repeat.

There are over 10,000 markers on the various linkage maps at CHLC, and most CHLC markers were genotyped in 15 CEPH pedigrees. The highest resolution CHLC maps have an average map distance of 1 -2 cM between markers. Some of the maps contain markers in well-supported unique positions along with other markers placed into intervals. Another set of genome-wide linkage maps was produced in 1996 by the group at Genethon. This group has identified and genotyped over 7,800 dinucleotide repeat markers and has produced maps containing only Genethon markers. These markers also have unique identifiers; each marker name has the symbols "AFM" at the beginning of the name. The Genethon map contains 5,264 genotyped in 8-20 CEPH pedigrees. These markers have been placed into 2,032 well-supported map positions, with an average map resolution of 2.2 cM. Because of homogeneity of their marker and linkage data and the RH and YAC-based mapping efforts at Genethon that incorporate many of their polymorphic markers, the Genethon map has become the most widely utilized human linkage map. The third and most recent set of human maps was produced at the Center for Medical Genetics at the Marshfield Medical Research Foundation.

This group has identified over 300 dinucleotide repeats and has constructed high-density maps using over 8,000 markers. Like the CHLC maps, the Marshfield maps include their own markers as well as others, such as markers from CHLC and Genethon. These maps have an average resolution of 2.3 cM per map interval. Markers developed at the Marshfield Foundation have an MFD identifier at the beginning of their names. The authors caution on their Web site that because only eight of the CEPH families were used for the map construction, the orders of some of the markers are not well determined. The Marshfield Web site provides a useful utility for dis-playing custom maps that contain user-specified subsets of markers. Two additional linkage maps have been developed exclusively for use in performing efficient large-scale and/or genome-wide genotyping.

The ABI PRISM linkage mapping sets are composed of dinucleotide repeat markers derived from the Genethon linkage map. The ABI marker sets are available at three different map resolutions (20, 10, and 5 cM), containing 811, 400, and 218 markers, respectively. The *Center for Inherited Disease Research* (CIDR), a joint program sponsored by The Johns Hopkins University and the National Institutes of Health, provides a genotyping service that uses 392 highly polymorphic tri- and tetranucleotide repeat markers spaced at an average resolution of 9 cM. The CIDR map is derived from the Weber v.9 marker set, with improved reverse primers and some additional markers added to fill gaps. Altnough each of these maps is extremely valuable, it can be very difficult to determine marker order and intermarker distance between markers that are not all represented on the same linkage map.

The MAP-O-MAT Web site at Rutgers University is a marker-based linkage map server that provides several map-specific queries. The server uses genotypes for over 12,000 markers (obtained from the CEPH database and from the Marshfield Foundation) and the CRI-MAP computer program to estimate map distances, perform two-point analyses, and assess statistical support for order for user-specified maps. Thus, rather than attempting to integrate markers from multiple maps by rough interpolation, likelihood analyses can be easily performed on any subset of markers from the CEPH database.

High-resolution linkage maps have also been constructed for many other species. These maps are often the most well-developed resource for animal species' whose genome projects are in early stages. The mouse and rat both have multiple genome-wide linkage maps other species with well-developed linkage maps include zebrafish, cat, dog, cow, pig, horse, sheep, goat, and chicken.

Hybrid Map

Radiation *hybrid maps* provide an intermediate level of resolution between linkage and physical maps. Therefore, they are helpful for sequence alignment and will aid in completion of the human genome sequencing project. Three human whole-genome panels have been prepared with different levels of X-irradiation and are available for purchase from Research Genetics. Three high-resolution genome-wide maps have been construeted using these panels, each primarily utilizing EST markers. Mapping servers for each of the three human RH panels are available on-line to allow users to place their own markers on these maps. RH score data are deposited to, and publicly available from, The Radiation Hybrid Database (RHdb). Although this section covers RH mapping in humans, many RH mapping efforts are also underway in other species. In general, lower-resolution panels are most useful formore widely spaced markers over longer chromosomal regions, whereas higher-resolution panels are best for localizing very densely spaced markers over small regions.

The lowest-resolution human RH panel is the Genebridge4 (GB4) panel. This panel contains 93 hybrids that were exposed to 3000 rads of irradiation. The maximum map resolution attainable by GB4 is 800-1,200 kb. An intermediate level panel was produced at the Stanford Human Genome Center (Stewart *et. al.*, 1997). The Stanford Generation 3 (G3) panel contains 83 hybrids exposed to 10,000 rads of irradiation. This panel can localize markers as close as 300—600 kb apart. The highest resolution panel was also developed at Stanford. The TNG panel has 90 hybrids exposed to 50,000 rads of irradiation and can localize markers as close as 50-100 kb. The Whitehead Institute/MIT Center for Genome Research constructed a map with approximately 6,000 markers using the GB4 panel.

Framework markers on this map were localized with odds ≥300:1, yielding a resolution of approximately 2.3 Mb between framework markers. Additional markers are localized to broader map intervals. A mapping server is provided for placing markers relative to the MIT maps. The Stanford group has constructed a genome-wide map using the G3 RH panel. This map contains 10,478 markers with an average resolution of 500 kb. Markers localized with odds = 1,000:1 are used to define "high-confidence bins," and additional markers are placed into these bins with lower odds. A mapping server is provided for placing markers scored in the G3 panel onto the SHGC G3 maps. A fourth RH map has been constructed using both the G3 and GB4 panels.

The combined map, the Transcript Map of the Human Genome was produced by the RH Consortium, an international collaboration between several groups. This map contains over 30,000 ESTs localized against a common framework of approximately 1,100 polymorphic Genethon markers. The markers were localized to the framework using the GB4 RH panel, the G3 panel, or both. The map includes the majority of human genes with known function. Most markers on the map represent transcribed sequences with unknown function. The order of the framework markers is well supported, but most ESTs are mapped relative to the framework with odds <1,000:1.

The majority of markers on the GeneMap have a lod score <2.0, and many are <1.0. Such markers are localized with relatively low support for local order, and their map positions should be confirmed by other means if critical. A mapping server for placing markers on GeneMap '99 is available at the Sanger Centre. The Radiation Hybrid Database (RHdb) is the central repository for all RH data. It is maintained at the European Bioinformatics Institute (EBI) in Cambridge, UK. RHdb is a sophisticated Web- and FTP-based searchable relational database that stores RH score data and RH maps. Data sub-mission and retrieval are completely open to the public. Data are available in multiple formats or as flatfiles. Release 18.0 contained over 126,000 RH entries for 100,000 different STSs scored on 15 RH panels in 5 different species, as well as 91 RH maps.

STS Cotent Maps

Many physical mapping techniques have been used to order genomic segments for regional mammalian genome mapping projects. However, only RH and STS content/large-insert clone mapping methods have yielded the high throughput and automation necessary for whole-genome analysis to date, although advances in sequencing technology and capacity have recently made sequence-based mapping feasible. Two land-mark achievements by the CEPH/Genethon and WICGR groups have mapped the entire human genome in YACs.

The most comprehensive human physical mapping project is the collection of overlapping BAC and PAC clones being identified for the human DNA sequencing project, along with the now complete draft sequence of the human genome. This information is being generated by many different labs, and informatics tools to utilize the data are rapidly evolving. The WICGR physical map is STS content based and contains more than 10,000 markers for which YAC clones have been identified, thus providing an average resolution of approximately 200 kb. This map has been integrated with the Genethon GL and the WICGR RH maps.

Together, the integration provides STS coverage of 150 kb, and approximately half the markers are expressed sequences also placed on GM99. The map was generated primarily by screening the CEPH Mega YAC library with primers specific for each marker and then by assembling the results by STS content analysis into sets of YAC contigs. Contigs are separately divided into *"single-linked"* and *"double-linked,"* depending on the minimum number of YACs (one or two) required to simultaneously link markers within a contig. Predictably, the double-linked contigs are shorter and much more reliable than the single-linked ones, largely because of the high chimeric rate of the MegaYAC library. Thus, some skill is required for proper interpretation of the YAC-based data. The WICGR Human Physical Mapping Project Home Page provides links to downloadable (but large) GIFs of the maps, a number of ways to search the maps, and access to raw data.

Maps can be searched by entering or selecting a marker name, keyword YAC, orYAC contig. Text-based displays of markers list marker-specific information, YACs containing the marker, and details of the associated contig. Contig displays summarize the markers contained within them, along with their coordinates on the GL and RH maps, which is a very useful feature for assessing contig integrity. Details of which YACs contain which markers and the nature and source of each STS/YAC hit are also shown. Clickable STS content maps are also provided from the homepage, and users have the option of viewing the content map alone or integrated with the GL and RH maps. Although there are numerous conflicts between the GL, RH, and STS content maps that often require clarification with other techniques, this resource is very informative once its complexities and limitations are understood, especially where BAC/PAC/ sequence coverage is not complete and in linking together BAC/PAC contigs.

The CEPH/Genethon YAC project is a similar resource to the WICGR project, also centered around screening of the CEPH MegaYAC library with a large set of STSs. Much of the CEPH YAC screening results have been incorporated into the WICGR data (those YAC/STS hits marked as C). However, the CEPH data includes YAC fingerprinting, hybridization of YACs to inter-Alu PCR products, and FISH localizations as complementary methods to confirm contig holdings.

As with WICGR, these data suffer from the high YAC chimerism rate; longrange contig builds should be interpreted with caution, and the data are best used only as a supplement to other genomic data. The CEPH YAC Web site includes a rudimentary text search engine for STSs and YACs that is integrated with the Ge-nethon GL map, and the entire data set can be downloaded and viewed using the associated QUTCKMAP application.

Much of the human draft sequence was determined from BAC libraries that have been whole-scale DNA fingerprinted and end sequenced.

To date, over 346,000 clones have been fingerprinted by Washington University Genome Sequencing Center (WUGSC), and the clone coverage is sufficient to assemble large contigs spanning almost the entire human euchromatin. The fingerprinting data can be searched by clone name at the WUGSC Web site and provides a list of clones overlapping the input clone, along with a probability score for the likelihood of each overlap. Alternatively, users can download the clone database and analyze the raw data using the Unix platform software tools IMAGE and FPC (for contig assembly), which are available from the Sanger Centre. ln parallel with the BAG fingerprinting, a joint project by The Institute for Genome Research (TIGR) and the University of Washington High-Throughput Sequencing Center (UWHTSC) has determined the insert-end sequences (STCs) of the WUGSC-fingerprinted clones (743,000 sequences). These data can be searched by entering a DNA sequence at the UWHTSC site or by entering a clone name at the TIGR site.

Together with the fingerprinting data, this is a convenient way to build and analyze maps in silico. The fingerprinting and STC data have been widely used for draft sequence ordering by the human sequencing centers, and the BAC/PAC contigs displayed by the NCBI Map Viewer are largely assembled from these data. Many human single-chromosome or regional physical maps are also available. Because other complex genome mapping projects are less well developed, the WICGR mouse YAC mapping project is the only whole-genome nonhuman physical map available. This map is arranged almost identically to its human counterpart and consists of 10,000 STSs screened against a mouse YAC library. However, whole-genome mouse fingerprinting and STC generation projects similar to their human counterparts are currently in production by TIGR/UWHTSC and the British Columbia Genome Sequence Centre (BCGSC), respectively.

DNA Sequence

As mentioned above, the existing human and forthcoming mouse draft genomic sequences are excellent sources for confirming mapping information, positioning and orienting localized markers, and bottom-up mapping of interesting genomic regions. NCBI tools like BLAST (Chapter 3) can be very powerful in finding marker/sequence links. NCBI's LocusLink lists all homologous sequences, including genomic sequences, for each known human gene *e*-PCR results showing all sequences containing a specific marker are available at the GM99, dbSTS, GDB, and eGenome Web sites, where each sequence and the exact base pair position of the marker in the sequence are listed.

Large sequence contigs can also be viewed schematically by NCBI's Entrez contig viewer and the Oakridge National Laboratory's Genome Channel web tool. As the mammalian sequencing projects progress, a "sequence first" approach to mapping becomes more feasible. As an example, a researcher can go to the NCBI's human genome sequencing page and click on the idiogram of the chromosome of interest or on the chromosome number at the top of the page. Clicking on the idiogram shows an expanded idiogram graphically depicting all sequence contigs relative to the chromosome. Clicking on the chromosome number instead displays a list of all sequence contigs listed in order by cytogenetic and RH-extrapolated positions. These contigs can then be further viewed for clone, sequence, and marker content, and links to the relevant GenBank and dbSTS records are provided.

Genomic Cataloguing

GDB's Comprehensive Maps provide an estimated position of all genes, markers, and clones in GDB on a megabase scale. This estimate is generated by sequential pairwise comparison of shared marker positions between all publicly available genome-wide maps. This results in a consensus linear order of markers. At the GDB Web site, the Web page for each genomic element lists one or more maps on which the element has been placed, with the estimated Mb position of the marker on each map:

Element	*Chromosome*	*Map*	*Coordinate*	*Units*	*EST MB*	*+/–*
D1S228	1	GeneMap '99	782.0000	cR	32.2	0.0

This example shows that marker D1S228 has been placed 782 cR from the Ip telomere on GM99, and this calculates to 32.2 Mb from the telomere with the GDB mapping algorithm. Well-mapped markers such as the Genethon microsatellites generally have more reliable calculated positions than those that are mapped only once and/or by low-resolution techniques such as standard karyotype-based FISH.

For chromosomes with complete DNA sequence available, the Mb estimates are very precise. LDB and UDB are two additional sites that infer physical positions of a large, heterogeneous set of markers from existing maps using algorithms analogous to GDB's. Both Web sites have query pages where a map region can be selected by Mb coordinates, cytogenetic band, or specific marker names. The query results show a text-based list of all markers in the region ordered by their most likely positions, along with an estimated physical distance in Mb from the p telomere. LDB also displays the type of mapping technique(s) used to determine the comprehensive position, the position of the marker in each underlying single-dimension map, and appropriate references. An added feature of the UDB site is its provision of marker-specific links to other genomic databases. At present, there are no graphical depictions for either map. Physical map positions derived from the computationally based algorithms used by GDB, LDB, and UDB are reliant on the accuracy and integrity of the underlying maps used to determine the position. Therefore, these estimates serve better as initial localization guides and as supportive ordering information rather than as a primary ordering mechanism. A researcher defining a disease locus to a chromosome band or between two flanking markers can utilize these databases to quickly collect virtually all mapped elements in the defined region, and the inferred physical positions serve as an approximate order of the markers. This information would then be supplanted by more precise ordering information present in single-dimension maps and/or from the researcher's own experimental data. The eGenome project uses a slightly different approach for creating integrated maps of the human genome.

All data from RHdb are used to generate an RH framework map of each chromosome by a process that maximizes the number of markers ordered with high confidence (1,000:1 odds). This extended, high-resolution RH framework is then used as the central map scale from which the high-confidence intervals for additional RH and GL markers are positioned. As with GDB, the absolute base pair positions of all markers are calculated for chromosomes that have been fully sequenced. eGenome also integrates UniGene EST clusters, large-insert clones, and DNA sequences associated with mapped markers, and it also infers cytogenetic positions for all markers. The eGenome search page allows querying by marker name or GenBank accession ID or by defining a region with cytogenetic band or flanking marker coordinates. The marker displays include the RH and GL (if applicable) positions, large-insert clones containing the marker, cytogenetic position, and representative DNA sequences and UniGene clusters.

Advantages of eGenome include the ability to view regions graphically using GDB's Mapview, exhaustive cataloguing of marker names, and an extensive collection of marker-specific hypertext links to related database sites. eGenome's maps are more conservative than GDB, LDB, and UDB as they show only the high-confidence locations of markers (often quite large intervals). Researchers determining a regional order *de novo* would be best advised to use a combination of these integrated resources for initial data collection and ordering. Because of the large number of primary data-sources available for human genome mapping, ensuring that the data collected for a specific region of interest are both current and all-inclusive is a significant task.

Genomic catalogues help in this regard, both to provide a single initial source containing most of the publicly available genomic information for a region and to make the task of monitoring new information easier. Human genomic catalogues include the NCBI, GDB, and eGenome Web sites. NCBI's wide array of genomic data sets and analysis tools are extremely well integrated, allowing a researcher to easily transition between marker, sequence, gene, and functional information. GDB's concentration on mapped genomic elements makes it the most extensive source of positional information, and its inclusion of most genomic maps provides a useful mechanism to collect information about a defined region. *e*Genome also has powerful *"query-by-position"* tools to allow rapid collection of regional information. No existing database is capable of effectively organizing and disseminating all available human genomic information. However, the eGenome, GDB, and NCBI Web sites faithfully serve as genomic Web portals by providing hyperlinks to the majority of data available for a given genomic locus. WICGR's mouse mapping project and the University of Wisconsin's Rat Genome Database have aligned the GL and RH maps for the respective species in a comparative manner. MGD's function as a central repository for mouse genomic information makes it useful as a mouse genomic catalogue, and, increasingly, RGD can be utilized as a rat catalogue. Unfortunately, other complex species' genome projects have not yet progressed to the point of offering true integrated maps or catalogues.

COMPARATIVE RESOURCES

Comparative maps provide extremely valuable tools for studying the evolution and relatedness of genes between species and finding disease genes through position-based orthology. There are several multispecies comparative mapping resources available that include various combinations of most animal species for which linkage maps are available. In addition, there are also many sequence-based comparative analysis resources. Each resource has different coverage and features. Presently, it is necessary to search multiple resources, as no single site contains all of the currently available homology information.

Only the most notable resources will be described here. A good starting point for homology information is NCBI's LocusLink database. The LocusLink reports include links to HomoloGene, a resource of curated and computed cross-species gene homologies. Currently, HomoloGene contains human, mouse, rat, and zebrafish homology data. For example, a LocusLink search of all organisms for the gene PMP22 (peripheral myelin protein) returns three entries, one each for human, mouse, and rat. At the top of the human PMP22 page is a link to HOMOL (HomoloGene). HomoloGene lists six homologous elements, including the rat and mouse Pmp22 genes, as well as additional mouse UniGene cluster and a weakly similar zebrafish UniGene cluster.

The availability of both curated and computed homology makes this a unique resource. However, lack of integrated corresponding homology maps is a disadvantage. The MGD does provide homology maps that simplify the task of studying conserved chromosome segments. Homologies are taken from the reported literature for mouse, human, rat, and 17 other species. Homology information can be obtained in one of three manners: searching for genes with homology information, building a comparative linkage map, or viewing an Oxford Grid. The simple search returns detailed information about homologous genes in other species, including map positions and codes for how the homology was identified, links to the relevant references, and links for viewing comparative maps of the surrounding regions in any two species. For example, a homology search for the Pmp22 gene returns a table listing homologous genes in cattle, dog, human, mouse, and rat. Figure 6.5 shows the mouse-human comparative map for the region surrounding Pmp22 in the mouse. A comparative map can also be obtained by using the linkage map-building tool to specify a region of the mouse as the index map and to select a second, comparison, species.

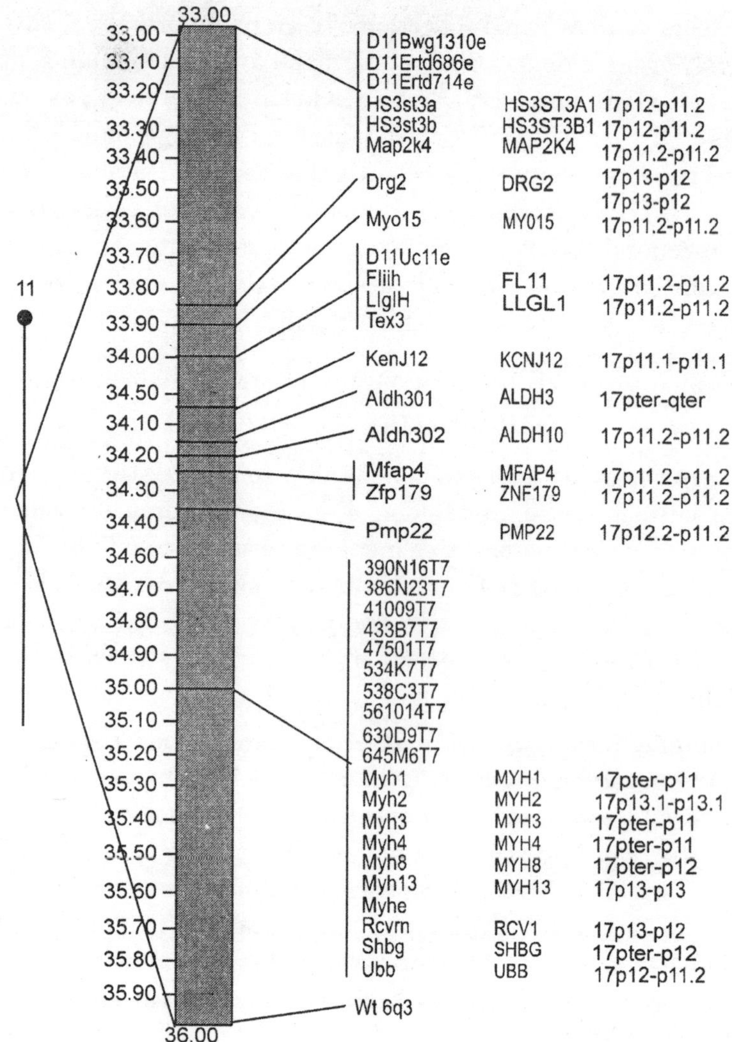

Fig. 7.1. MGD mouse-human comparative map of the region surrounding the mouse Pmp22 gene. Pmp22 is on mouse chromosome 11 at the position 34.5 cM on the mouse linkage map. As shown by the human genes displayed on the right, a segment of human chromosome 17 is homologous to this mouse region.

The resulting display is similar to that shown in Figure 7.1. An Oxford Grid can also be used to view a genome-wide matrix in which the number of gene homologies between each pair of chromosomes between two species is shown. This view is currently available for seven species. Further details on the gene homologies can be obtained via the links for each chromosome pair shown on the grid. The map-viewing feature of MGD is quite useful; however, the positions of homologous nonmouse genes are only cytogenetic, so confirmation of relative marker order within small regions is not possible. It is also possible to view MGD homology information using GDB. *In silica* mapping is proving to be a very valuable tool for comparative mapping. The Comparative Mapping by Annotation and Sequence Similarity (COMPASS) approach has been used by researchers studying the cattle genome to construct cattle-human comparative maps with 638 identified human orthologs.

Automated comparison of cattle and human DNA sequences, along with the available human mapping information, facilitated localization predictions for tens of thousands of unmapped cattle ESTs. The COMPASS approach has been shown to have 95% accuracy. The Bovine Genome Database displays the gene-based comparative maps, which also integrate mouse homologies. A similar approach is being used at the Bioinformatics Research Center at the Medical College of Wisconsin. Here, human, rat, and mouse radiation hybrid maps are coupled with theoretical gene assemblies based on EST and cDNA data (such as the UniGene set at NCBI) for all three species and provide the fundamental resources allowing for the creation of iteratively built comparative maps.

Homologies with uniformly mapped ESTs form the anchor points for the comparative maps. This work has, so far, identified 8,036 rat-human, 13,720 rat-mouse, and 9,745 mouse-human UniGene homologies, most mapped on one or all of the organisms. The creation of these comparative maps is an iterative exercise that is repeated as the radiation hybrid maps, ESTs, and UniGene rebuilds are developed. In addition, the algorithm predicts the placement of unmapped assemblies relative to the anchor information, providing a powerful environment for *"virtual mapping"* before radiation hybrid or other wet-lab methods are used to confirm the predictions. Another project utilizing electronic mapping has developed a high-resolution human/mouse comparative map for human chromosome 7. Recent efforts have greatly increased the number of identified gene homologies and have facilitated the construction of sequence-ready BAC -based physical maps of the corresponding mouse regions. An additional notable resource details homology relationships between human, mouse, and rat. Derived from a high-resolution RH maps, homologies for over 500 genes have been identified and are available in tabular format at a user-friendly Web site.

Regional Map Resources

Although whole-genome mapping resources are convenient for initial collection and characterization of a region of interest, data generated for only a single chromosome or a subchromosomal region are often important for fine mapping. In many cases, these regional maps contain more detailed, better integrated, and higher resolution data than the whole-genome maps can provide. There are numerous such data sets, databases, and maps available for each human chromosome, although little regional information is yet available on-line for other complex genomes.

Most published human chromosome maps are listed and can be viewed at GDB's Web site. Another excellent resource is the set of human chromosome-specific Web sites that have been created by various groups. Recently, the Human Genome Organization (HUGO) has developed individual human chromosome Web pages, each of which is maintained by the corresponding HUGO chromosome committees. Each page has links to chromosome-specific information from a variety of mapping sources, most of them being chromosome-specific subsets of data derived from whole-genome resources (such as the chromosome 7 GL map from Genethon). At the top of most HUGO chromosome pages are links to other chromosome pages designed by groups mapping the specific chromosome. These sites vary widely in their utility and content; some of the most useful are briefly mentioned below.

The sites offer a range of resources, including chromosome-and/or region-specific GL, RH, cytogenetic, and physical maps; DNA sequence data and sequencing progress, single chromosome databases and catalogues of markers, clones, and genomic elements; and links to related data and resources at other sites, single chromosome workshop reports, and chromosome E-mail lists and discussion forums. The major genome centers often include detailed mapping and sequence annotation for particular chromosomes at their sites. The Sanger Centre and the WUGSC have two of the most advanced collections of chromosome-specific genomic data, informatics tools, and resources.

Sanger has collected and generated most available mapping data and reagents for human chromosomes 1, 6, 9, 10, 13, 20, 22, and X. These data are stored and displayed using ACeDB, which can be utilized through a Web interface (WEBACE) at the Sanger Web site or, alternatively, downloaded onto a local machine (Unix OS). ACeDB is an object-oriented database that provides a convenient, relational organizational scheme for storing and managing genomic data, as well as for viewing the information in both text-based and graphical formats. ACeDB is the database of choice for most researchers tackling large genomic mapping projects. WUGSC has recently implemented single-chromosome ACeDB sequence and mapping databases for most human chromosomes, each of which has a Web interface. The Human Chromosome 1 Web site is an example of a community-based approach to genomic research. This site includes a repository for chromosome data generated by several labs, an extensive list of hyperlinks to chromosome 1 data, an E-mail list and discussion forum, a listing of chromosome 1 researchers and their interests, and several workshop reports.

The University of Texas at San Antonio's chromosome 3 site contains a database of large-insert clones and markers along with GL, RH, cytogenetic, and comparative maps. The University of California-Irvine has an on-line chromosome 5 ACeDB database, whereas the Joint Genome Institute (JGI) maintains chromosome 5 large-insert clone maps and some external Web links at their site. The University of Toronto chromosome 7 Web site includes a searchable comprehensive chromosome 7 database containing markers, clones, and cytogenetic information; this site also has a long list of chromosome links. Also, the National Human Genome Research Institute's chromosome 7 Web site contains a YAC/STS map, a list of ESTs, and integration with chromosome 7 sequence files. The University College London maintains a good comprehensive resource of chromosome 9 genomic links, an E-mail group, workshop reports, and a searchable chromosome 9 database.

Genome Therapeutics Corporation has developed an inclusive Web site for chromosome 10. This site has both GL/physical and integrated sequence-based maps, links to related data, and workshop reports. Imperial College maintains a searchable chromosome 11 database at their chromosome 11 Web site, whereas the chromosome 16 Web site at JGI contains restriction-mapped BAC and cosmid contigs and determined sequence, along with a list of chromosome 16 hyperlinks. A similar JGI resource for chromosome 19 includes a completely integrated physical map with sequence links and a list of external resources. The University of Colorado, the RIKEN Genomic Sciences Center, and the Max Planck Institute for Molecular Genetics (MPIMG) have an interconnected set of resouces that together integrate genomic clones, markers, and sequence for the completely sequenced chromosome 21. The Sanger Centre and the LDB have comprehensive resources for the viewing and analysis of chromosome 22. It is expected that additional recources for all completely sequenced chromosomes will be available soon.

The resources for the X chromosome are most impressive. The MPIMG has established a complete genomic catalogue of this chromosome that features integration of genomic mapping and sequence data derived from many sources and experimental techniques. These data can be viewed graphically with the powerful online Java application derBrowser. Finally, the sequenced and well-characterized mitochondria! genome is well displayed at Emory University, where a highly advanced catalogue encompassing both genomic and functional information has been established.

MAPPING RESOURCES

Potential applications of genomic data are numerous and, to a certain extent, depend on the creativity and imagination of the researcher. However, most researchers utilize genomic information in one of three ways: to find out what genomic elements— usually transcribed elements—are contained within a genomic region, to determine the order of defined elements within a region, or to determine the chromosomal position of a particular element. Each of these goals can be accomplished by various means, and the probability of efficient success is often enhanced by familiarity with many of the resources discussed in this chapter. It is prudent to follow a logical course when using genomic data.

During the initial data acquisition step, in which genomic data are either generated experimentally or retrieved from publicly available data sources, simultaneous evaluation of multiple data sets will ensure both higher resolution and greater confidence while increasing the likelihood that the genomic elements of interest are represented. Second, the interrelationships and limitations of the data sets must be sufficiently understood, as it is easy to overinterpret or under-represent the data. Finally, it is important to verify critical assignments independently, especially when using mapping data that are not ordered with high confidence. Below, we give some brief suggestions on how to approach specific map-related tasks, but many modifications or alternative approaches are also viable. The section is organized in a manner similar to a positional cloning project, starting with definition of the region's boundaries, determining the content and order of elements in the region, and defining a precise map position of the targeted element.

Genomic Region

A genomic region of interest is best defined by two flanking markers that are commonly used for mapping purposes, such as polymorphic Genethon markers in humans or MIT microsatellites in mice. Starting with a cytogenetically defined region is more difficult due to the subjective nature of defining chromosomal band boundaries. Conversion of cytogenetic boundaries to representative markers can be approximated by viewing the inferred cytogenetic positions of markers in comprehensive maps such as GDB's universal map, UDB, LDB, or eGenome. Because these cytogenetic positions are inferred and approximate, a conservative approach is recommended when using cytogenetic positions for region definition.

The choice of flanking markers will impact how precisely a region's size and exact boundary locations can be defined. Commonly used markers are often present on multiple, independently derived maps, so their *"position"* on the chromosome provides greater confidence for anchoring a regional endpoint. In contrast, the exact location of less commonly used markers is often locally ambiguous. These markers can sometimes be physically tethered to other markers if a large sequence tract that contains multiple markers can be found. This can be performed by BLASTing marker sequences against GenBank or by scanning e-PCR results in UniGene or eGenome for a particular marker.

Determining and Ordering the Contents of Defined Region

Once a region has been defined, there are a number of resources available for determining what lies within the region. A good way to start is to identify a map that contains both flanking markers, either from a chromosome-wide or genome-wide map from the sources listed above, from a genomic catalogue, or from a local map that has been generated by a laboratory interested in this particular region. For humans, GDB is the most inclusive map repository, although many regional maps have not been deposited in GDB but can be found with a literature search of the corresponding cytogenetic band or a gene known to map to the region. Many localized maps are physically based and are more accurate than their computationally derived, whole-chromosome counterparts. For other species, the number of maps to choose from is usually limited, so it is useful to first define flanking markers known to be contained in the available maps.

The map or maps containing the flanking markers can then be used to create a consensus integrated map of the region. This is often an inexact and tedious process. To begin, it is useful to identify from the available maps an index map that contains many markers, high map resolution, and good reliability. Integration of markers from additional maps relative to the index map proceeds by comparing the positions of markers placed on each map. For example, if an index map contains markers in the order A-B-C-D and a second map has markers in the order B-E-D, then marker E can be localized to the interval between markers B and D on the index map. Importantly, however, the relative position of marker E with respect to marker C usually cannot be accurately determined by this method. Repeated iterations of this process should allow localization of all markers from multiple maps relative to the index map. This process is of course significantly reinforced by experimental verification, such as with STS content mapping of large-insert clones identified for the region-specific markers or, ideally, by sequence-determined order.

Each marker represents some type of genomic element: a gene, an EST, a polymorphism, a large-insert clone end, or a random genomic stretch. In humans, identifying what a marker represents is relatively straightforward. Simply search for the marker name in GDB or eGenome, and, in most cases, the resulting Web display will provide a summary of what the marker represents, usually along with hyperlinks to relevant functional information. For mice, MGD provides a similar function to GDB. For other organisms, the best source is usually either dbSTS or, if present, Web sites or publications associated with the underlying maps. GenBank and dbSTS are alternatives for finding markers, but, because these repositories are passive (requiring researchers to submit their markers rather than actively collecting markers), many marker sets are not represented. If a marker is known to be expressed, UniGene, LocusLink, and dbEST are excellent sources of additional information. Many genes and some polymorphisms have been independently discovered and developed as markers multiple times, and creating a nonredundant set from a collection of markers is often challenging. GDB, eGenome, MGD, and (for genes) UniGene are good sources to use for finding whether two markers are considered equivalent but even more reliable is a DNA sequence or sequence contig containing both marker's primers. BLAST and the related BLAST2 are efficient for quickly determining sequence relatedness. Obviously, the most reliable tool for marker ordering is a DNA sequence or sequence contig. For expressed human markers, searching with the marker name in UniGene or Entrez Genomes returns a page stating where (or if) the marker has been mapped in GeneMap '99 and other maps, a list of mRNA, genomic, and EST sequences, and with Entrez Genomes, a Mapviewer-based graphical depiction of the maps, sequence-ready contigs, and available sequence of the region.

Similarly, GDB and *e*Genome show which DNA sequences contain each displayed marker. For other markers, the sequence from which the marker is derived, or alternatively one of the primer sequences, may be used to perform a BLAST search that can identify completely or nearly homologous sequences. The nonredundant, EST, GSS, and HTGS divisions of GenBank are all potentially relevant sources of matching sequence, depending on the aim of the project. Only long sequences are likely to have worthwhile marker-ordering capabilities. Finished genomic sequence tracts have at least some degree of annotation, and scanning the GenBank record for the large sequence will often yield an annotated list of what markers lie within the sequence and where they are. Keep in mind that such annotations vary considerably in their thoroughness and most are fixed in time; that is, they only recognize markers that were known at the time of the annotation. BLAST, BLAST2, or other sequence-alignment programs are helpful in identification or confirmation of what might lie in a large sequence.

The NCBI *e*-PCR Web interface can be used to identify all markers in dbSTS contained within a given sequence, and this program can be installed locally to query customized marker sets with DNA sequences. For genomes for which DNA sequencing is complete or is substantially underway, it may be possible to construct local clone or sequence contigs. Among higher organisms, this is currently possible only for the human and mouse genomes. Although individual clone sequences can be found in GenBank, larger sequence contigs —sequence tracts comprising more than one BAG or PAC—are more accessible using the Entrez Genomes Web site (see above). Here, by entering a marker or DNA accession number into the contigs search box, researchers can identify sequence contigs containing that marker or element. This site also provides a graphical view of all other markers contained in that sequence, the base pair position of the markers in the sequence, and, with the Mapviewer utility, graphical representations of clone contigs. This process can also be performed using BLAST or e-PCR, although it is somewhat more laborious. Once a sequence has been identified for markers in a given region, YAC clone, DNA fingerprinting, and STC data can be used to bridge gaps.

For humans and mice, the WICGR YAC data provide a mechanism for identifying YAC clones linking adjacent markers. However, caution should be exercised to rely mainly on double-linked contigs and/or to experimentally confirm YAC/marker links. Also for human genome regions, the UWHTSC and TIGR Web sites for identifying STCs from DNA sequence or BAC clones are very useful. For example, researchers with a sequence tract can go to the UWHTSC TSC search page, enter their sequence, and find STCs contained in the sequence. Any listed STC represents the end of a BAC clone whose insert contains a portion of the input sequence. The TIGR search tool is complementary to the UWHTSC search, as the TIGR site requires input of a large-insert clone name, which yields STC sequences. STCs represent large-insert clones that potentially extend a contig or link two adjacent, nonoverlapping contigs. Similarly, ~375,000 human BAG clones have been fingerprinted for rapid identification of overlapping clones (Marra *et. al.*, 1997). The fingerprinting data are available for searching at the Washington University Human Genome Sequencing Center (WUGSC). Combined use of Entrez, BLAST, the STC resources, and the BAC fingerprinting data can often provide quick and reliable contig assembly by in silico sequence and clone walking.

Map Position

Expressing the chromosomal position of a gene or genomic element in physical, RH, GL, or cytogenetic terms is not always straightforward. The first approach is to determine whether the element of interest has already been localized. The great majority of human transcripts are precisely mapped, and many genes have been well localized in other organisms as well. For species with advanced DNA sequencing projects, it is helpful to identify a large DNA sequence tract containing the genomic element of interest and then determine what markers it contains by looking at the sequence annotation record in GenBank or by *e*-PCR.

Identified human and mouse genes are catalogued in GDB and LocusLink or MGD, respectively, and searching UniGene with a marker name, mRNA or EST sequence accession number, or gene name will often provide a localization if one is known. Here again, nomenclature difficulties impede such searches, making it necessary to search each database with one or more alternate names in some cases. Another alternative is to determine if the genomic element is contained in a genomic sequence by a simple BLAST search. Most large genomic sequences have been cytogenetically localized, and this information is contained in the sequence annotation record. If gene-specific or closely linked markers have been used previously for mapping a position can usually be described in terms specific to the mapping method that was employed.

For example, if an unknown gene is found to map very close to a Genethon marker, then the gene position can be reported relative to the Genethon GL centiMorgan coordinates. Most human markers and many maps have been placed in GDB, so this is a good first step in determining whether a marker has been mapped. Simply search for the relevant marker and see where it has been placed on one or several maps listed under "cytogenetic localizations" and "other localizations." Inferred cytogenetic positions of human genes and markers are usually listed in GDB, UniGene, and eGenome if the elements have been previously mapped. If not, band or band range assignments can usually be approximated by finding the cytogenetic positions of flanking or closely linked markers and genes. Many sequenced large-insert clones have been assigned by FISH to a cytogenetic position; I. this information can usually be found in the sequence annotation or at the clone originator's Web site.

The process of determining whether a transcript or genomic element from another organism has been mapped varies somewhat due to the lack of extensive genomic catalogs, making it usually necessary to cross-reference a marker with the GL and/or RH maps available for the species. If no previous localization exists for a genomic element, some experimental work must be undertaken. For human and mouse markers, an efficient and precise way to map a sequence-based element is to develop and map an STS derived from the element by RH analysis. A set of primers should be designed that uniquely amplify a product in the species of interest, but not in the RH panel background genome. An STS is usually unique if at least one primer is intronic. Primers designed from an mRNA sequence will not amplify the correct-sized product in genomic DNA if they span an intron, but a good approximation is to use primers from the 3' untranslated region, as these stretches' only rarely contain introns and usually comprise sequences divergent enough from orthologous or paralogous sequences. However, beware of pseudogenes and repetitive sequences, and genomic sequence stretches are superior for primer design.

Suitable primers can then be used to type an appropriate RH panel; currently, human (G3, GB4, or TNG), mouse, rat, baboon, zebrafish, dog, horse, cow and pig panels are available commercially. After the relevant panel is typed, the resulting data can be submitted to a panel-specific on-line RH server (see above) for the human, mouse, rat, and zebrafish panels. For other species, isolation and FISH of a large-insert clone or GL mapping with an identified or known flanking polymorphism may be necessary.

8

Genotypic and Phenotypic Data

Phenotypic data is perhaps the least analysed form of bioinformatics data, but new laboratory techniques are providing more opportunities for genome-scale phenotype observations. The combination of phenotype data together with other sources of bioinformatics data and with whole system and network analysis will open new possibilities for understanding and modelling the emergent and complex properties of the cell.

PHENOTYPE

The *phenotype* of an organism is its observable characteristics. This can comprise a whole range of properties, from the obviously visible whole-organism characteristics such as flower colour, leaf length and wing size, through drug resistance, to measurements of metabolic flux.

Phenotype data is the oldest form of data available to biologists, preceding the current wave of *'omics'* by almost 150 years. As phenotype data is observable data, this form of data has been available for as long as anyone had the insight and patience to observe it. In 1866 Gregor Mendel published his famous work on the analysis of pea plant phenotypes. The mathematician and biologist monk carefully observed seven traits by growing and cross-breeding thousands of pea plants, the traits including whether the peas were smooth or wrinkled, the length of the stem, and the colour of the unripe pods. These observations gave birth to the science of genetics and the realization that different *hereditary factors* (genes) controlled different aspects of the plant. The scientific community took 34 years to catch up with and accept this work, which was ahead of its time. Mendel had discovered dominant and recessive traits along with the idea of alleles, and had paved the way for scientists in the early 1900s to experiment by crossing organisms and looking for changes or mutants. Now, in the computational and genomic world, we integrate phenotype data with a wide range of other omic data in order to better understand biology. In recent times computational biology and bioinformatics has mainly concentrated on the more reductionist omics: on genomics, transcriptomics and proteomics.

Now we are beginning to see a shift toward whole-organism understanding, and higher-level omic data, of which phenomics can occupy a variety of positions, from total physical appearance of an organism to specific traits. However, phenotype is still perhaps one of the most underexploited sources of information in bioinformatics. This is due in part to the recent enthusiasm for analysis of the new genomic data, and to the relative lack of large-scale data when compared to the other omics. Its usage is now growing with the introduction of new laboratory techniques for measurement. Phenotypic data generation is currently undergoing the same technology revolution as the rest of whole-genome biology, and is now available on a genome-wide scale for many of biology's model organisms. The phenotype of an organism is its observable characteristics, but what is observable depends on our current technology.

Hellerstein (2004) reviews recent methods for observing metabolic pathway fluxes in order to better define the phenotype of an organism. Genes and proteins give a static picture of an organism, and high-level observable characteristics such as growth measurements of mutants give a blunt picture of the phenotype, but metabolic fluxes can give a much more detailed picture of the properties of the organism. He states *"what we call a phenotype depends completely on the tools that we have for seeing. If we lack tools for measuring biochemical fluxes (i.e. biochemical 'motion detectors'), we cannot see or control phenotype in a complete way."*

The phenotype is influenced both by an organism's genotype and by its interaction with the environment. This relationship can be exploited to make models and predictions for gene function and organism behaviour. *Classical genetics,* or *forward genetics,* is the process of starting with a phenotype,. for example 'yellow coloured pea', and then searching for the gene(s) responsible for this phenotype. When phenotypes are quantitative in nature, such as *'flowering times',* this is known as QTL analysis (quantitative trait loci). The converse is *reverse genetics,* which begins with knowledge of the gene of interest, and then tries to look at the phenotype of the organism with this gene disrupted.

The following sections give an introduction to the different ways that phenotype data has been used together with other sources of bioinformatics data in computational biology recently. First we consider forward genetics and QTL analysis, then reverse genetics, and then look at how other sources of data are being used to predict phenotype. Section 6.5 describes how phenotype data can be used as part of systems biology, and Section 6.6 describes the current state of integration of phenotype data with bioinformatics databases. Finally, we draw some conclusions.

QTL ANALYSIS

Forward genetics makes use of knowledge of the phenotype in order to find the gene responsible. In the past, in Mendel 's era, this was the only knowledge that could be used, but now we have a wide variety of other sources of data to assist. In the genomic era, we can use knowledge of a phenotype that exists for many organisms, together with detailed knowledge of the genomes of those organisms in order to pinpoint the common genes. An example of this is the work of Jim *et. al.*, (2004), who integrate phenotypic information gathered from the literature with phylogenetic data in order to identify individual genes associated with that phenotype. Given several organisms exhibiting a phenotype, they look for proteins that are conserved across these organisms, using a BLAST search. They then calculate for each protein the ratio of the fraction of genomes with phenotype f containing that protein compared with the fraction of genomes containing that protein. This gives the propensity that a particular protein is associated with phenotype f. These scores are screened for statistical significance.

Phenotype assignments were derived from PubMed abstracts and research Web pages. The phenotypes they investigated were flagella, pili, thermophily and respiratory tract trophism, and many genes were identified as being associated with these phenotypes. For example, several of the genes they found to be associated with thermophily were annotated as being involved in DNA repair or ferredoxin oxidoreductases, both of which are known to be important to thermophiles and their ability to survive at high temperatures. Identification of genes involved in human disease phenotypes is a very important goal. New large-scale medical-genetic databases, often known as *biobanks,* collect data about human populations and their phenotypes in order to find the genetic factors responsible.

The UK Biobank will be a database created for the study of the contributions played by environment and genetics toward human health and disease. From 2005, half a million people will be followed in a study spanning ten years, and records will be made of their diet and lifestyle habits, medical history and biology. The data will enable researchers to compare genetic and environmental influence on health phenotypes. Similar initiatives have been considered or implemented in several other countries, including Iceland, Estonia, Tonga and Singapore. As with any other project of such importance, these projects are not without their controversy, in terms of privacy, scope, cost, consent, patent issues and exploitation by pharmaceutical industries. Many phenotypic characteristics of an organism are quantitative in nature. Examples of such characteristics (or 'traits') for the plant *Arabidopsis thaliana* include height of the plant, flowering time, seed weight and phosphate content, and it is hoped that such traits can be linked to specific genes.

Phenotypic analysis of gene deletion mutants may not be enough to locate the genes, as there may be many genes involved in creating a particular trait, at several different loci along the genome. These polygenes may also mask the expression of each other (epistasis), or interact with each other. *Molecular markers*, spread throughout the genome and used as a genetic map, can be used to map the traits to certain areas of the genome. QTL mapping is the process of relating the quantitative traits with the molecular markers. The use of such markers requires their map positions to be known accurately.

The markers are either RFLPs (restriction fragment length polymorphisms), PCR-based markers, AFLPs (amplified fragment length polymorphisms) or SNPs (single-nucleotide polymorphisms). RFLPs are reliable, but take time to analyse; PCR-based markers are much faster to analyse, but can be either unreliable or limited in number; AFLPs are efficient, reliable and effectively unlimited but SNPs are the currently favoured type of marker, due to the ability to genotype them in a high-throughput manner, and to their availability (every 500 base pairs in *Arabidopsis*. *QTL data* comprises data about the markers (their locations in the genome) and phenotypic and genotypic data for each of the *recombinant inbred lines* (RILs) that have been tested. The phenotypic data gives the measurements for the trait in question (such as leaf area), and the genotypic data gives the genotype of the RIL at each marker position (whether both alleles of the marker were the same, or the marker is heterozygous or missing). The earliest method of analysis was simply to see which markers were associated with the phenotypes. If, for a particular marker M, all individuals homozygous AA were significantly taller than those that were homozygous aa, then it could be inferred that this marker is associated with this trait. The problems with this were the following.

(*i*) The significance level must be corrected for the fact that many markers are being tested.

(*ii*) Due to linkage among the genes on a chromosome any one QTL may be associated with several markers.

(*iii*) The actual loci for this trait may not be at a marker, but between markers. The method cannot distinguish between strong association of a marker with a small effect and weak association of a marker with a strong effect.

Since then, other more appropriate techniques for analysis of QTL data have included interval mapping, least-squares regression, marker regression, composite and multiple QTL mapping, Bayesian models, genetic algorithms, Cockerham 's model and many more. The production of RILs and marker maps has increased, and the prospects for the future of QTL analysis in proving associations between phenotype and the genes responsible look rosy.

Glazier, Nadeau and Aitman (2002) give a summary of the stages that must be used in order to prove particular genes are responsible for a complex phenotypic trait using QTL analysis and give a discussion of several such genes in different organisms that have been found so far. Paran and Zamir (2003) look at the future of QTL analysis and conclude that, after more than 500 publications on mapping QTL in the past 10 years, the next stage will be a framework for understanding the statistical outputs of QTL analysis.

REVERSE GENETICS

As described above, *forward genetics* is the process of finding the genes that control a known phenotype, and as such is an important tool for tracking the causes of known characteristics, such as human disease. *Reverse genetics* provides the opposite: given our detailed knowledge of the human genome and genomes of other organisms, can we use knowledge of the genes to discover knowledge of phenotype. In phenotypic growth experiments, specific genes are disrupted or removed from an organism to create mutant strains.

These single-gene mutants, or 'knockout organisms grown under different environmental conditions, give us a picture of each gene's individual contribution to the phenotype, and can indicate the function of the missing gene. Current technology, such as disruptive insertion, gene deletion, mutation or RNA interference, provides single-gene disruption mutants on a genomewide scale for model organisms such as yeast *C.elegans* and *A. thaliana* and now even for mammalian cells. These new and comprehensive methods provide an enormously valuable source of information and complement the lower-level genome analysis in allowing the measurements of effects to which causes are sought. Organisms can have genetic redundancy (where more than one gene has the same role), or functional composition (where another gene can step in for a missing gene, but would not normally have played this role). This means that single-gene mutants do not always show a phenotype different from that of the wild type.

Double and even triple-gene mutants are now being generated in yeast on a large scale in order to identify genes whose deletion effects can be buffered by other genes. Lethal double mutants indicate genes involved in the same pathway, or functionally overlapping with each other. Tong *et al.,* compared pairs of genes shown by double mutants to interact and their Gene Ontology classes and found that over 27 per cent of interactions were between genes with a similar or identical GO annotation. Identification of gene function is a primary aim of computational biology, and reverse genetics via knockout mutants can be an invaluable tool for this aim. Single-gene deletion phenotypic growth data have been used to make predictions for gene function. This work used decision trees in order to predict. MIPS functional annotation for unannotated yeast genes, from the phenotype data.

Phenotype growth data can still be sparse across the genome, due to experimental constraints. The techniques for producing mutants are new and improving, but have in the past been labour intensive in identifying the gene responsible and prone to residual gene activity. The phenotypes of mutants can often show no obvious differences from that of the wild type because of buffering effects from other genes. Furthermore, growth under a wide variety of conditions can be time consuming and expensive. Therefore this research used bootstrap techniques and a multilabel learning method to make better use of the sparse phenotypic data set. The decision trees produced prediction rules that could be easily comprehended. Several rules were analysed and shown to be consistent with known biology, for example that sensitivity or resistance to calcofluor white meant that the knocked-out gene was likely to be involved in cell wall biogenesis.

PREDICTION OF PHENOTYPE

Both forward and reverse genetics make use of phenotype data to learn about the genetics of an organism. Conversely, phenotype can be predicted from other sources of data. Phenotype prediction is useful when the actual growth and observation experiments would be costly or time consuming, or difficult to do. King *et. al.*, (2003) have made predictions of phenotype from functional annotations. Decision trees were used to learn from Gene Ontology (GO) annotations, and results were shown to be generally supported by literature searches and experimental phenotype assays.

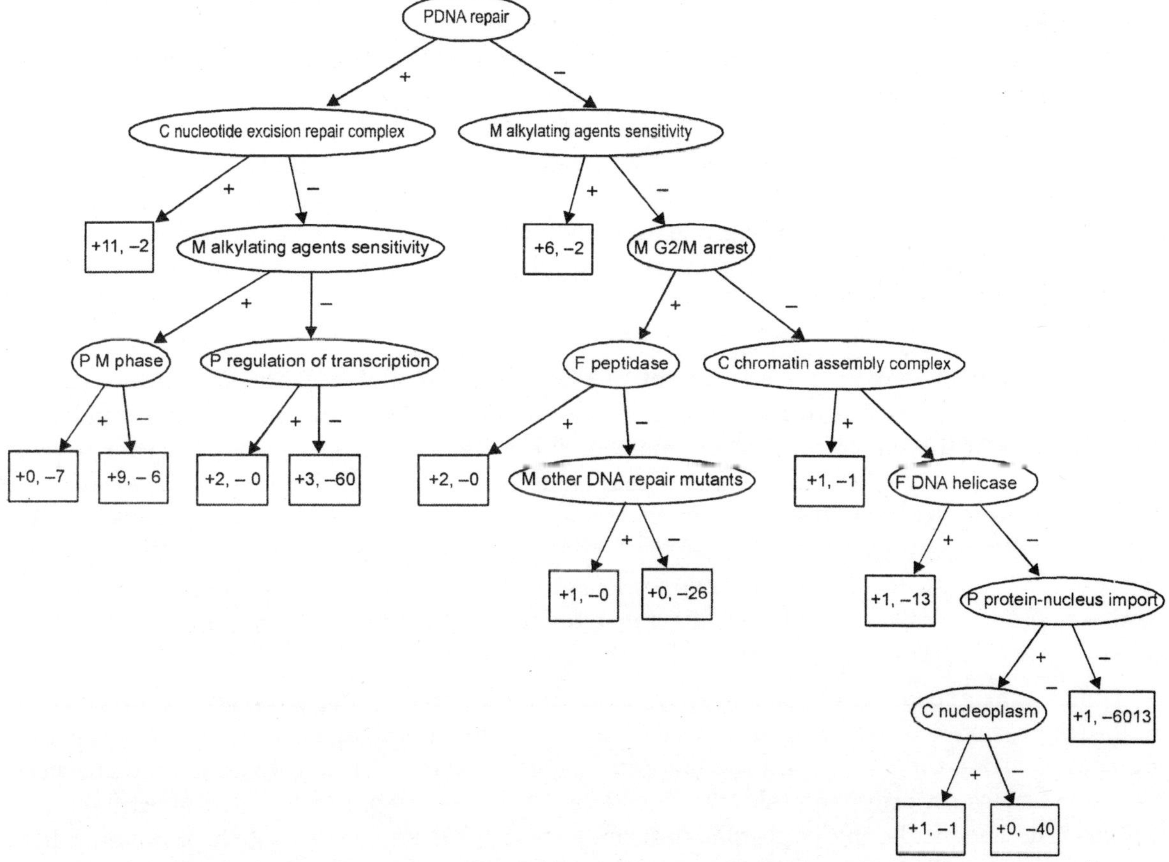

Fig. 8.1 : A decision tree used to predict whether or not a gene had the MIPS 'UV light sensitivity' phenotype, based on other MIPS phenotype annotations and on GO annotations. The letters P, C, F and M that precede the names of the attributes in the internal nodes indicate that the attributes are from the GO biological process branch, GO cellular component branch, GO molecular function branch, or MIPS yeast phenotype catalogue, respectively.

An example of one of their decision trees is given in Figure 8.1. They note the close relationship between function annotation and phenotype annotation. GO annotations with evidence codes such as IMP (inferred from mutant phenotype) are based on phenotype annotation. When integrating these sources of data, it must be realized that they are not necessarily independent and care must be taken to avoid circular data. In this case, King *et al.* removed genes with GO annotations with IMP, TAS, NAS, IC and NR evidence codes from their data. Understanding of the relationships in the data that is to be integrated is vital to ensure that results are not biased unfairly. If the data is not independent then care must be taken not to use machine learning methods that assume independence of attributes, such as naive Bayes. If the data is strongly relational, then multi-relational data-mining tools can be applied instead, such as those described in the July 2003 special edition of the *ACM SIGKDD* Explorations Newsletter. Complex and interdependent data sets are now a common feature in bioinformatics. Combining a variety of sources of data in order to predict a result is a challenge, and requires a variety of data-mining techniques.

Biological data sets present real-world problems for computer scientists specializing in data-mining, such as sparse data, noisy data and a range of different data types and formats. In acknowledgement of this, the data-mining community has used gene deletion experiments as part of the Eighth ACM SIGKDD International Conference on Knowledge Discovery and Data Mining (KDD 2002) Cup Challenge. This required participants to integrate and mine data from protein interactions, localization, function annotation and MEDLINE abstracts, to predict the activity of the AHR signalling pathway in yeast single-gene deletion strains. Krogel and Scheffer (2003) describe their experiences in analysing this combination of data, including text mining the scientific abstracts, and propositionalizing the relational data. Phenotype prediction is particularly useful in cases where mutants are difficult or impossible to obtain, such as in the pharmaceutical fields; *e.g.*, how will a person respond to a drug? Beerenwinkel *et al.* (2003) predict drug resistance phenotype by use of genomic sequence data.

Some HIV virus variants are resistant to some of the available drugs. Resistance can be determined through activity in the presence or absence of the drug (phenotype), or by analysing the sequences of the enzymes in the virus that the drug is known to target in order to look for mutations (genotypic testing). Genotyping is faster and cheaper, and hence Beerenwinkel *et al.* show how decision tree and support vector machines can be used to predict the relationship between genotype and phenotype, and to support the interpretation of the genotype. Parsons *et. al.*, (2004) use phenotypic experiments to test sensitivity of yeast single-gene mutant strains to different drugs, building a 'chemical-genetic profile' for each drug, indicating which genes interact with the drug and can buffer the drug target.

Genes that appear in the profiles for more than one drug can be said to be involved in multi-drug resistance, and they identified 65 genes involved in drug resistance to at least four compounds. They then went on to make double-gene mutants for selected genes, and scored these for fitness, in order to create profiles that could be compared to the chemical-genetic profiles. The chemical-genetic profiles indicate gene-drug interaction, and the double-mutant genetic profiles indicate gene-gene interaction. Similarities between profiles could be used to identify target pathways. For example, 75 genes showed sensitivity to the drug flucanzole. ERG11 is a target of flucanzole, and making double mutants with this gene showed that 13 of the 27 genes that interacted with ERG 11 also showed sensitivity to flucanzole.

DATA WITH SYSTEMS BIOLOGY

Prediction of phenotype is useful where the growth experiments themselves are expensive to perform. With the increasing capability of laboratory technology, the efficiency and reliability of these experiments improves and this method of data generation becomes more cost effective. Laboratory robots now provide the means to produce accurate and reproducible phenotype growth experiments, under the control of computer software. This opens new opportunities for automation in science.

The Robot Scientist is a project that uses a liquid handling laboratory robot and artificial intelligence software to automatically design and execute phenotype growth experiments in order to analyse the functions of gene products involved in a specific yeast metabolic pathway. In this work the results of phenotypic growth experiments of yeast knockout mutants are combined with data about the ORFs, enzymes and metabolites involved in the metabolic pathway within a logical model. The model is encoded in the logic-based programming language Prolog. Then a machine learning system creates hypotheses pertaining to the possible roles of the gene products within the pathway.

The Robot Scientist then chooses the best phenotype experiments to carry out next, choosing growth media to add to reinstate the pathway from the gene that was knocked out. In this case the 'best' experiment is the most discriminatory between the competing hypotheses of gene-enzyme pairings, while minimizing the cost ot growth media. Figure 8.2 shows the architecture of the Robot Scientist. In this way, the Robot Scientist closes the whole scientific loop constructing hypotheses, devising the experiments, conducting the experiments and using the results to construct the next round of hypotheses. All this is integrated into an automatic system, with no human input in the design of experiments or to interpret results (only to move trays in and out of incubators, and provide supplies of media). In this project, the phenotype experiments and resulting data have been integrated as a part of the whole process, from experiment construction to analysis, and provide a vital part of the evidence. If phenotype can be predicted from other sources of information, then this information too can be included within a system such as the Robot Scientist, and this will improve the choice of the next round of growth experiments.

Testing for protein essentiality, or growth/non-growth of a mutant strain, is a very important phenotype. Despite the availability of systematic gene deletion strains, many genes have still not been tested even for this most basic of phenotypes. Jeong *et. al.*, (2003) use statistics and network analysis to integrate function annotation, mRNA expression data and protein interaction data in order to predict essentiality in yeast. Dezso, Oltvai and Barabasi (2003) conclude from analysis of the combination of protein interaction data, expression data, cellular localization and function annotation that a gene's deletion essentiality phenotype depends upon the role that it plays within protein complexes. They discover essential and non-essential complexes, and note that the deletion of a gene whose protein is involved in the core of an essential complex will be lethal, whereas if the complex is non-essential the protein will be too. Thus the essentiality phenotype is a property of the complex rather than the individual protein.

The proteins in the cores of complexes share common functions, localizations and expression as well as essentiality. Then there are proteins surrounding the cores that do not share common essentiality, functions, localizations or expression and these are considered to be just temporary attachments to the complexes. Network analysis of metabolic networks and systems biology promise to open a whole new way of integrating data and understanding biology at a more comprehensive level. The current vision statement of the BBSRC (the Biotechnology and Biological Sciences Research Council in the UK) expects the realistic modelling of the cell by computer within the next ten years.

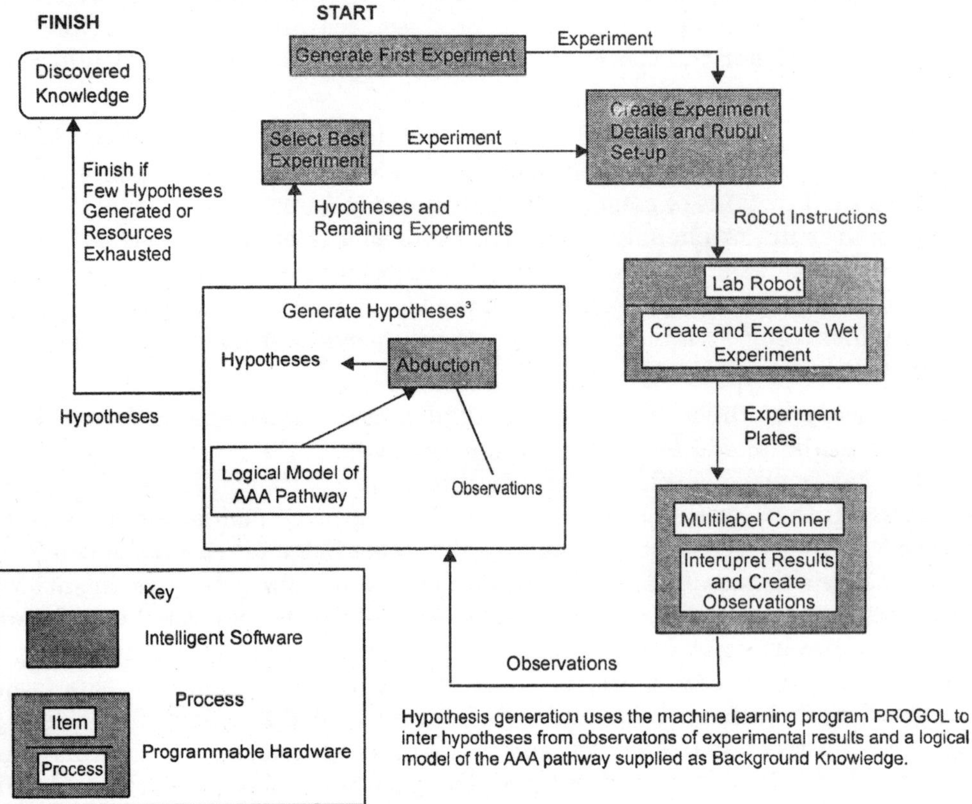

Fig. 8.2 : The architecture of the Robot Scientist (courtesy Ken Whelan, University of Wales, Aberystwyth).

Most phenotypic behaviours are a result of multiple interactions and effects caused by multiple components within a cell, and as such we need to use integrated data and models at the level of systems biology if we are to truly understand the cause of phenotypes. Famili *et. al.*, (2003) use a metabolic model of yeast to analyse phenotypic behaviour. A model of the metabolic network for yeast was constructed using genomic, biochemical and physiological information. Then the network was analysed by computer to discover the range of metabolic activity that could be displayed, adding constraints to eliminate invalid behaviours.

By placing demands on the network for growth and maintenance, predicted phenotypic behaviour can be calculated, and compared with experimental results. In some cases the model agreed with experimental results. Where the model behaved differently, this led to the researchers adding further constraints in order to improve the model. The metabolic model and the phenotype experiments then become part of an iterative process of improvement of knowledge of metabolism. The model was also used to predict phenotypic behaviour of mutant single-gene deletion strains of yeast, grown on complete media.

The predictions agreed with the experimental results from the SGD database in 81.5 per cent of the cases. Hellerstein (2004) in his analysis of new technology for measuring metabolic fluxes, concludes that recent tools will allow increasingly more detailed measurement and control. These tools will contribute still further to systems biology understanding by allowing measurement of dynamic phenotypes and whole processes, rather than the destructive analysis of parts after an experiment.

PHENOTYPE DATA IN DATABASES

If phenotypic data sets are to be effectively integrated with other sources of data, they will need to be stored as part of the existing system of databases for biological data. MIPS provides the Comprehensive Yeast Genome Database (CYGD), which now lists phenotypic information along with the entry for each gene.

The results of a search for a yeast gene in the CYGD will describe not only its function, physical features, localization and literature references but also its known disruption phenotypes. Then there are a wide range of specifically phenotype-oriented databases, which tend to specialize in some way, on either particular organisms or particular phenotypes. For example, the GDPC (Genomic Diversity and Phenotype Connection) database aims to collect phenotype data from different sources and to integrate it with genomic diversity (*e.g.*, SNPs or molecular markers such as AFLPs or RFLPs). This will be a resource for the plant community, and will allow the data to be made publicly available in a standardized format, using XML and SOAP and providing a Java API and a browsing facility.

In a similar way, the PharmGKB database (Klein *et. al.*, 2001) serves the human community, by storing data about clinical phenotypes, and associating these with genetic information. Data about gene polymorphisms, phenotype variabilities, environmental factors and treatment protocols is stored in the database, and access is provided through a Web front end, a Java API and XML formats. Privacy and confidentiality of the data is assured and no patient-identifying data is stored. A small of the variety of existing databases storing phenotype data can be seen in Table 8.1.

Table 8.1. A small sample of the wide variety of available online phenotype databases. All accessed 18 August 2004

Name	*Type*	*URL*
GDPC	Plant	http://www.maizegenetics.net/gdpc/
PharmGKB	Clinical	http://www.pharmgkb.org
Remedis	Rare metabolic diseases	http ://www. remedis.de
Mouse Phenome Database	Mouse	http://www.jax.org/phenome
Maize Phenotype Database	Maize	http://www.mutransposon.org/project/ RescueMu/zmdb/phenotypeDB/
RMA/DB	*C. elegans*	http://nematoda.bio.nyu.edu/
Chinese Gene Variation Databas	Human (Chinese)	http://ww.cgvdb.org.tw/
TRIPLES	*S. cerevisiae*	http://ygac.med.yale.edu/triples/ triples. htm
Database of Essential Genes	Multiple organisms, essential genes	http://tubic.tju.edu.cn/deg/

Phenotype data sets are currently stored in many disparate databases around the world, each database often describing a single organism, each with different representations, access and levels of detail. On the other hand, genome and proteome data have highly organized and well established databases microarray experiments have MIAME standards and metabolomics is in the process of developing standards. Phenotype data will in the future need to make use of controlled vocabularies and metadata standards in order to describe the experimental conditions and results fully and unambiguously. The use of multiple existing ontologies and vocabularies to describe phenotype are a step in the right direction for a review of the current use of ontologies, particularly in describing phenotype data), but serious thought is needed to ensure that future descriptions will be complete and useful. Phenotype data will also require well maintained databases to provide standardized access and cross-references to existing bioinformatics data.

CONCLUSION

Observations of phenotype are growing both in terms of genome-wide coverage and in terms of level of detail. Phenotype data will have much to contribute to the integrated analysis of systems biology. Integration of phenotypic data sources with other sources of data has already begun for investigation of gene function and prediction of phenotype. Integration of phenotypic experiments together with whole-system and network analysis will open new possibilities for understanding and modelling the emergent and complex properties of the cell.

Integration of phenotype data into the standard databases will allow the data to be used as part of automated processes, and introduction of standards for phenotype metadata will enable confidence in the data and results. As the biological technology for phenotype measurement and high-throughput experiments improves, bioinformatics becomes the ever-closer integration of biology and computer science. 'Wet-lab' experiments become an ongoing part of computational analysis and model construction and the whole scientific loop of hypothesis-experiment-interpretation. Integrated data sources permit a whole-system approach to understanding the cell, and making use of phenotype data is an important part of this process.

9

Genomics of *Escherichia coli*

The first genome sequence from a free-living organism was completed in 1995. We do not know yet which will be the first genome to be completely deciphered, but *Escherichia coli* is a natural candidate for such a feat. Certainly *E. coli* has been the most studied cell since the early days of molecular biology. Fred Neidhardt, a well-known microbiologist, said once that every biologist knows about at least two cells: the one he is studying and working with, and—even if he is not aware of it—*E. coli*. Nonetheless, there is a large amount of information still missing about the genome of *E. coli*. For instance, currently there is functional information for about half of its 4290 genes. While we know the transcription unit (TU) organization of a third of such genes (around 1400 genes), only about 600 promoters have been experimentally mapped. Despite being incomplete, the information gathered on *E. coli,* in a sense, offers a benchmark for computational biology.

Consider the annotation process of a new genome. This process consists in associating the known and predicted biological properties of segments of DNA, and their products, in the case of genes, with the corresponding region in the genomic sequence. A first step is guessing genes by methods that take into account the statistics of small oligonucleotide distributions based on an initial set of genes. This *"seed"* is either a group of experimentally known genes or a set of large ORFs (open reading frame, a continuous region of DNA coding for a protein or a fraction of it) found in the genome that by some evidence are likely to be genes. Typically the seed is obtained by the use of a computer program designed to translate the whole DNA sequence into all possible proteins, then compare them against the database or repository of experimentally known or predicted protein products.

It is here that E. *coli* contributes a large number of experimentally characterized genes and gene products. This chapter shows that in bioinformatics the knowledge accumulated on *E. coli* extends to other goals in the in *silico* description of the biology of gene regulation. As a result of an initial review of transcriptional regulation in *E. coli* our laboratory has for years been gathering data from the literature and organizing it into a database, RegulonDB. RegulonDB contains information on transcription initiation as well as operon organization in *E. coli* K-12. Thus, the information it contains can be mapped with its precise coordinate in the chromosome. Using the information gathered in RegulonDB, we have implemented several methods to enlarge the known cases and complement them with computational predictions. In this way a complete description of genes in TUs and their associated regulatory elements can be proposed.

In this chapter we illustrate how this information can be used in the analysis of global gene expression profiles in *E. coli* grown in different conditions. This implies generating predictions of steady states of a given regulatory network structure. Furthermore, some methods are illustrated that use the experimentally determined transcriptome profiles to generate predictions of the structure of the network. Certainly, we do not yet know all of the interacting molecules defining the network and its dynamics.

Finally, we discuss how these methods can be extended to analyze other microbial genomes. In the first part of this chapter we describe the methods used to predict the structure of the regulatory network using sequence information. We also discuss the use of Bayesian methods to enrich some of these predictions. Next we give some examples of the use of the information in the database in the analysis of transcriptome experiments. Finally, we discuss how these methods are expected to change with the increasing number of related genomes sequenced, and discuss some ideas on the evolutionary origin of transcriptional regulation.

Table 9.1 : Contents of RegulonDB: Number of Elements and Increase from Previous Years.

Object	*1997*	*1998*	*1999*	*2000*[a]	*2001*[b]
Regulons	99	83	83	165	166
Regulatory Interactions	533	433	433	642	935
Sites			406	469	750
Products:			4405	4405	4405
Regulatory polypeptides			83	165	318[c]
RNAs			115	115	115
Other polypeptides			4207	3976	3972
Protein complexes	99	83	83	165	166[d]
Genes	542	456	4405	4405	4405
Transcription units	292	230	374	528	657
Promoters	300	239	432	624	746
Effectors	35	36	36	36	66
External References	2050	2011	4394	4704	4943
Synonyms		681	3525	3525	3525
Terminators			40	86	106
RBSs			59	98	133
Conformations				83	201
Conditions				10	10

a Number of elements as of 1 October 2000 in RegulonDB.
[b]Number of elements as of 7 November 2001 in RegulonDB.
[c]Total of 318 transcriptional DNA-binding regulators, of which 165 have experimental evidence and the rest have been predicted based on their helix-turn-helix DNA binding motif.
[d]The number of protein complexes decreased from 165 to 163 due to the constant verification of the data. It was found that two pairs of regulatory polypeptides, with experimental evidence, actually bind together to create two single protein complexes. See the glossary for the definition of regulon, operon, and transcription unit.

COMPUTING GENOMIC ELEMENTS

Promoters and Regulatory Sites

Transcriptional regulatory proteins affect the binding of the RNA polymerase (RNAP) to the promoter, that is, the binding sites of RNAP located upstream of genes. By means of various mechanisms, regulatory proteins can alter any step in the transition involving the binding and the subsequent steps leading to the initiation of a stable elongation complex. Once the RNAP binds to the DNA, it undergoes conformational changes from the so-called closed complex to a conformation that separates the DNA strands into an open complex.

Transcription of the first two to seven nucleotides initiates, in a process that can be aborted, causing the unbinding of RNAP from the DNA and the liberation of short oligonucleotides. If transcription continues, RNAP forms a stable elongation complex that will proceed to termination of an mRNA transcript containing one or several genes. The amount of mRNA produced depends on the regulation of all possible biochemical transitions involved in the complete process. We limit the discussion here to the regulation at the level of transcription initiation, which depends on both intrinsic factors—strength of RNAP binding to the promoter, speed of transcription initiation—and extrinsic factors. The latter basically consist of regulatory proteins that bind to DNA around the promoter region and either facilitate or prevent transcription initiation. Transcription initiation is highly specific. The RNAP holoenzyme (RNAP core plus a sigma factor) recognizes the precise site of transcription initiation. In the absence of the sigma factor, the RNAP binds to DNA in a much less specific manner. While molecules are capable of such exact recognition, the challenge is to generate mathematical and computational methods that can mimic this high resolution of recognition. We are far from anything similar.

Computational methods to predict promoters and other protein-binding sites, such as operator sites, are known to generate a high number of false positives. Although one can solve this problem by limiting the search to a threshold equal to the average score of known sites, or to the average of such scores minus one or two standard deviations, this solution eliminates weaker sites that could potentially be relevant. Figure 9.1 shows the weight matrices of the two conserved elements of *E. coli* sigma 70 promoters, the -10 and the –35 boxes. The average scores of the boxes are 3.12 and 2.17 bits, respectively, which add up to 5.28 bits. This means that this signal might be found randomly every $2^{5.28}$, or around once every 60 base pairs (*bp*). The computational ability to identify similar putative promoter sites depends on the universe to be searched.

A	12	42	27	19	00	05	57	49	43
C	36	01	40	07	16	18	36	26	09
G	28	32	15	14	19	55	00	00	27
T	30	31	24	66	71	28	13	31	27
Consensus	C	A	C	T	T	G	A	A	A

-35 Matrix [15 .... 21] -10 Matrix

A	16	21	07	103	28	52	40	05	19
C	20	21	14	01	13	21	27	00	15
G	44	21	06	02	17	11	19	00	45
T	26	43	79	00	48	22	20	101	27
Consensus	G	T	T	A	T	A	A	T	G

Fig. 9.1 : Weight matrices of the –10 and -35 boxes of E. coli sigma 70 promoter. Consensus matrices were generated using the wconsensus program, This –10 consensus pattern agrees with that reported in the literature. TTGACA is the classic consensus on –35; our –35 differs from that in one base pair. We did not use the classic patterns as seeds to make the matrices, as had been done in previous analyses.

Lukashin et al. (1989) claim that they can identify 99% of promoters, but their search is limited to oligonucleotides of the same size as the promoters. This is useless in searching for promoters in a genome. The significant genomic universe to search for promoters can be estimated by looking at the distance distribution of promoters in relation to the beginning of genes. Figure 9.2 shows that 90% of all sigma 70 promoters are located within 250 *bp* upstream of the ATG of the beginning of the gene.

These 496 promoters are located in 392 regulatory regions; in other words, 81.6% of known TUs have a single promoter, 13.1% have two promoters, and 5.3% have three or more. Thus, the interesting challenge for computational genomics is that of identifying promoters, usually one, in regions of 200 to 300 bp. When identifying a promoter in a region of 250 *bp*, the efficiency can drop to 20%. For instance, searching in the whole *E. coli* genome for thresholds equal to the average of scores of experimentally known promoters generates 1341 possible promoters. If we search using the weight matrices of figure 9.1, using a threshold low enough to make sure we identify 90% of a set of known promoters (commonly the average minus three standard deviations), we get an average of 55 putative promoters per upstream region 250 bp long. That is, there are on average more than 50 additional apparent signals—or noise—mixed with each true signal. In fact, more than 50% of these upstream regions show a putative promoter with a better score than the true known promoter.

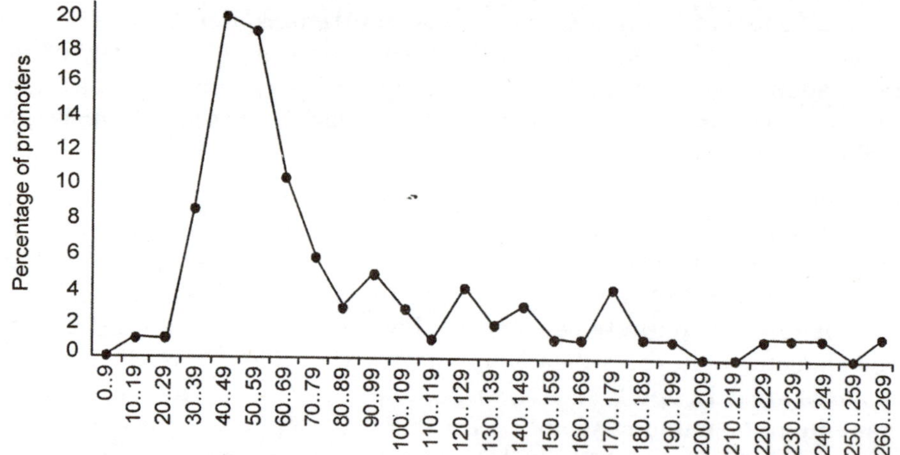

Fig. 9.2 : Distance distribution of promoters, relative to the beginning of the gene in E. coli. Distribution of 496 sigma 70 promoters in relation to the beginning of the downstream gene. Each promoter was positioned based on its –10 box. The highest fraction of promoters falls around 50 bp upstream from the beginning of the gene.

Figure 9.3*a* shows the curves of specificity and sensitivity as a function of the threshold. The curve will never rise above 50% sensitivity, since, as just mentioned, more than 50% of regulatory regions show a signal stronger than the true promoter. No doubt the statistics summarized in a weight matrix are a weak reflection of the molecular ability of RNAP to recognize its binding sites. We have implemented a method that improves the specificity of recognition by an order of magnitude.

The method consists of two steps. The first is to obtain a large population of potential candidates, using a low threashold. The second is a "competition" process that selects a few putative promoters based on a comparison of all candidates within the regulatory region. The competition involves the *"intrinsic"* strength (the added scores of the -10 and –35 boxes, plus the distance separating them) as well as information from the context (*i.e.*, the distance to the beginning of the gene).

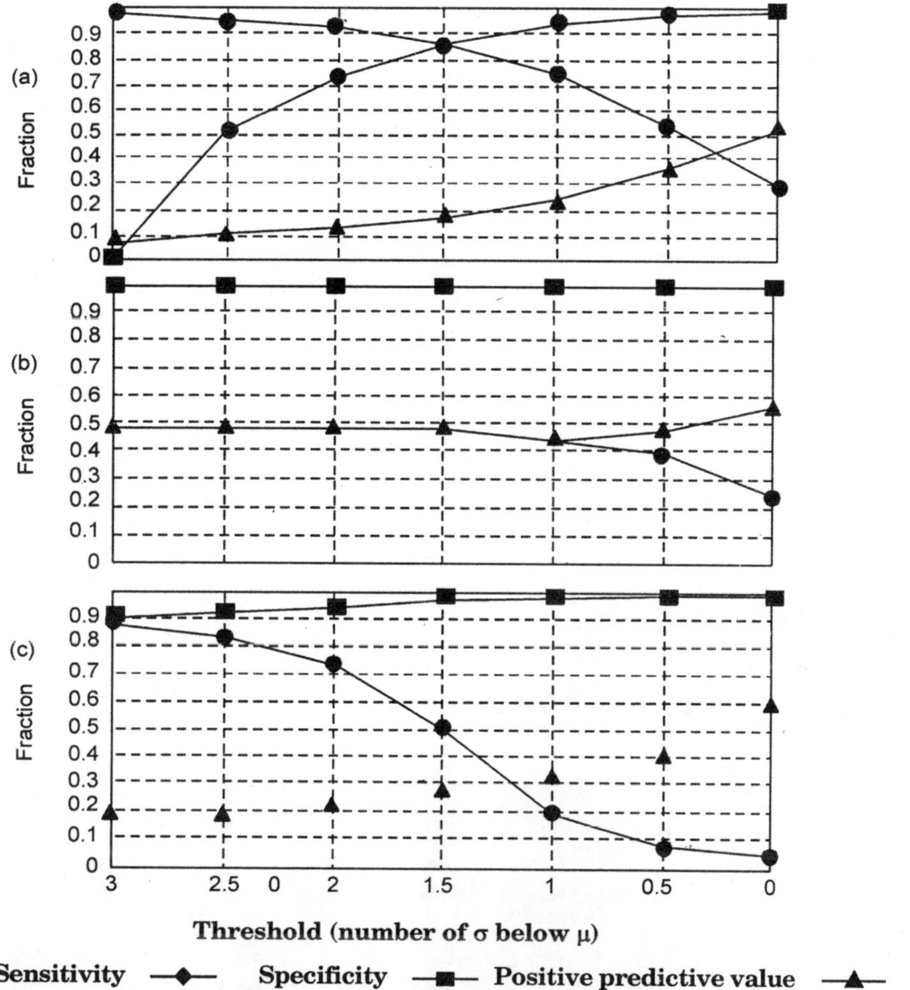

Fig. 9.3 : Evaluation of detection of promoters by several methods. Promoters were searched in a region 250 bp ustream of the gene, (a) Classic method using consensus / patser. (b) Selecting the strongest scoring site in the region, (c) Using the covering algorithm of competition among putative promoters as described in the text. Sensitivity is defined as the ratio of true positives divided by all true sites (the sum of true positives and false negatives). Specificity is defined as the ratio of true negatives divided by the sum of true negatives plus false positives. The positive predictive value is the ratio of true positives divided by all the sites reported by the method (true positives plus false positives). True sites are those experimentally reported, and positive sites are those found by the method.

Interestingly, the method works better when this competition is performed by splitting all signals into groups of candidate promoters with boxes separated by a given distance. All candidate signals are split into seven groups (those whose boxes are separated by a distance of 15 bp, up to those separated by 21 bp), and each group contributes one best candidate. Although analyses splitting promoters in this way have already been performed we initially had no idea of why this independent competition performs better.

A plausible explanation is that by means of this mechanism, we are conserving at the end signals that can overlap (since only promoters with different in-between distances can overlap). This would imply that the precise recognition of a promoter by RNAP is achieved by multiple promoter signals located close together, overall providing a higher energy of binding. Our method yields a true promoter out of six candidates on average (with 80% efficiency), compared with one out of 50 if selected at random. The computational analyses of operator-binding sites for specific transcriptional regulators around the promoter region follows the same basic strategy of weight matrices that is applied to upstream regions in a genome, but the number of known sites is fewer.'RegulonDB currently accounts for around 160 DNA-binding transcriptional regulators with experimental evidence. As of May 2001, we have been able to gather experimental information on known binding sites for 62 proteins.

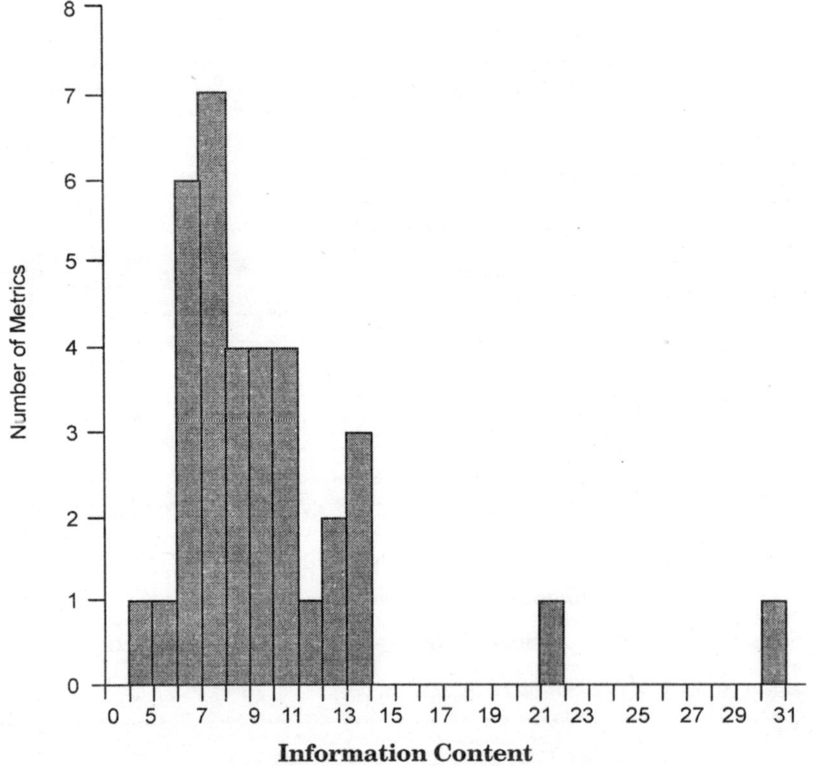

Fig. 9.4 : Information content of weight matrices for specific regulators in RegulonDB. The information content was obtained using consensus with the reported binding sites of 35 regulatory proteins. The values were adjusted against the sample size. For instance, two of the matrices obtained had less than 6 bits of information, and only one, corresponding to ArcA, has 30 bits of information.

We can construct specific weight matrices (with at least four sites) for 35 regulators and search for potential new sites in upstream regions in the genome. This time, given the distribution of sites located around 200 bp upstream from the promoter initiation site we have to search in a region of around 400 bp upstream from the beginning of the gene. This interval is obtained by adding the distance from the promoter to the beginning of the gene, and the distance from the sites to the promoter. We did not consider promoter prediction to be reliable enough to limit the search for sites to the vicinity of the predicted promoters. In fact, if one looks at the information content of all regulatory proteins, the gradient goes from around 6 to 28 bits. This indicates that the promoter site (–35 plus –10), with around six bits, is on the lower range of information content, compared with sites for transcriptional regulators.

Assuming that this information content reflects the energy of binding, and therefore the equilibrium constant of the association between the protein and the DNA sites, it would make sense to see a lower constant for RNAP, given the greater abundance of RNAP compared with that of specific regulators. Searching in the set of upstream regions of all genes in the genome, we find more than 44,000 putative CRP sites, whereas for LexA we find only 160 sites. These numbers illustrate the problem of high false positives or poor specificity in the search for operator sites for transcriptional regulators. Furthermore, as mentioned before, we have sites for only around one sixth of all transcriptional regulators.

A way to considerably diminish the false positives is by restricting the predicted sites to those located at positions relative to the beginning of transcription similar to positions of known sites. This can be achieved by detailed analysis of the combination of sites into modules or *"phrases"*. So far we have focused on methods that deal with sensors that are built from a set of known binding sites for a given protein.

The availability of transcriptome data has motivated the implementation of different types of methods aimed at identifying common motifs in sets of regulatory regions of genes observed to be coexpressed. Although programs in this direction were implemented years ago the recent boom of transcriptome and chip methodologies has motivated new methodologies such as Gibbs sampling and others. We have implemented a method that searches exhaustively in all the space of sequences, and identifies *"words"* or *oligonucleotides* up to seven bases long that are overrepresented in a given family of upstream regions. Its performance has been similar to several other methods when analyzing a set of coexpressed genes in yeast.

Later, a variation of this basic idea was implemented that allows for the search for words with internal symmetry, with *direct* or *inverted* repeats or *palindromes*. This program is called *dyad-detector.* More recently we have tested the performance of dyad-detector against sets of coregulated genes obtained from RegulonDB. An interesting observation in the first results enabled us to develop a strategy that improves the accuracy of the method in detecting genes regulated by a given protein, and to identify the binding site. This method starts with a family of upstream regions containing the set of binding sites for the known or unknown protein (if the set comes from coexpressed genes, for instance). We then use the dyad-detector to identify the overrepresented words. Most of these significant words accumulate near true binding sites.

Focusing on islands of overlapping words or motifs offers an additional step to increase the specificity of the method. With additional information coming from other sources that define groups of genes or operons as plausibly coregulated sets, we may be able to predict, in principle, the binding sites for a set close to the complete collection of 314 transcriptional regulators in *E. coli*.

This will be an important step forward in the characterization of the structure of the network of transcriptional interactions of the whole cell. We can imagine having 314 sets of binding sites distributed in the upstream regions of the genome, and defining the corresponding weight matrices that expand the currently known 50 or so weight matrices. We turn now to the analysis of transcriptional activators and repressers that affect transcription initiation. Helix-turn-helix (HTH) is the dominant motif used by these specific regulators to bind to DNA in bacteria, although other motifs, such as beta-sheet and zinc fingers, are found in a few cases. According to our estimates, more than 90% of bacterial regulators involve an HTH motif.

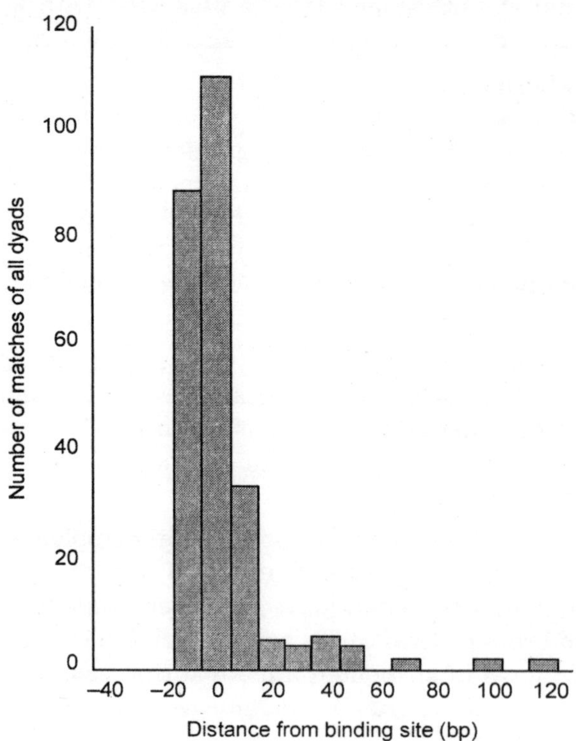

Fig. 9.5 : Distances from binding site of all dyads detected at the upstream regions of genes regulated by LexA. Dyads with a significance higher than 0 were obtained with the dyad-detector program (van Helden *et. al.*, 2000). Their location in relation to the true LexA sites is shown, with negative distances corresponding to those that overlap true sites.

Similar methods, based on weight matrices (profiles) and regular expressions (pattern search, *i.e.*, prosite patterns), can be used to predict transcriptional regulators identifying the conserved HTH motif by means of sequence similarity. In this way the set of 160 known regulators expands to a total of 314 transcriptional regulators in *E. coli.* These regulators are grouped in 24 different families, which are assumed to be groups of proteins with a common evolutionary origin. The functional roles of these families can be inferred from the observation that repressors have their HTH motif located predominantly at the N-terminus, and activators have it at the *C*-terminus, of the protein. This is another useful piece of information that contributes to the solution of the problem of predicting the whole network of regulatory interactions involved in transcription initiation. Assume that we have identified an equal number of families of protein-binding motifs as we have of transcriptional regulators. The problem then is to identify which protein binds to which set of sites, and thus to identify which protein regulates which set of genes.

Ideally, we would like to directly identify the sequence of the operator site from the sequence of the DNA-binding motif of the regulator, as has been done for the GalS protein. However, with the data in RegulonDB we have verified that this striking observation is indeed an exceptional case. A genomic approach putting together several pieces of information seems reasonable. Some of these pieces and their contribution to limiting the association of the regulator and its corresponding regulatory sites and regulated genes are the following:

(*i*) Regulators are predicted as activators or repressors, and are also predicted as members of a particular family of evolutionarily related proteins. We have observed that families tend to regulate genes of relatively similar function.

(*ii*) The position of the binding site in relation to the regulated promoter can corroborate if it is an activator or a repressor. Even in the case of dual proteins, we have observed that for a given specific regulator, when it works as an activator, its sites are closer to the promoter, compared with the positions of the sites used to repress.

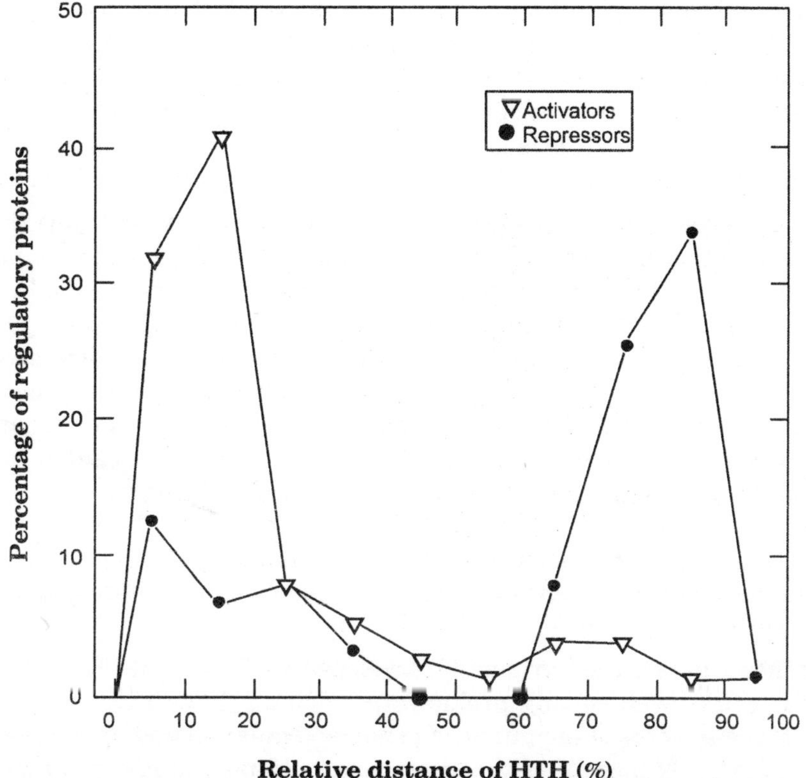

Fig. 9.6 : Relative location of helix-turn-helix (HTH) motifs within regulatory proteins. The total length in amino acids of each regulatory protein was normalized to 100, and the relative position of the center of the HTH motif for each protein was obtained.

(*iii*) Transcriptome snapshots of a regulated transition would in principle enable the search for correlations in the expression profiles of regulatory and regulated genes, and in this way contribute to restricting their identification.

(*iv*) Sets of genes subject to common regulation, or regulons, can be predicted on the basis of phylogenetic profiles and on the basis of expanding the method of operon prediction described below.

(*v*) We know that regulatory proteins are autoregulated in a good number of cases. Thus, finding a motif present upstream of a regulatory protein gene would suggest that all genes having similar sites are targets of the regulator. A clustering approach can be implemented that uses all this information and that is capable of grouping genes on the basis of both knowledge from the database and data from transcriptome experiments.

Operons

In bacteria, genes are clustered in *monocistronic* and *polycistronic* (*operons*) TUs. The information available in RegulonDB for more than 400 TUs with experimental evidence enabled us to implement a method that can predict operons. This method is based on the clearly different distribution of intergenic distances of pairs of genes in the same operon, as opposed to pairs at the boundaries of TUs. It also uses the conservation of functional class of genes within an operon to predict operons in the rest of the genome. The organization of all genes into TUs has motivated the definition of tne set of *"minimal upstream regions"* (MURs) of the genome as the set of upstream regions of all TUs in the genome. This clustering simplifies the problems of Jhgjroblemsof motif searching and of the architecture of the network of interactions.

Forty-two hundred genes and ORFs cluster in around 2600 TUs, and thus there are 2600 MURs. We have shown that the set of TUs in a given regulon can be expanded by the use of comparative genomics combined with evidence of upstream regulatory motifs. In this example we show that *"orthologue TUs,"* found in *E. coli* and *Haemophilus influenzae,* can be added to a regulon if such TUs have a similar binding site. An expansion of this idea, as suggested by Galperin and Koonin (2000), takes advantage of the high frequency of rearrangement of operons. It is expected that such rearrangements of the gene elements of operons occur among related TUs (regulons). Starting with a set of genes known to be part of a regulon in *E. coli,* one would need to find the orthologues in other genomes. If such orthologues are associated in an operon with another gene, and that gene is also present in E. *coli,* there would be a chance that this gene is part of the same regulon. Some evidence that the rearrangements occur among related operons has been published. A complete analysis would need TU predictions, in different genomes, to start with.

We have expanded the method originally developed in *E. coli* and have found evidence suggesting that it works well in any prokaryotic genome. It will be very interesting to extrapolate some of these pieces of the puzzle of genomic regulation and apply them to different bacterial genomes. Certainly the role of *E. coli* as an oasis of annotations, compared with much less studied related microorganisms, must be extrapolated further.

Transcriptome Analyses

All the accumulated experimental knowledge and computational predictions of gene clustering in operons and gene regulation provide valuable information for analyzing expression profiles from transcriptome experiments in *E. coli.* The analysis of any given condition requires the identification of the function of the set of affected—induced or repressed—genes, the organization of such genes into operons, their regulatory regions with promoters and regulatory sites; that is, the gathering of all known information on such genes.

There is also a need to have at hand methods that can organize and manipulate the information. We have implemented interfaces that integrate outputs of programs that search in RegulonDB and display integrated information about the regulation of gene expression of a given set of genes. Several interesting questions can be raised when comparing the data that have accumulated through years of experimentation with data obtained in a single transcriptome. One question is to evaluate and understand how congruent the observed profile is to what we know about the network of interactions. At a first, broad level of analysis, statistical evaluations of such congruence can be obtained. Genes grouped in operons tend to be subject to a common mechanism of regulation. Thus, most of the time one would expect that under any growth condition, all genes within an operon will be either induced or repressed. We define such operons as congruent. We could count the fraction of congruent operons as a function of the threshold of gene expression. If the low expression values get mixed with the noise of detection, one would expect a curve of increasing congruence as the threshold rises. This is what we observe in several transcriptome experiments, one of which is illustrated in figure 9.7.

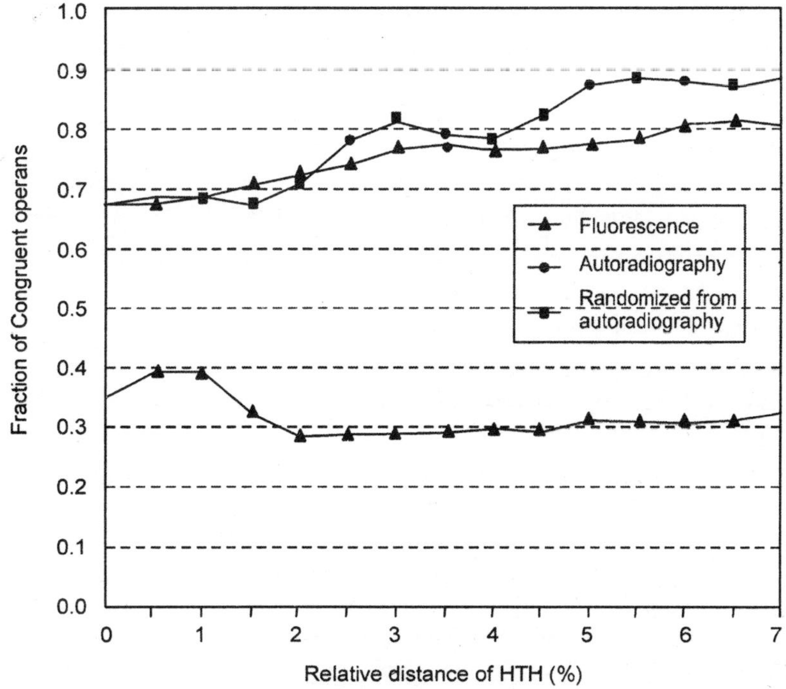

Fig. 9.7 : Congruence of operons in transcriptome experiments. Evaluation of congruence of transcriptome experiments as measured by the number of operons completely expressed or repressed by the number of operons represented at different levels of induction or repression (here shown as absolute value). The example shows the heat shock experiments measured by fluorescence and by autoradiography.

Although such a congruent behaviour is not surprising, it is a useful index, based on the biology of *E. coli* genes and operons, to use as an *"internal control"* of experiments across different conditions. More interestingly, if the high region of congruence of this curve has some biological meaning, it suggests that for a large number of changes in growth conditions, *E. coli* adapts through the regulation of an important number of genes, no fewer than 100 or 150. This clearly suggests mechanisms for concerted regulation of a number of genes that is well beyond the average number of genes in a single known regulon. There is enough information to analyse the congruence of cell behaviour in more detail and from a different perspective.

Snapshots of the expression flexibility of the cell are like photographs taken from a boat at high speed—the picture can easily be blurred. The set of regulatory interactions in a cell forms a network in which the level of expression of one gene can easily affect that of others, either directly or indirectly. In order to evaluate the congruence of genes within such networks, we need to consider all known interactions. This type of analysis, at a given moment in time, can be called *"internal congruence"* or *"synchronous congruence"*— as opposed to the analysis of a transition of the cell when changing the conditions of growth. We investigated whether groups of genes known to be precisely coregulated show homogeneous patterns of expression in transcriptome data. We concentrated on the study of absolute levels of expression of *E. coli* grown at 37 °C in minimal medium, since we have data from the experiment performed four times.

These experiments were performed in the laboratory of Fred Blattner at the University of Wisconsin, Madison. Defining expression levels as either *on* or *off,* based on their absolute values, we can identify subsets of coregulated genes that behave in a homogeneous way—either *on* or *off.* A binomial analysis with a priori probabilities p and $(1 - p)$ for *on* and *off,* based on their frequency in the complete set of genes in the experiments, shows the statistical significance of these results. A set of genes coregulated by a single protein and with a statistically significant behaviour is shown in table 9.2. This may not be the case for all coregulated groups. More elaborated analyses will surely be required to deal with the flexible response of regulons.

In order to perform similar analyses, this time comparing transcriptome experiments on E *coli* in different conditions, we need to employ a computational machinery that will simulate direct and indirect effects and their state transitions.

In a way, the ideas presented here are rather simple in terms of simulation of networks. The recent literature addressing the analysis of transcriptome data and modeling of regulatory networks is growing very fast.

Table 9.2 : Internal congruence in gene expression of coregulated gene.

	Data						
	Experiments						
Co-regulating proteins	*Gene*	*1*	*2*	*3*	*4*	*Off*	*On*
CsgD CsgG OmpR	csgD	*off*	*off*	*off*	*off*	4	0
CsgD CsgG OmpR	csgE	*off*	*off*	*off*	*off*	4	0
CsgD CsgG OmpR	csgF	*off*	*off*	*off*	*off*	4	0
GsgD CsgG OmpR	csgG	*off*	*off*	*off*	*off*	4	0
Totals *on*		0	0	0	0		
Totals off		4	4	4	4	16	0
Summary							
%genes *off*	1.00						
%genes *on*	0.00						
Probability	0.01						

An example of a group of genes known to be precisely coregulated by three regulatory proteins: CsgD, CsgG, OmpR. The group shows a homogeneous pattern of expression in repetitions (experiments 1, 2, 3, and 4) of transcriptome data. The expression levels are described as either on or off pattern, for which we analyzed the statistical significance using a binomial distribution model.

COMPARATIVE GENOMICS

Since the work by Zuckerkandl and Pauling (1963), we have been aware of the historical content within macromolecular sequences. Conservation in the course of evolution is becoming a powerful source of information for a new generation of computational methods that use full genomic sequence and information. For instance, predicting regulatory elements in eukaryotic genomes was recognized until very recently as a difficult computational problem*Although their position in relation to the beginning of the genes is much more flexible than in prokaryotes and allows remote positions, one would expect, when comparing two related organisms, that their functional sites are visible as small, conserved regions as opposed to the rest of noncoding DNA. This identification is what is called a *"phylogenetic footprint"*. In an analogy to chemical footprinting of operator sites. We have mentioned that regulatory proteins can be grouped into families with a common evolutionary origin.

The interesting correlation between the position of the HTH motif and the role in regulation of the protein can be either the result of a convergent process in evolution due to physical constraints, or a trace of the common origin of these proteins. To address this issue, we expanded our data set of transcriptional regulators, and compared them with all ORFs in sequenced bacterial and archaeal sequenced genomes. Our accumulated evidence points to a common origin of this positional restriction. Certainly, when the HTH conserved sequence within a family is analyzed, it expands beyond the strict functional, characterized HTH motif, suggesting that a trace of similar sequence with no binding task is still visible. Second, not only evolutionary families tend to have conserved positions; the distribution of families across the different microbial genomes is such that we can estimate the point in evolution when they emerged.

A quite interesting observation in this analysis is a set of four repressor families that show homologues in archaea, indicating that this motif existed before the divergence, around three billion years ago between the current bacteria and the archaea. The evolutionary trace in operon organization is also rather striking in the microbial world. The organization of operons in *E. coli* has shown, as already mentioned, genes within operons with a strong preference for short intergenic distances as opposed to a flat, almost equiprobable distribution for a range of distances between genes at the boundaries of TUs. This is a strict genomic observation in the sense of a property that can be analyzed only when complete genomes are available.

The intergenic distances aid in predicting functional relationships between genes—because genes in the same operon tend to participate in the same function—without any need for homology detection among proteins. In order to evaluate whether the same method can be applied in predicting operons in bacteria other than *E. coli,* one needs known sets of experimentally characterized operons in other bacteria. Fortunately, a collection of experimentally characterized operons is available for *Bacillus subtilis,* a gram-positive bacterium.

To our surprise, the predictive accuracy of our method for this set is as high as that for the operons in RegulonDB, using the parameters defined with the *E. coli* collection. Still, evidence is necessary to show that the method is applicable to any prokaryotic genome.

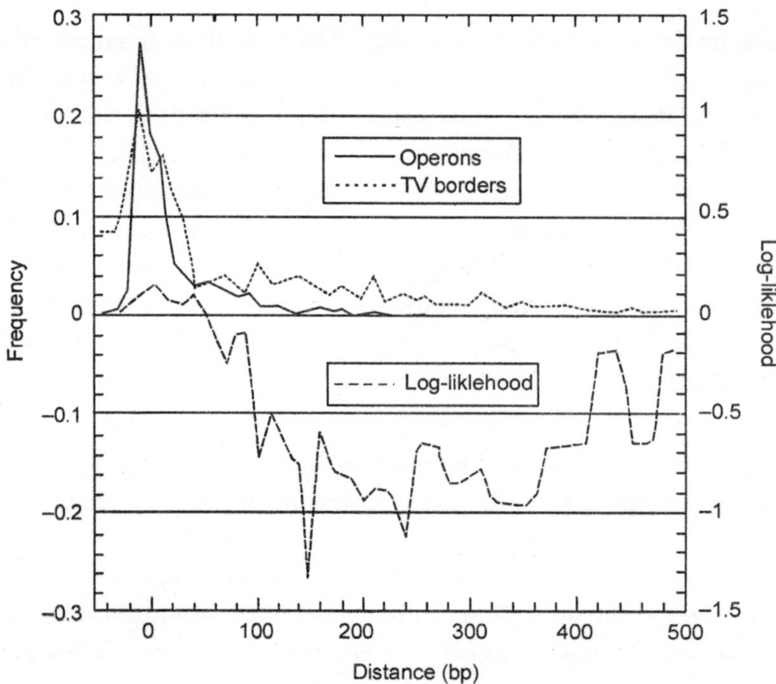

Fig. 9.8 : Intergenic distances of genes within operons and at TU borders. Distribution of intergenic distances of genes within operons (continuous line), and of genes at the borders of transcription units (discontinuous line, upper part). The figure also displays the log-likelihoods derived from the comparison of the frequency of pairs of genes within operons separated by a given distance versus that among genes at TU borders. Intergenic distances were obtained from known operons, and distances at borders were obtained from transcription units (i.e., operons as well as genes transcribed in isolation).

Another observation about the properties of genes within experimentally known operons, compared against genes at TU boundaries, is that they have a higher tendency to be conserved as neighbours in other genomes. We tested to see if this holds true for predicted genes within operons and predicted boundaries. This independent test gave the expected results, that is, the method derives predicted genes within operons that are conserved as neighbours more frequently than predicted boundaries. Other tests confirm the applicability of the method, such as the conservation of the peak observed in the distance distributions of all genes transcribed in the same direction.

The peak is always located at the same place, and it is a reflection of the abundance of operons of each genome. This information on abundance can be used to adjust the predictions for greatest accuracy, but in principle the method performs surprisingly well in any prokaryotic genome. In summary, we started with an expected observation: that pairs of genes in the same operon tend to be close to each other, in contrast to pairs of genes at the boundaries between two neighbouring TUs.

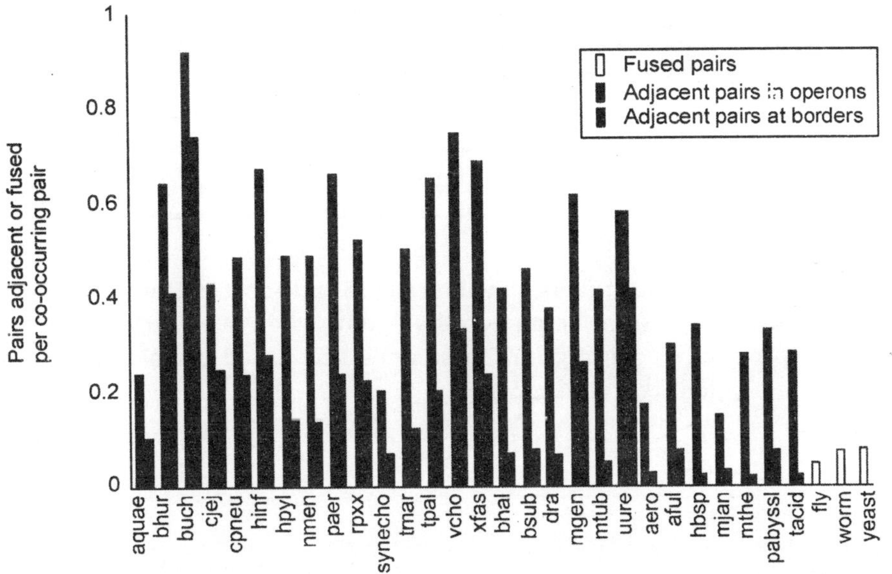

Fig. 9.9 : Conservation of E. coli pairs of genes inside operons and at their border. The first column in each organism represents pairs in operons, while the second represents genes at the borders of transcription units. The column representing pairs in operons is always higher, and fusions occur only among conserved genes corresponding to operons. The labels mostly correspond to the file names at GENBANK: aquae, Aquifex aeolicus; bbur, Borrelia burgdorferi; buch, Buchnera sp. APS; cjej, Campylobacter jejuni; cpneu, Chlamydia pneumoniae; hinf, Haemophilus influenzae Rd; hpyl: Helicobacter pylori 26695; nmen, Neisseria meningitidis MC58, paer, Pseudomonas aeruginosa; rpxx, Rickettsia prowazekii, synecho, Synechocystis PCC6803; tmar, Thermotoga maritima; tpal, Treponema patidum; vcho, Vibrio cholerae; xfas, Xylella fastidiosa; bhal, Bacillus halodurans; bsub, Bacillus subtilis; dra, Deinococcus radiodurans; mgen, Mycoplasma genitalium; mtub, Mycobacterium tuberculosis; uure, Ureaplasma urealyticum; aero, Aeropyrum pernix; aful, Archaeoglobus fulgidus; hbsp, Halobacterium sp.; mjan, Methanococcus jannaschii; mthe, Methanobacterium thermoautotrophicum; pabyssi, Pyrococcus abyssi; tacid, Thermoplasma acidophilum; fly, Drosophila melano-gaster; worm, Caenorhabditis elegans; yeast, Saccharomyces cerevisiae.

The difference in the distribution is so clear that this single parameter suffices to predict operons with an accuracy higher than 80% in *E. coli.* A similar accuracy in the set of known operons in *Bacillus* is rather surprising, given the evolutionary distance between these organisms. Also important is the different evolutionary conservation of pairs of genes in operons, as opposed to genes at the boundaries of TUs. The resulting methodology enables a quite accurate prediction of TUs in all microbial genomes. This computational ability is based on a universal conservation of the basic architecture of operons within the microbial world.

CONCLUDING REMARKS

Many questions are open, and some of them may remain mysteries for a long time. Transcription regulation as a network of interactions clearly illustrates the complexity of biological architectures. We do not know how deeply we will be able to understand such networks with current scientific approaches. There is no clear quantitative method to address the emergent properties of complex biological systems. These limitations bring a healthy skepticism to what we will be able to do.

However, the defined phenotypes of individual mutations in complex differentiation pathways, such as aging, suggests that complex as biology may be, it is decomposable and subject to analysis. Transcriptional regulation and operon organization in bacteria bring us to a different point in the history of the biological complexity, the one of the origin of life, or at least the origin of cellular life and its early speciation in what became the archaeal, bacterial, and eukaryotic domains of life. *E. coli* or another bacterium may be the first completely annotated genome in the near future. It is true that in several aspects bacteria are simpler systems, compared to higher organisms.

But this simplicity does not prevent novel methods and representations that can integrate the biology of a bacterial cell from being equally useful when studying eukaryotic cells. One further step will be to study gene regulation and the associated metabolic capabilities of the regulated genes in a single integrated model.

10

Sequencing Genome

The first complete genome sequences of cellular life forms have become available in just the last several years. In 1995, the genomes of the first two bacteria, *Haemophilus influenzae* and *Mycoplasma genitalium,* were reported. One year later, the first archaeal *(Methanococcus jannaschii)* and the first eukaryotic (yeast *Saccharomyces cerevisiae)* genomes were completed. 1997 was marked by a landmark achievement—the sequencing of the genomes of the two best-studied bacteria, *Es-cherichia coli* and *Bacillus subtilis.* Many more bacterial and archaeal genomes, as well as the first genome of a multicellular eukaryote, the nematode *Caenorhabiditis elegans,* have since been sequenced providing ample material for comparative analysis. A notable (and perhaps disappointing to many biologists) outcome of these first genome projects is that at least one-third of the genes encoded in each genome had no known or predictable function; for many of the remaining genes, only a general functional prediction appeared possible.

The depth of our ignorance becomes particularly obvious on examination of the genome of *Escherichia coli* K12, arguably the most extensively studied organism among both prokaryotes and eukaryotes. Even in this all-time favorite model organism of molecular biologists, at least 40% of the genes have unknown function. On the other hand, it turned out that the level of evolutionary conservation of microbial proteins is rather uniform, with ~70% of gene products from each of the sequenced genomes having homologous in distant genomes. Thus, the functions of many of these genes can be predicted simply by comparing different genomes and by transferring functional annotation of proteins from better-studied organisms to their orthologous from lesser-studied organisms. This makes comparative genomics a powerful approach for achieving a better understanding of the genomes and, subsequently, of the biology of the respective organisms. Here, we describe databases that store genomic information and bioinformatics tools that are used in the computational analysis of complete genomes.

The subject of comparative genomics includes a number of distinct aspects, and it is unrealistic to cover them all in a brief chapter. We limit the discussion to the analysis of protein sets from completely sequenced genomes. Because most of the latter are from prokaryotes, there is an inevitable focus on prokaryotic biology in the presentation. Furthermore, in our choice of the genome analysis tools to discuss in detail, we decided to concentrate largely on Web-based ones that are readily accessible to any user, as opposed to stand-alone software that has more limited applicability.

PROGRESS IN GENOME SEQUENCING

By the beginning of 2000, genomes of 23 different unicellular organisms (5 archaeal, 17 bacterial, and 1 eukaryotic) had been completely sequenced. At least 70 more microbial genomes were in different stages of completion with respect to sequencing. Periodically updated lists of both finished and unfinished publicly funded genome sequencing projects are available in the GenBank Entrez Genomes division and at the site maintained by *The Institute for Genome Research* (TIGR) and at Integrated Genomics.

A complete list of sequencing centers world-wide can be found at the *NHGRI* Web site. One can retrieve the actual sequence data from the *NCBI FTP* site or from the *FTP* sites of each individual sequencing center. A convenient sequence retrieval system is maintained also at the DNA Data Bank of Japan. In the framework of the *Reference Sequences* (RefSeq) project, *NCBI* has recently started to supplement the lists of gene products with some valuable sequence analysis information, such as the lists of best hits in different taxa, predicted functions for uncharacterized gene products, frame-shifted proteins, and the like. On the other hand, sequencing centers like TIGR have been updating their sequence data, correcting some of the sequencing errors and, accordingly, their sites may contain more recent data on unfinished genome sequences.

General Purpose Databases for Comparative Genomics

Because the World Wide Web makes genome sequences available to anyone with Internet access, there exists a variety of databases that offer more or less convenient access to basically the same sequence data. However, several research groups, specializing in genome analysis, maintain databases that provide important additional information, such as operon organization, functional predictions, three-dimensional structure, and metabolic reconstructions.

Pedant: This useful Web resource provides answers to most standard questions in genome comparison. PEDANT provides an easy way to ask simple questions, such as finding out how many proteins in *H. pylori* have known (or confidently predicted) three-dimensional structures or how many NAD^+-dependent alcohol dehydrogenases (EC 1.1.1.1) are encoded in the *C. elegans* genome. The list of standard *PEDANT*, queries includes EC numbers, PROSITE patterns, Pfam domains, *BLOCKS*, and *SCOP* domains, as well as PIR keywords and PIR superfamilies. Although *PEDANT* does not allow the users to enter their own queries, the variety of data available at this Web site makes it a convenient entry point into the field of comparative genome analysis.

COGs : The Clusters of Orthologous Groups (COGs) database has been designed to simplify evolutionary studies of complete genomes and improve functional assignments of individual proteins. It consists of ~2,800 conserved families of proteins (*COGs*) from each of the completely sequenced genomes. Each *COG* contains orthologous sets of proteins from at least three phylogenetic lineages, which are assumed to have evolved from an individual ancestral protein. By definition, *orthologous* are genes that are connected by vertical evolutionary descent (the *"same"* gene in different species) as opposed to *paralogs*—genes related by duplication *within* a genome. Because *orthologous* typically perform the same function in all organisms, delineation of orthologous families from diverse species allows the transfer of functional annotation from better-studied organisms to the lesser-studied ones.

The protein families in the *COG* database are separated into 17 functional groups that include a group of un-characterized yet conserved proteins, as well as a group of proteins for which only a general functional assignment appeared appropriate. This site is particularly useful for functional predictions in borderline cases, where protein similarity levels are fairly low. Due to the diversity of proteins in *COGs*. sequence similarity searches against the *COG*.database can often suggest a possible function for a protein that otherwise has no clear database hits. This database also offers some convenient tools for a comparative analysis of complete genomes as will be described below.

***KEGG* :** The *Kyoto Encyclopedia of Genes and Genomes* (KEGG) centers around cellular metabolism (Kanehisa and Goto, 2000). This Web site presents a comprehensive set of metabolic pathway charts, both general and specific, for each of the completely-sequenced genomes, as well as for *Schizosaccharomyces pombe, Arabidopsis thaliana, Drosophila melanogaster,* mouse, and human. Enzymes that are already identified in a particular organism are color-coded, so that one can easily trace the pathways that are likely to be present or absent in a given organism.

For the metabolic pathways covered in *KEGG*, lists of orthologous genes that code for the enzymes participating in these pathways are also provided. It is also indicated whenever these genes are adjacent, forming likely operons. A very convenient search tool allows the user to compare two complete genomes and identify all cases in which conserved genes in both organisms are adjacent or located relatively close (within 5 genes) to each other. The *KEGG* site is continuously updated and serves as an ultimate source of data for the analysis of metabolism in various organisms.

***MBGD* :** The *Microbial Genome Database* (MBGD) at the University of Tokyo offers another convenient tool for searching for likely homologous among all sequenced microbial genomes. In contrast to *COGs, MBGD*, assigns homology relationships based solely on sequence similarity (*BLASTP* values of 10^{-2} or less). *MBGD* allows the user to submit several sequences at once (up to 2,000 residues) for searching against all of the complete genomes available, displays colour-coded functions of the detected homologous, and shows their location on a circular genome map. The output of *MBGD's BLAST* search also shows the degree of overlap between the query and target sequences, which could help in discerning multidomain proteins. For each sequenced genome, *MBGD* provides convenient lists of all recognized genes that are involved in a particular function, *e.g.*, the biosynthesis of branched-chain amino acids or the degradation of aromatic hydrocarbons.

***WIT* :** The WIT ("What Is There?") database, like KEGG, aims at metabolic reconstruction for completely sequenced genomes. The distinguishing features of the WIT approach are that it (1) considers as a pathway any sequence of reactions between two bifurcations and *(2)* includes proteins from many partially-sequenced genomes. These features allow WIT to offer many more sequences of the same enzymes from different organisms than any other database, which significantly facilitates the recognition of additional members of these enzyme families. On the other hand, complex pathways like glycolysis or the TCA cycle have been split into many separate reactions, which sometimes makes pathway analysis unnecessarily complicated. However, anyone who overcomes the initial difficulties in using the WIT system will be rewarded by the ability to easily predict metabolic pathways in many organisms with complete and still unfinished genomes.

Organism-Specific Databases

In addition to general genomics databases, there exists a variety of databases that center around a particular organism or a group of organisms. Although all of them are useful for specific purposes, those devoted to *E. coli, B. subtilis,* and yeast are probably the ones most widely used for functional assignments in other, less studied organisms. Following are short descriptions of the most frequently used of these databases.

Escherichia Coli : The importance of *E. coli* for molecular biology is reflected in the large number of databases dedicated to this organism. Two of these are maintained at the University of Wisconsin-Madison and at the Nara Institute of Technology, the research groups that carried out the actual sequencing of the *E. coli* genome. Because the Wisconsin group is now involved in sequencing the enteropathogenic *E. coli* O157:H7 and other enterobacteria, their database is most useful for analysis of enteric pathogens.

The group at the Nara Institute of Technology is primarily interested in resolving the functions of still-unannotated *E. coli* genes and strives to create an ultimate resource for further studies of *E. coli.* Their site provides a convenient link of genomic data to the Kohara restriction map of *E. coli* and allows one to search for Kohara clones that cover the region of interest. Another useful database on *E. coli,* EcoCyc, lists all experimentally studied *E. coli* genes; it also provides exhaustive coverage of the metabolic pathways identified in *E. coli.* The goal of another *E. coli* database, EcoGene, is to provide curated sequences of the *E. coli* proteins.

This is a good source for frame-shifted and potentially mistranslated proteins. Finally, Colibri and RegulonDB are the databases of choice for those interested in regulatory networks of *E. coli.* The *E. coli* Genetic Stock Center (CGSC) Web site also provides gene linkage and function information; it also lists the mutations available at the *CGSC.*

Mycoplasma Genitalium : *Mycoplasma* has the smallest genome of all known cellular life forms, which offers some clues as to what is the lower limit of genes necessary to sustain life (the *"minimal genome"*). Its comparison to the second smallest known genome, that of *Mycoplasma pneumoniae,* is available online. Recent data from TIGR provides insight into the range of *M. pneumoniae* and *M. genitalium* genes that can be mutated without loss of viability. From both computational analysis and mutagenesis studies, it appears that 250—300 genes are absolutely essential for the survival of mycoplasmas.

Bacillus Subtills : The *B. subtilis* genome also attracts considerable attention from biologists and, like that of *E. coli,* is being actively studied from the functional perspective. The Subtilist Web site, maintained at the Institute Pasteur, is constantly updated to include the most recent results on functions of new *B. subtilis* genes. In addition, a convenient index of *B. subtilis* sporulation genes is maintained at the Royal Holloway University of London.

Saccharomyces Cerevisiae : The major databases specifically devoted to the functional analysis of yeast *S. cerevisiae* genome are the *Saccharomyces Genome Database* (SGD) at Stanford University, the Yeast Database at *Munich Institute for Protein Sequences* (MIPS), and *Yeast Protein Database* (YPD) at Proteome, Inc. All three databases provide periodically updated lists of yeast proteins with known or predicted functions, appropriate references, and mutant phenotypes and reflect the ongoing efforts aimed at complete characterization of all yeast proteins. SGD is probably the largest and most comprehensive source of information on the current status of the yeast genome analysis and includes the *Saccharomyces* Gene Registry.

The *MIPS* database provides most of the same data and serves as a resource for new results coming from the multinational *EUROFAN* project. *YPD* is a curated database that is an useful resource for current information on the function of yeast proteins. *YPD* now allows free access for academic researchers using the database for non-commercial purposes. Other useful sites for yeast genome analysis include *Saccharomyces cerevisiae* Promoter Database, listing known regulatory elements and transcriptional factors in yeast; Transposon-Insertion Phenotypes, Localization, and Expression in *Saccharomyces (TRIPLES)* database, which tracks the expression of transposon-induced mutants and the cellular localization of transposon-tagged proteins, and the *Saccharomyces* Cell Cycle Expression Database, presenting the first results on changes in mRNA transcript levels during the yeast cell cycle.

ANNOTATION OF GENOME

With recent progress in rapid, genome-scale sequencing, sequence analysis and annotation of complete genomes have become the limiting steps in most genome projects. This task is particularly daunting given the paucity of functional information for a large fraction of genes even in the best-understood model organisms, let alone poorly-studied ones such as those from Archaeal species. The standard steps involved in the structural-functional annotation of uncharacterized proteins includes (1) sequence similarity searches using programs such as BLAST, FASTA, or the Smith-Waterman algorithm; (2) identifying functional motifs and structural domains by comparing the protein sequence against *PROSITE, BLOCKS, SMART* or Pfam; (3) predicting structural features of the protein, such as likely signal peptides, transmem-brane segments, coiled-coil regions, and other regions of low sequence complexity; and (4) generating a secondary (and, if possible, tertiary) structure prediction.

All these steps have been automated in several software packages, such as GeneQuiz, *MAGPIE, PEDANT, Imagene*, and others. Of these, however, *MAGPIE* and *PEDANT* do not allow outside users to submit their own sequences for analysis and display only the authors' own results. GeneQuiz offers a limited number of searches (up to 100 a day) to general users but is still a good entry point for comparative genome analysis. However, GeneQuiz relies on unrealistically high cutoff scores to infer homology, which inevitably results in relatively low sensitivity. In some cases, the user may be better off by simply using the same tools that are packaged in the aforementioned programs separately. To perform sequence analysis on a large scale, it is frequently desirable to run the requisite software locally, in batch mode. One such package that is currently available for free downloading is *SEALS*, developed at *NCBI*. It consists of a number of *UNIX*-based tools for retrieving sequences from GenBank, running database search programs such as *BLAST* and *MOST* viewing and parsing search outputs, searching for sequence motifs, and predicting protein structural features.

Prediction of Protein Functions

Analysis of the first several bacterial, archaeal, and eukaryotic genomes to be sequenced showed that the sequence comparison methods mentioned above failed to predict protein function for at least one-third of gene products in any given genome. In these cases, other approaches can be used that take into consideration all other available data, putting them into "genome context". By taking advantage of the availability of multiple complete genomes, these approaches offer new opportunities for predicting gene functions in each of these genomes. All these approaches rely on the same basic premise, that the organization of the genetic information in each particular genome reflects a long history of mutations, gene duplications, gene rearrangements, gene function divergence, and gene acquisition and loss that has produced organisms uniquely adapted to their environment and capable of regulating their metabolism in accordance with the environmental conditions. This means that cross-genome similarities can be viewed as meaningful in the *evolutionary* sense and thus are potentially useful for functional analysis.

The most promising comparative methods—specifically employ information derived from multiple genomes to achieve robustness and sensitivity that are not easily attainable with standard tools. It seems that they are indeed the tools for the "new genomics," whose impact will grow with the increase in the amount and diversity of genome information available. Here, some of these new approaches are briefly reviewed using for illustration, whenever possible, examples provided by currently available Web-based tools. A disproportionate number of these examples are from the *COG* system. This should not be construed as a claim that this is, in any sense, the best tool for genome annotation; rather, it reflects a degree of flexibility in for-mulating queries that is provided by the *COGs* as well as the subjective factor of the authors' familiarity with the organization of this system.

Transfer of Functional Information : The simplest and by far the most common way to utilize the information embedded in multiple genomes (at least at this time) is the transfer of functional information from well-characterized genomes to poorly-studied ones. Implicitly, this is done whenever a prediction is made for a newly sequenced gene on the basis of a database hit(*s*). There are, however, many pitfalls that tend to hamper accurate functional prediction on the basis of such hits. Perhaps the most important ones relate to the lack of sufficient sensitivity, error propagation because of reliance on incorrect or imprecise annotations already present in the general-purpose databases, and the difficulty in distinguishing orthologous from paralogous.

The issue of orthology vs. paralogy is critical because transfer of functional information is likely to be reliable for orthologous (direct evolutionary counterparts) but may be quite misleading if paralogous (products of gene duplications) are involved. All these problems are, in part, obviated in the COG system, which consists of carefully annotated sets of likely orthologous and does not rely on arbitrary cutoffs for assigning new proteins to them. The COGs can be employed for annotation of newly-sequenced genomes using the COGNITOR program.

This program assigns new proteins to *COGs*, by comparing them to protein sequences from all genomes included in the *COG* database and detecting genome-specific best hits (BeTs). When three or more BeTs fall into the same *COG,* the query protein is considered a likely new *COG* member. The reasoning is that it is extremely unlikely that such coherence occurs by chance, even if the observed sequence similarity *per se* is not statistically significant. The requirement of multiple BeTs for a protein to be assigned to a *COG* serves, to some extent, as a safeguard against the propagation of errors that might be present in the *COG* database itself. Indeed, if a *COG* contains one or even two false-positives, this will not result in a false assignment by *COGNITOR* under the three-BeT cutoff rule. Figure 10.4 shows two examples of the *COGNITOR* application to proteins from the bacterium *Deinococcus rodiodurons* and the archaeon *Aeropyrum pernix* that have not been assigned a function in the original genome annotation.

Phylogenetic Patterns (Profiles) : The *COG*-type analysis applied to multiple genomes provides for the derivation of *phylogenetic patterns,* which are potentially useful in many aspects of genome analysis and annotation. Similar concepts have been introduced by others in the form of phylogenetic profiles. The phylogenetic pattern for each protein family (*COG*) is defined as the set of genomes in which the family is represented. The *COG* database is accompanied by a pattern search tool that allows the user to select *COGs* with a particular pattern. Predictably, genes that are functionally related (*e.g.*, those that encode different subunits of the same enzyme or participate in consecutive steps of the same metabolic pathway) tend to have the same phylogenetic pattern.

In a complementary fashion, closely related species tend to co-occur in *COGs* Because of these features, phylogenetic patterns can be used to improve functional predictions in complete genomes. When a particular genome is represented in the *COGs* for a subset of components of a particular complex or pathway but is missing in the *COGs* for other components, a focused search for the latter is justified. The same applies to cases in which a gene is found in one of two closely related genomes, but not the other, particularly if it is conserved in a broad range of other genomes. There are several reasons why unexpectedly incomplete phylogenetic patterns may be observed.

In the simplest case, certain proteins, typically small ones, could have been missed in genome translation. Thus, the apparent absence of *secE* genes in the genomes of *Aquifex aeolicus* and *Helicobacter pylori* that encoded all the other components of the Sec protein translocation machinery suggests that the *secE* genes could have been missed in the original genome annotation for these two bacteria. Indeed, these genes are easily recognized by searching the six-frame translation of the respective genomes using *TBLASTN*. Examination of the *COGs* that miss one representative from a group of close species similarly may result in the identification of otherwise undetected genes. For example, only one COG (COG 1546) contained an *M. genitalium* protein but not an *M. pneumoniae* protein. A search for a possible missing *M. pneumoniae* counterpart identified a candidate, whose inclusion into this *COG* was subsequently verified by *COGNITOR* and sequence alignment.

An unexpected absence of a species in a phylogenetic pattern also may indicate that the given species encodes a highly diverged member of the respective orthologous family. For example, the presence of easily-recognizable A, B, D, and I subunits of the archaeal type H^+-ATPase in *Borrelia burgdorferi, Treponema pallidum,* and both chlamydia (COGs 1155, 1156, 1394, and 1269) immediately suggests that membrane-bound subunits of this enzyme should also be encoded in these genomes. Indeed, genes for the E and K subunits of the H+-ATPase could be recognized in these genomes (COGs 1390 and 0636) despite their low sequence similarity to the corresponding subunits of the archaeal enzymes.

The gene for the *F* subunit however, has been identified so far only in *T. pallidum* but not in the three other bacterial species whereas the gene for the *C* subunit (COG 1527) has not been recognized in any of them. Finally, unexpected *"holes"* in phylogenetic patterns and differences between components of the same complex or pathway may be the manifestation of a phenomenon termed *nonorthologous gene displacement,* in which unrelated or distantly related proteins are responsible for the same function in different organisms. When essential functions are involved, this tends to result in phylogenetic patterns that are perfectly or partially complementary, together spanning the entire range of genomes. Note that, in each case, the complementarity is not perfect because certain genomes encode both forms of the respective enzyme.

In the case of lysyl-tRNA synthetases the two forms of the enzyme are completely unrelated, whereas the two fructose-biphosphate aldolases are distantly related, but are not orthologous. During the analysis of new genomes, it is possible to focus on families with complementary phylogenetic patterns to identify candidates for missing components of complexes and pathways.

Use of Phylogenetic Patterns for Differential Genome Display

The phylogenetic pattern approach and, specifically, the pattern search tool associated with the COGs can be used in a systematic fashion to perform formal logical operations (AND, OR, NOT) on gene sets—an approach suitably dubbed "differential genome display". Figure 10.7 shows examples of such analyses. This type of genome comparison allows a researcher to delineate subsets of gene products that are likely to contribute to the specific lifestyles of the respective organisms, for example, thermophily. The use of this approach to identify candidate drug targets in pathogenic bacteria is perhaps of special interest. It seems logical to look for such targets among those genes that are shared by several pathogenic organisms, but are missing in eukaryotes.

Simple exercises in this direction show, however, that this is not a straightforward strategy. It is tempting to suggest that the best targets for new broad-spectrum antimicrobial agents would be genes that are shared by all pathogenic microbes, but not by any other organisms. The trouble is that such genes do not seem to exist, even if one allows for those that are missing in mycoplasmas, which have by far the smallest genomes. Furthermore, even when the conditions are relaxed so that it is only required that the genes be present in all pathogenic bacteria (except possibly mycoplasmas) and absent in yeast and *E. coli* (the dominant component of the normal gut microbial population), the net comes back empty. It seems therefore that the best one can do to search for such potentially universal antimicrobial agents is to isolate the genes that are present in all pathogens, possibly in other bacteria and archaea but not in eukaryotes. This results in a list of 35 families, most of which are, in fact, represented in all bacteria it seems likely that some of these proteins are indeed good candidates for drug targets. More specifically directed searches can be easily set up; for example, searching for families that are represented in two species of chlamydia, possibly other pathogenic bacteria, but not any other genomes produces just two COGs. These could be of interest for a detailed experimental study aimed at the development of new agents that could be active against both chlamydia and spirochetes.

Domain Fusions: Another recently developed comparative genomic approach involves systematic analysis of protein and *domain fusion* (and fission). The basic assumption is that fusion would be maintained by selection only when it facilitates functional interaction between proteins, for example, kinetic coupling of consecutive enzymes in a pathway. Thus, proteins that are fused in some species can be expected to interact, perhaps physically or at least functionally, in other organisms. A straightforward example of functional inferences that can be drawn from domain fusion is seen in the histidine biosynthesis pathway, which in *E. coli* and *H. Influenzae* includes two two-domain proteins, His I and His B.

The two domains of His I catalyze two sequential steps of histidine biosynthesis and thus represent subunits that are likely to physically interact even when produced as separate proteins; this correlates with the predominance of the domain fusion among these enzymes. In contrast, the two domains of His B catalyze the seventh and ninth steps of the pathway and hence are not likely to physically interact, which is compatible with the relative rarity of the fusion. The COG database includes about 700 distinct multidomain architectures that have stand-alone counterparts.

Thus using domain fusion for functional prediction has considerable heuristic potential although this approach will not work for *"promiscuous"* domains such as, for example, the DNA-binding helix-turn-helix domain, which can be found in combination with a wide variety of other domains. In addition, several databases (with accompanying search tools) have recently been developed for detecting domains and exploring architecture of multidomain proteins: Pfam, ProDom and SMART. Although not comprehensive as of this writing, SMART seems to be the most advanced of these systems, combining high sensitivity of domain detection with accuracy, high speed, and extremely informative presentation of domain architectures.

Rapid searches for protein domains, based on a modification of the PSI-BLAST program is now available through the Conserved Domains Database (CDD) at NCBI. It seems worth considering an example of a complex multidomain protein analysis in some detail, to see how assigning functions to various domains of a multi-domain protein helps one understand its likely cellular role(s). The *M. tuberculosis* protein Rvl364c consists of 653 amino acid residues. Its annotation in GenBank correctly indicates that it has statistically significant similarity to the *B. subtilis* sigma factor regulation protein, RSBU_BACSU, which, however, is only 335 amino acids long. The region of similarity between these two proteins is said to be even shorter, 244 amino acid residues. Thus, in addition to the portion apparently homologous to RsbU, Rvl364c probably contains other domains. Submitting Rvl364c for a Pfam search gives an unexpected result: Pfam search identifies a SpoIIAA(RsbV)-like domain in the 550 – 652 region of the protein. The confidence level is not particularly high (E = 0.0049), but examination of the alignment shows conservation of the phosphorylatable serine residue and the surrounding motif, as well as conservation of the secondary structure elements (not shown). This suggests that Rvl364c actually contains *four* domains, the second and the fourth of which correspond to *B. subtilis* RsbU and RsbV proteins.

The structures and functions of the first (residues 1-155) and the third (334—550) domains remain to be analyzed. This could be done by PSI-BLAST analysis of individual segments encompassing the presumptive domains, but this route would involve careful examination of the complex search outputs. In contrast, SMART provides a one-step solution. The SMART output indicates that the protein under analysis contains N-terminal PAS and PAC domains, which are ligand-binding sensor domains present in many histidine kinases and other signal-transduction proteins, a PP2C-like phosphatase domain (the actual biochemical activity of the RsbU domain) and a histidine kinase-type ATPase domain (residues 433—527). For the latter domain, however, the statistical significance of the hit is low (E = 0.16) and the assignment needs further verification. A PSI-BLAST search started with the segment of Rvl364c identified by SMART as the ATPase domain reveals similarity to the RsbW proteins, a distinct group of serine kinases of the histidine kinase fold involved in the anti-sigma regulatory systems (hence, the low-significance hit to the general profile for this domain in SMART). SMART does not detect the *C*-terminal SpoIIAA domain of Rvl364c, which has been identified by Pfam, emphasizing the importance of complementary methods for complete assignments of domains and functions.

The domain organization of the protein can also be probed using the COG database. Entering Rvl364c as a COGNITOR query assigns its domains to four COGs (where the protein already belongs since it originates from a completely sequenced genome): (1) Rvl364c_l-COG2202 "PAS/PAC domain," (2) Rvl364c_2-COG2208 "Serine phosphatase RsbU, regulator of sigma subunit," (3) Rvl364c_3-COG2172 *"Anti-sigma regulatory factor* (Ser/Thr protein kinase)," and (4) Rvl364c_4-COG1366 "*Anti-anti-sigma regulatory factor* (antagonist of antisigma factor)." Thus, taken together, the results obtained using different methods converge on an unprecedented four-domain architecture for Rvl364c that juxtaposes the sensor PAS/PAC domain with all three components of the anti-sigma regulatory system fused within a single protein.

The PAS/PAC domain is most likely involved in sensing the energetic state of the cell, similarly to the recently characterized Aer protein of *E. coli*, whereas the phosphatase, kinase, and phosphorylatable adapter domains are expected to efficiently transmit this information to the downstream signal response machinery. Thus, we can tentatively annotate Rvl364c protein as a complex regulator of sigma factor activity; the exact implications of the unusual domain fusion remain to be investigated experimentally.

Analysis of Conserve Genes brings (operons) : An approach that is conceptually similar to the analysis of gene fusions, but is more general, if less definitive, involves systematic analysis of gene *"neighbourhoods"* in genomes. Because functionally linked genes frequently form operons in bacteria and archaea, gene adjacency may provide important functional hints. Of course, many functionally related genes never form operons, and, in many instances, adjacent genes are not connected in any way. However, due to the lack of overall conservation of gene order in prokaryotes, the presence of a pair of adjacent orthologous genes in three or more genomes or the presence of three orthologous in a row in two genomes can be considered a statistically meaningful event and can be used to infer potential functional interaction for the products of these genes.

The simplest current tool for identification of conserved gene strings in any two genomes is available as part of *KEGC*. It allows the user to select any two complete genomes (*e.g.*, *B. burgdorferi* and *R. prowazekii)* and look for all genes whose products are similar to each other (*e.g.*, have *BLAST* scores greater than 100) and are located within a certain distance from each other (that is, separated by 0-5 genes). The results are displayed in a graphical format illustrating the gene order and the presumed functions of gene products. In the example shown in Fig. 10.1 the uncharacterized conserved protein BB0788 from *B. burgdorferi* is similar to the RP042 protein of *R. prowazekii,* and BB0789 is similar to RP043. These pairs of genes indeed have been identified as orthologous in the COGs (COG0037 and COG0465, respectively), and examination of the relative genomic locations of other members of these COGs shows that orthologous gene strings are present in the genomes of *C. trachomatis* (CT840-CT841), *C. pneumoniae* (CPn0997-CPn0998), and *T. maritima* (TM0579-TM0580), whereas in *B. subtilis* the corresponding genes (*yacA* and *ftsH*) are one gene apart. This conservation of gene juxtaposition in phylogenetically distant bacteria is suggestive of a functional connection.

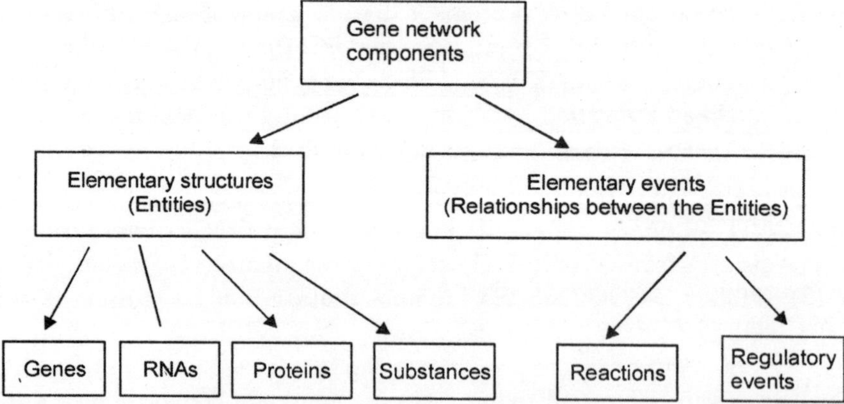

Fig. 10.1 : Elucidation of a protein's domain architecture using SMART. The SMART output for the Rv1364c protein is shown. The additional bar above the line shows the location of the SpollAA(RbsV) domain that is not recognized by SMART.

The functions of one of the genes in all these pairs are well known, as they are clear orthologous of the *E. coli ftsH* (*HflB*) gene. This gene encodes an ATP-dependent metalloprotease, which is responsible for the degradation of short-lived cytoplasmic proteins and a distinct class of membrane proteins. By association, BB0788 and its orthologous may be expected to also play some, perhaps regulatory, role in the degradation of specific protein classes. The *E. coli* orthologous of BB0788, MesJ, is annotated in the SWISS-PROT database as a putative cell cycle protein. We have been unable to identify the source of this information. Nevertheless, it is compatible with a functional (and possibly also physical) interaction between MesJ and FtsH, which also has been implicated in cell division on the basis of genetic data.

In the COG database, MesJ and its orthologous (COG0037) are annotated as "Predicted ATPases of the PP-loop superfamily." All these enzymes share a diagnostic sequence motif, which has been discovered previously in a number of ATP pyrophosphatases. This motif is clearly seen in the multiple alignment accompanying the COG and in the ProDom entry PD000352. Therefore, by combining operon information with sequence-based prediction of the biochemical activity, we hypothesize that MesJ and its orthologous are ATP-pyrophosphatases involved in the regulation of FtsH-mediated proteolysis of specific bacterial proteins, which may be important for cell division. Because a PP-loop superfamily ATPase would comprise a novel class of cell division regulators, experimental verification of this hypothesis will be of considerable interest.

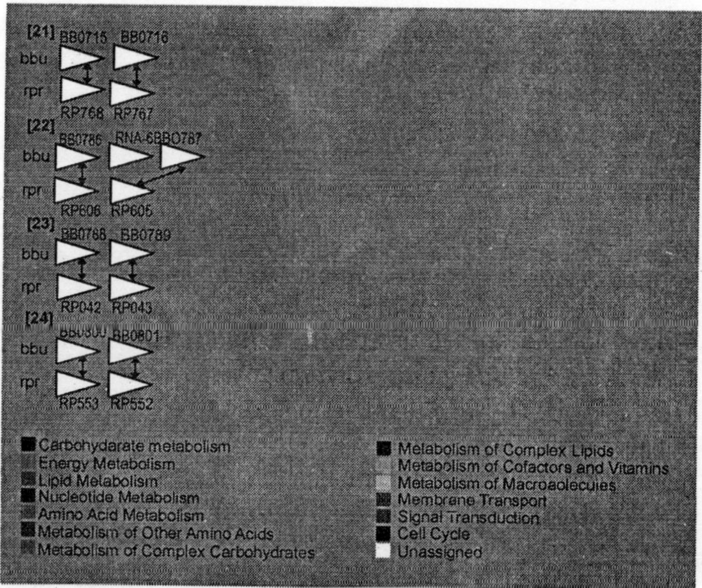

Fig. 10.2 : The results of a search for conserved gene strings (operons) in Borrelia burgdorferi (upper arrows) and Rickettsia prowazekii (lower arrows) using the KEGG gene cluster tool. The arrows are colour-coded according to the gene function. Genes without an assigned function (all of the genes in this page) are shown in white.

RECONSTRUCTION OF METABOLIC PATHWAYS

To succinctly recap the genome analysis tools discussed above, we present here a reconstruction of the glycolytic pathway in the archaeon *Methanococcus jannaschii*. Metabolic reconstruction is one of the indispensable final steps of all genome analyses and a natural convergence point for the data produced by different methods. Glycolysis is perhaps *the* central pathway of cellular biochemistry as it becomes obvious from a cursory exploration of the general scheme of biochemical pathways, available in the interactive form on the ExPASy Web site. The names of all the enzymes and metabolites on this map are hyperlinked and searchable. Entering *"glycolysis"* as the search term finds three fields, B5, F5 and U6, the first two of which indicate the border areas of glycolysis, which actually stretches from C5 to E5.

The enzyme names are hyperlinked to the ENZYME database, which is now the official site of the Enzyme Commission (IUPAC-IUBMB Joint Commission for Biochemical Nomenclature). The ENZYME database lists names and catalyzed reactions for all the enzymes that have been assigned official *Enzyme Commission* (EC) numbers, whether or not their protein sequences are known. Thus, clicking on the name "phosphoglucomutase" will bring up the corresponding page in the *ENZYME database*, which will also indicate that the official name of this enzyme is glucose-6-phosphate isomerase (EC 5.3.1.9).

Process of Glycolysis

1. *Glucose-6-Phosphate Isomerase :* Each page in the *ENZYME* database lists all enzymes with the corresponding EC number that are included in the current release of *SWISS-PROT*. There is, however, no entry for *M. jannoschii* under glucose-6-phosphate isomerase; therefore, we will instead use the *KEGG*, database. On entering *KEGG*, one can go to Open *KEGG*, then to Metabolic Pathways, then to Glycolysis. This takes one to an easy-to-navigate chart of compounds and enzymes that participate in glycolysis and gluconeogenesis. The pull-down menu in the upper left corner allows one to select the organism of choice. When *Methanococcus jannaschii* is selected, the boxes that indicate enzymes already recognized in this organism become shaded green. One can see that the box 5.3.1.9 is highlighted. Clicking on this box shows *M. jannaschii* protein *MJ1605.*, which indeed can be confidently identified as phosphoglucomutase (glucose-6-phosphate isomerase).

In the *SWISS-PROT/TREMBL* database, this protein (Q59000) is currently annotated as "similar to prokariota glucose-6-phosphate isomerase." To verify that *MJ1605* is indeed the orthologous of known glucose-6-phosphate isomerases, one can use the *COGs, WIT*, or *MBGD* systems. For example, in the *COG* database, *MJ1605* is the only member of *COG0166* from *M. jannaschii;* since this *COG* includes glucose-6-phosphate isomerases from a number of species, identification of the *M. jannaschii* protein appears to be reliable. This can be confirmed by examination of the multiple sequence alignment associated with this *COG*, which shows a particularly high similarity between *MJ1605* and phosphoglucoisomerases from *Thermotoga maritima* and *B. subtilis*. The *WIT* database will show that *MJ1605* is even more similar to the (predicted) glucose-6-phosphate isomerases of *Campylobacter jejuni* and *Streptococcus pyogenes*. Collectively, this evidence leaves no doubt that we have identified the correct *M. jannaschii* protein.

2. *Phosphofructokinase :* The next glycolytic enzyme, phosphofructokinase (EC 2.7.1.11), illustrates the opportunities and limitations of metabolic reconstruction based on comparative genomics. The most common version of this enzyme, PfkA, uses either ATP (in bacteria and many eukaryotes) or pyrophosphate (primarily in plants) as the phosphate donor. In addition, *E. coli* encodes a second version of this enzyme, PfkB, which is unrelated to PfkA and instead belongs to the ribokinase family of carbohydrate kinases. All the databases we can use agree that there is no readily identifiable candidate for this activity in *M. jannaschii*. Indeed, the

KEGG chart for *M. jannaschii* does not show this enzyme as predicted, WIT does not suggest any candidates for this function, and the COG database does not show any archaeal members in its COG0205 (6-phofructokinase) or COG 1005 [fructose-1-phosphate kinase and related fructose-6-phosphate kinase (PfkB)] entries.

Thermophilic archaea possess a distinct, ADP-dependent phosphofructokinase, the gene(*s*) for which has been recently identified in *Pyrococcus furiosus* (*Juininga et. al., 1999*). An ortholog of this protein is readily identifiable in other Pyrococci and in *M. jannaschii* (MJ1604) but not in any other archaeal or bacterial species; its sequence shows no detectable similarity to other kinases. This is a clear case of nonorthologous gene displacement; due to the limited phylogenetic distribution of this novel Pfk, the candidate protein in *M. jannaschii* could not have been detected by computational means until its ortholog had been experimentally characterized in *P. furiosus.*

3. *Aldolase :* The next glycolytic enzyme, fructose- 1,6-bisphosphate aldolase (EC 4.1.2.13), is found in two substantially different variants, namely, metal-independent (class I) and metal-dependent (class II) aldolases in bacteria and eukaryotes. There is no ortholog of either of these in *M. jannaschii.* Instead, predicted archaeal aldolases comprise COG1830 ("DhnA-type fructose- 1,6-bisphosphate aldolase and related enzymes"), which includes two *M. jannaschii* proteins, MJ0400 and MJ1585. This COG includes orthologous of the recently described class-I aldolase of *E. coli,* which is only very distantly related to the typical class-I enzymes. As indicated above, the phylogenetic patterns for typical bacterial class-II aldolase (COG0191) and this DhnA-type aldolase (COG1830) complement each other, with the exception that both types of aldolases are encoded by *E. coli* and *A. aeolicus.* This complementarity allows one to predict that the DhnA-type enzyme functions as the only fructose- 1,6-bisphosphate aldolase in chlamydiae and archaea, including *M. jannaschii.*

4-6. *Triosephosphate Isomerase, Glyceraldehyde-3-Phosphate Dehydrogenase, and Phosphoglycerate Kinase :* These enzymes catalyze the next three steps of glycolysis and are nearly-uniformly represented in all organisms. The *M. jannaschii* candidates for these activities (MJ1528, MJ1146, and MJ0641, respectively) can be easily identified by a BLAST search. Accordingly, all major databases converge on these functional assignments.

7. *Phosphoglycerate Mutase :* The activity of phosphoglycerate mutase (EC 5.4.2.1) has been experimentally demonstrated in a close relative *of M. jannaschii,* but there are no obvious candidate proteins to carry out this function. As a result, KEGG does not show this enzyme as encoded in the M. *jannaschii* genome. WIT does suggest a candidate protein (MJ1612 or RMJ05975 in WIT) but annotates it as "phosphonopyruvate decarboxylase." Indeed, WIT shows only limited sequence similarity of this protein to the phosphoglycerate mutases from the mycoplasmas and *H. pylori,* whereas all close homologous of MJ1612 seem to be phosphonopyruvate decarboxylases.

Finally, a search of the COG database for "phosphoglycerate mutase" retrieves three COGs, only one of which, COG1015 (phosphopentomutase/ predicted phosphoglycerate mutase and related enzymes), contains archaeal members, including two *M. jannaschii* proteins, MJ1612 and 0010. A detailed analysis shows that both of them could possess phosphoglycerate mutase activity. The definitive identification of the phosphoglycerate mutase in *M. jannaschii* awaits direct biochemical studies.

8-9. *Enolase and Pyruvate Kinase :* In *M. jannaschii,* these enzymes are readily identified through strong sequence similarity to the corresponding bacterial orthologous. As a result, enzymes has been experimentally characterized in *M. jannaschii,* computational analysis provides for the identification of all of them, with uncertainty remaining with regard to just one step.

Glycolysis in *H. Pylori?*

A Cautionary Note. Metabolic reconstruction based on genome analysis requires considerable caution to be exerted with regard to the plausibility of functional predictions in the general context of the biology of the organism in question. So as to not depart from glycolysis, consider the case of phosphofructokinase in the gastric ulcer-causing bacterium, *Helicobacter pylori.* This organism lacks a homolog of PfkA but does encode a close homolog of PfkB. Accordingly, WIT suggests this PfkB homolog, HP0858, as a candidate for the phosphofructokinase function in *H. pylori.* Perusal of the COG database, however, suggests a different solution, since HP0858 consists of two distinct domains, one of which indeed belongs with PfkB and other sugar kinases (COG0524), whereas the other one is predicted to be a nucleotidylyltransferase (COG0615).

Given this domain architecture, it appears most likely that this protein is an ADP-heptose synthetase, an enzyme of peptidoglycan biosynthesis. A simple analysis of the biology of *H. pylori* as an acid-tolerant bacterium shows that it is likely to use gluconeogenesis, but not glycolysis, which makes phosphofructokinase unnecessary for this organism. Thus, the assignment of this activity to HP0858, which appeared to be statistically supported, is, in all likelihood, biologically irrelevant. Examination of this protein's domain architecture could be the first indication of its role in a process other than glycolysis, with biological considerations further supporting this interpretation.

AVOIDING COMMON PROBLEMS IN GENOME ANNOTATION

Due to its intrinsic complexity, genome annotation defies full automation and is inherently errorprone. Accidental error rate can be minimized only through further development of the semiautomated annotation systems and the appropriate training of annotators. There are, however, several sources of systematic error that plague genome analysis. Awareness of these could help improve the quality of genome annotation.

Error Propagation and Incomplete Information in Database

Sequence databases are prone to error propagation, whereby erroneous annotation of one protein causes multiple errors as it is used for annotation of new genomes. Furthermore, database searches have the potential for noise amplification, so that the original annotation could have involved a minor inaccuracy or incompleteness, but its transfer on the basis of sequence similarity aggravates the problem and eventually results in outright false functional assignments. These aspects of current sequence databases make the common practice of assigning gene function on the basis of the annotation of the best database hit (or even a group of hits with compatible annotations) highly error-prone. Time- and labour-consuming as this may be, adequate genome annotation requires that each gene be considered in the context of both its phylogenetic relationships and the biology of the respective organism, hence the rather disappointing performance of automated systems for genome annotation. There are numerous reasons why functional annotation may be wrong in the first place, but two main groups of problems have to do with database search methods and with the complexity and diversity of the genomes themselves.

Database Searches

It is customary in genome annotation to use a cutoff for *"statistically significant"* database hits. It can be expressed in terms of the false-positive expectation *(E)* value for the *BLAST* searches and is set routinely at values such as $E = 0.001$ or $E = 10^5$. The problem with this approach is that the distribution of similarity scores for evolutionarily and functionally relevant sequence alignments is very broad and that a considerable fraction of them fail the E-value cutoff, resulting in undetected relationships and missed opportunities for functional prediction (false negatives). Conversely, spurious hits may have E-values lower than the cutoff, resulting in false positives. The latter is most frequently caused by compositional bias (low-complexity regions) in the query sequence and in the database sequences.

Clearly, there is a trade-off between *sensitivity* (false-negative rate) and *selectivity* (false-positive rate) in all database searches, and it is particularly difficult to optimize the process in genome-wide analyses. There is no single recipe to circumvent these problems. To minimize the false-positive rate, appropriate procedures for filtering low-complexity sequences are critical. Filtering using the SEG program is the default for Web-based BLAST searches, but additional filtering is justified for certain types of proteins. For example, filtering of predicted nonglobular domains using SEG with specifically adjusted parameters and filtering for coiled-coil domains using the COILS2 program is one way to minimize the false positive rate. Minimizing the false-negative rate (that is, maximizing sensitivity) is an open-ended problem.

It should be kept in mind that a standard database search (*e.g.*, using *BLAST*) with the protein sequences encoded in the given genome as queries is insufficient for an adequate annotation. To increase the sensitivity of genome analysis, it should be supplemented by other, more powerful methods such as screening the set of protein sequences from the given genome with preformed profile libraries.

Protein, and Organismal Context

As discussed above, protein domain architecture, genomic context and an organism's biology may serve as sources of important, even if indirect, functional information. However, those same context features, if misinterpreted, may become one of the major sources of error and confusion in genome annotation. Standard database search programs are not equipped with the means to explicitly address the implications of the multidomain organization of proteins. Therefore, unless specialized tools such as *SMART* or *COGs* are employed and/or the search output is carefully examined, assignment of the function of a single-domain protein to a multidomain homolog and vice versa becomes frequent in genome annotation.

Promiscuous, mobile domains are particularly likely to wreak havoc in the annotation process, as demonstrated, for example, by the proliferation of "IMP-dehydrogenase-related proteins" in several genomes. In reality, most or all of these proteins (depending on the genome) share with IMP dehydrogenase the mobile CBS domain but not the enzymatic part. As discussed above, it is also critical for reliable genome annotation that the biological context of the given organism is taken into account. In a simplistic example, it is undesirable to annotate archaeal gene products as nucleolar proteins, even if their eukaryotic homologs are correctly described as such. As a general guide to functional annotation, it should be kept in mind that current methods for genome analysis, even the most powerful and sophisticated of them, facilitate, but do not supplant the work of a human expert.

CONCLUDING REMARKS

With an increasing number of complete genome sequences becoming available and specialized tools for genome comparison being developed, the comparative approach is becoming the most powerful strategy for genome analysis. It seems that the future should belong to databases and tools that consistently organize the genomic data according to phylogenetic, functional, or structural principles and explicitly take advantage of the diversity of genomes to increase the resolution power and robustness of the analysis. Many tasks in genome analysis can be automated, and, given the rapidly growing amount of data, automation is critical for the progress of genomics. This being said, the ultimate success of comparative genome analysis and annotation critically depends on complex decisions based on a variety of inputs, including the unique biology of each organism. Therefore, the process of genome analysis and annotation taken as a whole is, at least at this time, not automatable, and human expertise is necessary for avoiding errors and extracting the maximum possible information from the genome sequences.

11

GeneNet System

The molecular genetic systems that control the processes occurring in organisms on the basis of the hereditary information contained in their genomes are called *gene networks*. Numerous biological, biochemical, and physiological molecular processes occur simultaneously in humans, animals, and plants, as well as in prokaryotes, eukaryotes, and archaea. Cells divide and differentiate, tissues and organs are formed. Organisms enter into complex interactions with the environment while consuming matter, energy, and information flows during their growth, development, and reproduction.

All of these diverse processes are regulated genetically. Gene networks, with groups of genes functioning in concert as their central elements, are the backbone of this regulation. Theoretical studies of gene networks commenced in the 1960s. They considered general organization patterns of molecular genetic systems controlling functions of prokaryotes and described the dynamics of gene networks within the simplest logic schemes. Further studies involved approaches based on differential equations and stochastic models. Numerous methods for mathematical simulation have been developed, including (1) Boolean networks allowing gene networks to be reconstructed from experimental data (2) the logical approach (3) Petri nets and (4) threshold models. However, insufficient experimental data limited the development of the theory of gene networks until the mid-1980s.

During the 1990s, the appearance of efficient methods for studying molecular mechanisms that regulate gene expression and successful research in structural-functional organizations of various genomes triggered an explosive accumulation of experimental data on functions of gene networks. This accumulation led to a wide diversity of databases on various features of the gene network functions, genetic regulation of metabolic processes, morphogenesis, development, and so on.

It was a powerful stimulus for developing both new and earlier methods, approaches, and algorithms designed to detect functional regularities of biological systems at all levels of organization. The search for regularities based on contextual analysis of nucleotide sequences is considered by McGuire and Church in chapter 6 of this volume. In chapter 8 of this volume, Huang introduces the application of logical approaches to the development of multicellular organisms. In this chapter we consider the problems connected with regularities of gene network function and point out two aspects of the question: (1) developing methods for storing and preserving the information accumulated both in experiments and by means of numeric analysis of mathematical models; (2) working out the methods of compiling and analyzing mathematical models of gene network functioning. The databases listed here and many other databases are an important source of information for both the experimental study and the computer analysis of gene networks, genetically controlled metabolic processes, physiological systems, and other items.

The role and significance of such databases will grow with the amount of experimental information on functions of gene networks and genetically controlled systems and processes. Consequently development of efficient technologies for accumulating this information in computer databases is of the utmost importance. Detailed here is the technology we have developed for computer-assisted description of gene networks and the database GeneNet, constructed using this technology. Analysis of a variety of actual gene networks suggested us two important methodological principles for development of the technology in question:

1. The function of any gene network in either a unicellular or a multicellular organism involves a limited set of elementary structures and events of their interactions at different hierarchical levels of organization (genes, cell nucleus, cytoplasm, nuclear membrane, intercellular space, tissue, or organ). The specific combination of elementary units and events generates a tremendous diversity of gene networks with typical patterns of structural-functional organization and functional modes.
2. In any genetically controlled organism system, it is impossible to separate in a pure form the genetic component itself (*i.e.*, the component that performs the control) and the controlled component (which provides for a particular biochemical, physiological, or other elementary function). This means that description of a gene network implies simultaneous consideration of these components.

This chapter details the technology of gene network description; the GeneNet database created using this technology and containing the information on more than 20 gene networks of multicellular organisms; and basic principles of the gene network organization and functions. We also describe mathematical simulation of biological systems and illustrate the method used by particular models of gene network dynamics.

DESCRIBING GENE NETWORKS

Object-Oriented Approach

In the object-oriented approach the components of gene networks in the GeneNet database are divided into elementary structures (entities) and elementary events (relationships between the entities).

Elementary Structures : We consider the following elementary structures to be significant for the function of gene networks: gene, RNA, protein, and nonproteinaceous substanc. This list can be extended, if necessary, with other elementary structures. Each object class is described in a separate table, using a specialized dataj-representation format that takes into account the peculiarities of each class.

The database contains the following tables: GENE, RNA, *PROTEIN*, and *SUBSTANCE*.

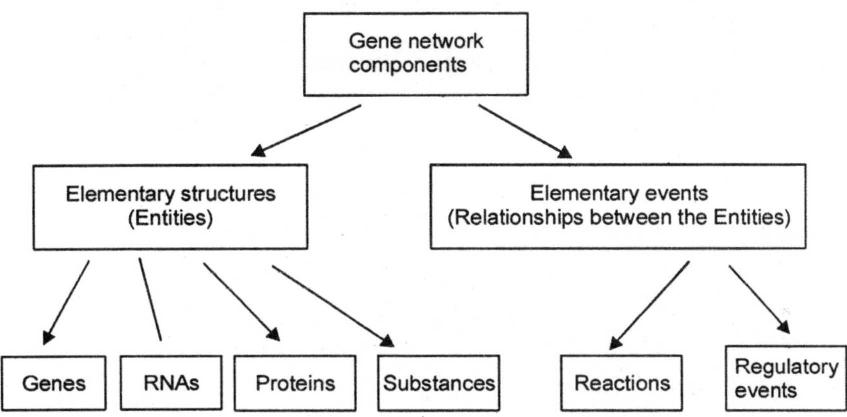

Fig. 11.1 : Class hierarchy in the GeneNet database.

Elementary Events : In the GeneNet database we consider two types of elementary events: reaction and regulatory event.

Chemical notation is the basis for describing the elementary events. Thus, any event is described as follows:

$$\begin{array}{c} C_1, \;\ldots,\; C_k \\ \downarrow \\ A_1 + \ldots A_n \;\Rightarrow\; B_1 + \ldots B_m, \end{array}$$

where *A's* are the elementary entities in the interaction; *C's* are the entities modulating this interaction; and *B's* are the products of the interaction. Based on this model, we consider two types of interactions between gene network components:

1. *Reaction* the interaction between entities that leads to generation of a new entity (assembly or disassembly of a multimeric complex, expression of a protein, secretion of certain substances, protein modifications, etc.). In some cases such a reaction corresponds to a single biochemical reaction (*e.g., protein phosphorylation*), while in other cases it corresponds to a series (cascade) of successive biochemical reactions (e.g., expression of a protein). Interactions of the former type are designated in the GeneNet database as direct reactions; of the latter type, as indirect reactions.
2. *Regulatory event* the effect of an entity (it may be a catalyst or an inhibitor) on a certain reaction. Regulatory events of four types are distinguished, depending on their effect on the reaction: switching on, switching off, positive effect, and negative effect. Regulatory events described in the GeneNet database are induction of gene expression, repression of gene expression, activation of signal transduction pathway by effector molecules, enzymatic catalysis etc.

Levels of Gene Network

The GeneNet system allows the fact that gene network components may be distributed in various organs, tissues, cells, and cell compartments to be taken into account. As a first approximation, three main hierarchical levels (organism, cell, and gene) are considered in the description of the gene network.

Organism Level

The entities described at this level are organs, tissues, particular types of cells, and secreted proteins and substances affecting other organs, tissues, and cells. The description at this level enables the spatial order of gene network components in the organism to be considered.

Cell Level

The entities described at this level are cell compartments (for example, cytoplasm, nucleus, and mitochondria), proteins, RNAs, genes, and substances (*e.g.*, steroids, lipids, energy-stored molecules, metabolites). The description at this level enables distribution of the gene network components throughout cell compartments to be considered.

Gene Level

Regulation of gene expression is described in detail at this level employing the information from the TRRD database (Kolcha nov *et.al.*, 2000).

FORMAT FOR DESCRIBING ELEMENTARY STRUCTURES AND ELEMENTARY EVENTS

Elementary Structures

The elementary structures essential for the function of gene networks in the GeneNet database are described in the tables *GENE, PROTEIN, RNA* and *SUBSTANCE*. The data format used for describing the elementary structures involves several information fields.

```
ID Gg: GATA1 ¯p
DT 09.10.97; Ananko E.; created.
DT 19.6.1999.; Podkolodnaya O.A.; updated.
OS Callus gallus (chicken) .
SN GATA1
NM transcription factor GATA1
SY NF-E1
FN active
MM no data
MD phosphorylated
GN Gg:GATA-1
DR TFFACTOR; T00267;
DR SWISSPROT; P17678;
DR EMBL; M26209;
DR PIR; A32993;
SO GgEryth Prog
RF Briegel K., et al. 1996
//
```

Fig. 11.2 : Description of the elementary structure in the table PROTEIN of GeneNet database, exemplified by GATAl transcription factor, which is essential for the function of the gene network of erythrocyte differentiation and maturation.

For example, the entry shown in figure 11.2 indicates that chicken protein *GATA1-p* is a transcription factor (*NM*) that is phosphorylated (*MD*); exists in an active state (*EN*); has a synonymous name *NF-E1* (*SY*); is described in databases *TFFACTOR, SWISS-PROT, EMBL,* and *PIR* (*DR*); and is isolated from erythroid progenitor cells (*SO*); and the information stored was obtained from a paper by K. Briegel et al. (1996) (*RF*).

Elementary Events

Two types of interactions are considered when describing the gene networks in the GeneNet database: reactions and regulatory events (see above). In the GeneNet database, elementary events are described in the table *RELATION*.

(a)

```
ID <protein>Hs :preSREBP^cytoplasm ->
<protein>Hs : SREBP1^Acytoplasm
DT 02.8.1999.; Ignatieva E.V.; created.
EF direct
RF Wang X. et al., 1994
//
```

(b)

```
ID <substance>Cholesterol^Acytoplasm -»
<protein>Hs:SRP^Acytoplasm -» <protein>Hs :preSREBP1^Acytoplasm
-> <protein>Hs:SREBP1Acytoplasm
DT 02.8.1999.; Ignatieva E.V.; created.
AT decrease
EF indirect
RF Wang X. et al., 1994
\\
```

Fig. 11.3 : Description of relationships in the GeneNet database, (a) reaction — transformation of the inactive precursor transcription factor preSREBPl into the mature factor SREBP1; (b) regulatory event: inhibition of SRP protease activity by cholesterol (gene network of lipid metabolism).

Reactions **:** The data format used for describing reactions involves several information fields.

Description of a reaction within the gene network of lipid metabolism, includes transformation of transcription factor pre*SREBPl* into active transcription factor *SREBP1*, the key regulator of this gene network. The record means that the human protein pre*SREBP1* transforms into human protein *SREBPE1*. the reaction takes place in the cytoplasm (*ID*), and the interaction described is direct (*EF*). As is described below, this reaction requires sterol-regulated protease (*SRP*). The activity of this protease decreases with the increase in cell cholesterol level, thus closing the negative feedback circuit controlling the cholesterol level in the cell.

Description of a regulatory event significant for the function of the gene network of lipid metabolism (*i.e.*, inhibition of SRP by cholesterol) is shown in Fig. 11.3*b*. The *SRP protease* provides for activation of the inert transcription factor *preSREBP1* into the operative *SREBP1*. The record shown means that cholesterol inhibits (*AT*) the reaction transforming pre*SEBP1* into active *SREBP1*, which requires *SRP*(*ID*). The reaction takes place in the cytoplasm and is described as indirect (*EF*).

VISUALIZATION OF GENE NETWORK

Since the language described is rather complex, the user deals only with a specially developed graphic interface that generates and interprets the code of the language. With the help of this program, the user can input the data into the . GeneNet database, operating with concepts of molecular biology associated with the expression of genes.

The input interface automatically translates the input information into a standard GeneNet format, examples of which were considered above. A specialized *JAVA* program, GeneNet Viewer, processes the formalized data accumulated in the GeneNet database and presents it to the user as a graphic diagram. The Viewer allows the GeneNet database to be explored and visualized through the Internet and includes tools for automated generation of gene network diagrams, a system of filters, tools for data navigation, on-line help, interactive cross-references within the GeneNet database, and references to other databases. A standard set of images corresponding to elementary structures and events is used for gene network visualization.

GENE NETWORKS IN GENENET DATABASE

Analysis of the information contained in the GeneNet database and the available literature suggests several basic types of gene networks.

1. Gene networks controlling cyclic processes, such as the cell cycle and the cycle of heart muscle contraction.
2. Gene networks underlying cell growth and differentiation, morphogenesis of tissues and organs, growth and development of the organism.
3. Gene networks maintaining homeostases of biochemical and physiological parameters of the organism.
4. Gene networks providing for responses of the organism to changes in the environment (e.g., stress response).

We will now consider certain typical patterns of gene networks

Homeostasis

Negative feedback regulation plays an important role in the operation of gene networks, providing for maintenance of a parameter within a certain range around its optimal level. An example is the gene network regulating intracellular cholesterol concentration Cholesterol synthesis is implemented with the involvement of mevalonate pathway enzymes. Increased transcription of the genes coding for these enzymes raises the intracellular cholesterol concentration. The key regulators ot this pathway are transcription factors of the SREBP subfamily. They activate the transcription of a cassette of genes coding for many enzymes of the mevalonate pathway, since regulatory regions of all these genes contain binding sites for these factors. SREBP1 is formed through proteolytic cleavage of its inactive precursor

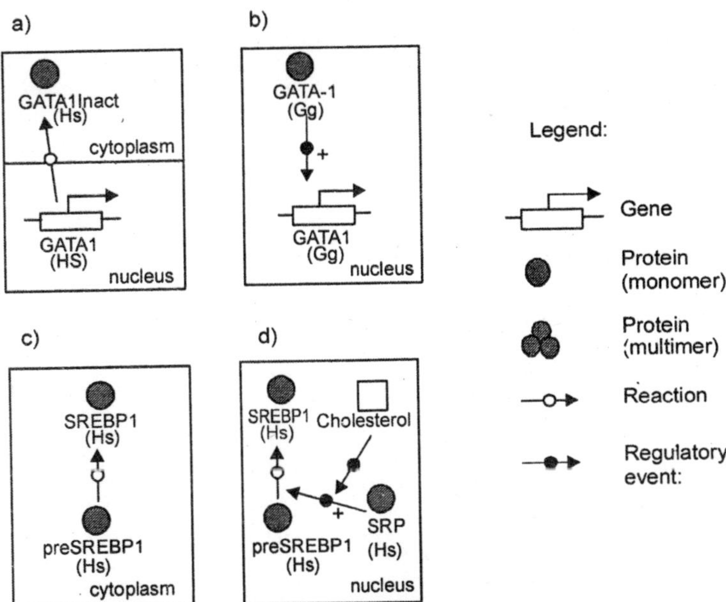

Fig. 11.4 : Graphical representation of elementary structures and events essential for the gene network function, (a) The GATAl gene expression resulting in the emergence of the inactive precursor of the transcription factor GATAlinact in the cytoplasm, (b) Autoactivation of GATA-1 gene transcription by factor GATAl encoded by this gene (gene network of erythrocyte differentiation and maturation), (c) Transformation of the transcription factor preSREBPl inactive precursor into the mature factor SREBP1. (d) Inhibition of SRP protease activity by cholesterol (gene network of lipid metabolism).

(preSREBPl), with a molecular weight of 125 kDa, performed by a sterol-regulated protease, resulting in the active factor, with a molecular weight of 68 kDa. An increase in cholesterol level suppresses SRP activity, slowing the transition of preSREBPl mto the active form and decreasing the level of active SREBP1 and the transcriptional activities of the genes of the mevalonate pathway.

In turn, decreases in the levels of the mevalonate pathway enzymes reduce the rate of cholesterol synthesis, thus normalizing its level in the cell. The intensity of cholesterol transport across the cell membrane from the intercellular space plays an important role in the maintenance of the intracellular level of cholesterol.

The transport involves low-density lipoprotein receptor (LDLR). At a decreased cholesterol concentration, the concentration of active SREBP1 is increased, activating transcription of the gene encoding LDLR.

With an increased intracellular cholesterol level, the activities of sterol-regulated proteases, and therefore the concentration of active SREBP, decrease. This in turn decreases the transcription activity of the LDLR gene and cholesterol transport into the cell. A characteristic feature of the gene network being considered is its activation by the decrease in the level of the parameter it adjusts (cholesterol concentration), and its halting when the level of this parameter exceeds the optimal value. Such mechanisms are present in virtually all the homeostatic gene networks.

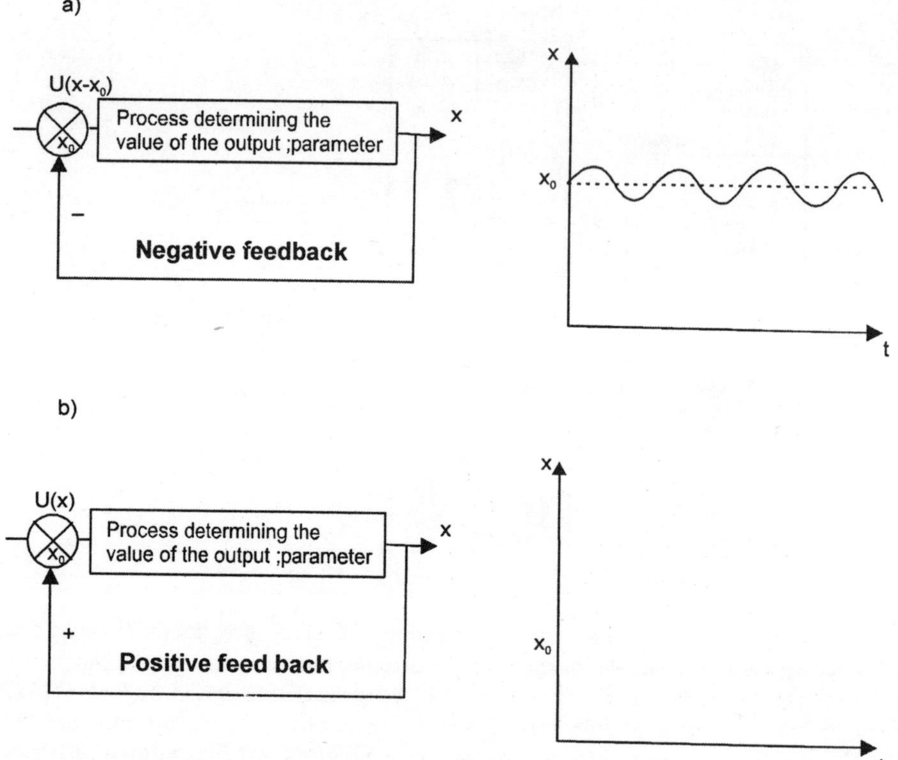

Fig. 11.5 : Two basic types of regulatory circuits in the gene networks, (a) Negative feedback, (b) Positive feedback.

Irreversible Processes

Cell differentiation, the morphogenesis of tissues and organs, and the growth and development of organisms are examples of gene network-controlled irreversible processes. Typically the gene networks controlling irreversible processes (1) are triggered by an external signal and (2) contain positive feedback circuits, which essentially boost the external signal and thereby trigger the irreversible processes of the gene network function. Such circuits provide for the maximal effective deviation of a parameter from its current value. An example of the gene network providing for differentiation and maturation of erythrocytes.

The key external regulator of this gene network is erythropoietin (*EPO*). Binding of erythropoietin to the membrane receptor (*EPOR*) results in dimerization of the latter, triggering the signal transduction pathway implemented by cell protein kinases. As a result, certain transcription factors are phosphorylated, transferred to the nucleus, and acetylated. They activate transcription of several genes, including the gene coding for transcription factor *GATA1*. Transcription factor *GATAL1* is the key regulator of erythrocyte differentiation and maturation.

The presence of *GATA1* binding sites in regulatory regions of virtually all the genes constituting this gene network underlies its key regulatory role. The transcription factor *GATA1* provides two positive feedback circuits that boost the initial differentiation signal received by the eryth-ropoietin receptor through its interaction with erythropoietin, and conveyed to the cell nucleus by the signal transduction pathway. The frirst circuit functions as follows. Expression of the gene *GATA1* results in the appearance of the inactive transcription factor precursor GATAlinact in the cytoplasm. This precursor enters the cell nucleus and acts there in a phosphorylated form (GATAl-p) with its binding site in the promoter of the *GATA1* gene.

The presence of GATA1 binding sites in the promoter of its own gene results in a rapid self-enhancement of its transcription according to the positive feedback mechanism. This positive feedback is very efficient and rapid because no other mediator genes are involved in its function. The second positive feedback circuit, enhancing the *GATA1* gene transcription, functions in the following way. The GATA1 binding site is present in the promoter of the erythropoietin receptor (EPOR) gene. Transcription activation of this gene increases the number of erythropoietin receptor molecules on the cell membrane and, correspondingly, the intensity of signal transduction from erythropoietin via its receptor to the gene *GATA1*. Thus, this positive feedback circuit is closed, providing for additional self-enhancement of the *GATA1* gene transcription. When this positive feedback brings the level of transcription factor GATAl-p to a threshold value, transcription of a cassette of genes is activated simultaneously. This cassette includes genes coding for (1) α-and β-subunits of hemoglobin and (2) heme biosynthesis enzymes. Transcription of the cassette of genes encoding cell surface antigens (GPD, GPB, and GPC) is also activated. Finally, transcription of the cassette of genes encoding transcription factors particularly important for switching to differentiation, is activated. These factors provide for an additional stimulation of erythroid-specific genes. In this way a cascade of processes ensuring the terminal differentiation and maturation of the erythrocyte is triggered. The cassette-like activation of genes, positive regulatory effects, and the key activators are observed in virtually all the gene networks described in the GeneNet database that trigger irreversible processes in the organism.

FUNCTIONAL DYNAMICS

Before considering the main body of results, some introductory remarks on specifics of the objects simulated are necessary because they essentially determined our selection of the simulation tools. First, even the simplest gene networks comprise dozens of physical components and interactions between them. Typically, gene networks are composed of hundreds of elements and more.

Structural-functional organizations of many gene networks have been clarified in sufficient detail. Of the utmost importance is the fact that, despite their great diversity, the structural-functional organization of gene networks permits them to be directly and graphically represented by means of chemical kinetic description of elementary events constituting the gene network function. Second, gene networks as actual objects exist simultaneously in a variety of different physical (material) forms due to such natural phenomena as gene polyallelism, genetic rearrangements, and performance of similar functions in different organisms. The more complex a gene network is, the wider the permissible diversity range of its variants.

The number of variants is increased considerably by construction of artificial systems (such as expression vectors). In addition, gene networks themselves are elements of more complex biological systems. Thus, the methods used for formalization should allow not only analysis of the initial gene networks and their minor modifications (mutations), but also calculation of the dynamics of virtually any genetic variants constructed from them.

In other words, any model of a gene network should potentially contain models of many gene networks, including those drastically different from the initial one. A generalized chemical kinetic method, developed earlier and already applied to simulation of various gene networks meets all these requirements. This method provides a precise and effective simulation of the gene network performance patterns because its computer realization adequately reflects the basic properties of biological systems.

CHEMICAL KINETIC SIMULATION METHOD

The generalized chemical kinetic simulation method is oriented to a formalized, primarily portrait, description of the performance patterns of arbitrary biological systems. Formalization is performed according to a block principle: a simulated system is divided into elementary subsystems in order to describe each subsystem individually. Elementary subsystems are described in terms of formal blocks. A formal block is uniquely characterized by an ordered list of formal dynamic variables $\bar{X}$, an ordered list of formal parameters $\bar{P}$, and the law of information transformation $\bar{F}$.

The law of information transformation $\bar{F}$ can be uninterrupted, discrete, logical, or stochastic. The choice of elementary process description is determined by the nature of a biological process, the task to be solved, and the preference of a model's author. Therefore the models worked out within the generalized kinetic simulation method are generally hybrid (*i.e.*, they can have uninterrupted, discrete, logical, and other blocks). The models that belong to the systems of common differential equations with a rational right-hand side are an important class of models for successful descriptions of gene network functioning. Such models appear when elementary blocks with differential equations expressing the immediate rate of concentration changes through immediate values of component concentration changes (mRNA, proteins, nonproteinaceous substrates, their complexes, etc.) in a simulated system are used as a law of information transformation.

1. *Bimolecular reversible reaction :* $x_1 + x_2 \underset{k_2}{\overset{k_1}{\Leftrightarrow}} x_3$

$$\bar{X} = (x_1, x_2, x_3), P(k_1, k_2),$$

$$F : \frac{dx_1}{dt} = k_2 \cdot k_3 - k_1 \cdot x_1, \frac{dx_1}{dt} = \frac{dx_2}{dt} = -\frac{dx_3}{dt}.$$

2. *Monomolecular irreversible reaction :* $x \xrightarrow{x} y$

$$\bar{X} = (x, y), \bar{P} = (k),$$

$$\bar{F} : \frac{dx}{dt} = \frac{dy}{dt} = -k \cdot x.$$

3. *Michoelis – Menten scheme*

$$X = (e, s, p), P = (k_o, k_m),$$

$$\bar{F} : \frac{ds}{dt} = Z, \frac{dp}{dt} = Z$$

$$Z = \frac{k_o es}{k_m + s}.$$

4. *Universal scheme*

$$\bar{X} = (x_1, ..., x_m, ..., y_n),$$

$$\bar{P} = (s_1, ..., s_l, a_1, ..., a_m, b_1, ..., b_n), m \geq 1, n \geq 0,$$

$$\bar{F} = \begin{cases} \frac{dx_j}{dt} = -a_j \cdot Z, j = 1,...,m, \\ \frac{dy_l}{dt} = b_l \cdot Z, l = 1,...,n \end{cases}$$

where

$$Z = \frac{R(x_1,...,x_m,s_1,...,s_l)}{Q(x_1,...,x_m,s_1,...,s_l)}.$$

R,Q–polynomes from variable values ***m***.

Fig. 11.8 : Blocks of the Generalized Chemical Kinetic Simulation Method (GCKSM): $\bar{X}$*, ordered list of formal variables;* $\bar{P}$*, ordered list of formal parameters; and* $\bar{F}$*, the law of transformation of information.*

The chemical-kinetic approach is a methodological base for developing this class of models. The reason for its application is that biochemical reactions are the base of the processes controlled by gene networks. Mono- and bimolecular reactions are the base of the simplest biochemical reactions. The rate of changes in component concentration values in the reactions proceeding in a perfect mixture can be described with the systems of common and autonomous equations shown in figure 11.8, nos 1 and 2.

General view of an equation setting the regularity of alteration of

variable s rate

$dx_i/dt=F_i(x_1, ..., x_i, ..., x_n, k_1, ..., k_j, ..., k_m)$,

$i=1,$,n;x i⁻model dynamic variables, k_j-constants.

1 reaction	2 reaction
A + B<=>C	C=> A + P
$da/dt= \{-k_1ab+k_2c\}$,	$dc/dt= \{-k_3c\}$,
$db/dt= \{-k_1ab+k_2c\}$,	$da/dt= \{-k_3c\}$,
$dc/dt=-\{-k_1ab+k_2c\}$	$dp/dt=-\{-k_3c\}$,

Model

$x_1=a$: $da/dt= \{-k_1ab+k_2c] - \{-k_3c\} = F_1$,

$x_2=b$: $db/dt= \{-k_1ab+k_2c\} =F_2$,

$x_3=c$: $dc/dt=-\{-k_1ab+k_2c\}+ \{-k_3c\} =F_3$,

$x_4=p$: $dp/dt = -\{-k_3c\} =F_4$,

Fig. 11.9. Rule of summing immediate velocities in elementary process.

It is not complicated to formulate a general system of equations even when several mono- and bimolecular biochemical reactions proceed in a medium simultaneously. It is necessary to use the rule of summing immediate velocities of the simplest reactions: the product of changes in immediate concentration velocities of the agent involved in several reactions is the sum of immediate velocity changes of the given agent in these reactions. As can be seen, the equation systems are quite simple when described only with mono- and bimolecular reactions. Velocities of concentration changes are formulated by means of bilinear expressions from the same concentrations. In principle, two reaction types (bi- and monomolecular) are quite sufficient to describe arbitrary gene networks. However, such models will have a great many variables and parameters.

In addition it is necessary to conduct bio-system decomposition on mono- and bimolecular reactions, although it is not always justified due to the lack of knowledge about concrete mechanisms of a particular gene network-controlled process. The natural way out of the deadlock is to consider more complicated processes as elementary and describe the laws of their proceeding in the form of autonomous equation systems with rational right-hand sides. The Michaelis-Menten equation, frequently used in approximate descriptions of enzyme synthesis, is one of the most famous examples. More complicated reactions will be described with more complicated equations. The right-hand sides can be arbitrarily rational in a general case. Since the rule ot summing up immediate velocities does not depend on the internal complexity of elementary processes, it does not make the procedure of formulating a complete equation system more complicated.

The total of agent concentration velocity changes is still the sum of the velocity changes in all elementary processes. It allows us to formulate a common approach to the description of regularities in the functioning of biological systems, particularly gene networks. The simulated system splits into simpler parts that will be considered elementary. Further on, elementary processes will be the building blocks used to develop the models not only of initial systems but also of their various modifications. A great number of other models can be developed out of elementary processes in general, since any integration of an elementary process is a potential mode.

In particular it is possible to construct the models of practically arbitrary genetic variants of the initial network. There is no need to construct all the models at once; each particular model is constructed as required. Our method in its most general realization implies the possibility of simulation not only of trans-interactions but also of cis-interactions. The latter should be taken into account in cases where patterns of gene network performance depend not only on the functions, but also on relative location, of the genes involved.

REGULATION OF CHOLESTEROL BIOSYNTHESIS IN THE CELL

Cholesterol, an amphiphilic lipid, is an essential structural component of the cell membrane and the outer layer of the lipoproteins of blood plasma. Simultaneously it is a precursor of corticosteroids, sex hormones, bile acids, and vitamin D. Cholesterol is synthesized from acetyl-CoA, and its major fraction in the blood plasma is in the low-density lipoproteins (LDL). Cholesterol is removed from tissues with involvement of high-density lipoproteins (HDL) to be transported to the liver and transformed there into bile acids. In pathology, cholesterol is a factor causing *atherosclerosis* of vital cerebral arteries, heart muscle, and other organs.

A high ratio of LDL cholesterol to HDL cholesterol in the plasma is observed in *coronary atherosclerosis*. This reveals the great biomedical and applied importance of studying cholesterol turnover in the organism. The gene network regulating intracellular cholesterol biosynthesis has now been studied in sufficient detail. Data on its performance patterns are accumulated in the GeneNet database. Acetyl-CoA is the source of all the carbon atoms of the cholesterol molecule. The cholesterol biosynthesis pathway has numerous stages and is controlled by a variety of enzymes, including *HMG-CoA reductase, farnesyl diphosphate synthetase*, and *squalene synthetase*. Syntheses of these enzymes are activated by *SREBP*. The activity of *SREBP* depends, in turn, on the intracellular cholesterol concentration in a negative feedback mode: the lower the concentration of metabolically active cholesterol in the cell, the higher the SREBP activity. We have developed a model of functional dynamics of this gene network. It describes all the stages of cholesterol biosynthesis shown in figure 11.4*a* as edges with adjacent nodes.

In addition, the model describes the mechanisms underlying the interchange of intracellular and blood cholesterol. Negative feedbacks whereby cholesterol controls its own synthesis and the synthesis of LDL receptors at the transcription level are also considered. The model fragment that consists of three equations describing the cycle of molecules acetyl-CoA, acetoacetyl-CoA, and HMG-CoA—three cholesterol precursors—is presented in figure 11.10.

The equations are based on considering six processes in which they are agents. Right-hand equation members corresponding to one process are in braces and accompanied by indexes, displayed as subscripts, following the right brace: 1, entry of acetyl-CoA into the medium of the gene network functioning; 2, tiolase-cata-lyzed synthesis of acetoacetyl-CoA from two molecules of acetyl-CoA; 3 and 4, withdrawal of acetyl-CoA and acetoacetyl-CoA from the medium; 5, HMG-CoA synthase (HMGCS)-catalyzed synthesis of HMG-CoA from acetoacetyl-CoA and acetyl-CoA; 6, HMG reductase (HMGR)-catalyzed synthesis of mevalonic acid from HMG-CoA. In totally, the model comprises 65 elementary processes. The model of performance dynamics of this gene network described in the Gene-Net database contains 40 products (dynamic variables) and 93 constants.

Values of a number of constants were assessed using the relevant published data. The rest of the parameters were determined through numerical experiments using quantitative and qualitative characteristics known in the literature as criteria of their adequacies. This model allows the equilibrium state of the biosystem to be calculated. The equilibrium persists while the environmental conditions remain constant. If they change (*e.g.*, the content of LDL particles in blood plasma increases twofold), the system.

$$\begin{cases} d[Acetyl-CoA]/dt = \{k_{s,1}\} + \{-k_{o,5}\} HMG-CoA\ synthase] [Acetyl-CoA] [Acetoacetyl-CoA]/ \\ (K_{m,5} + [HMG-CoA\ \text{synthase}] [Acetyl-CoA] + [Acetyl-CoA] [Acetoacety-CoA] + \\ [HMG-CoA\ Synthase] [Acetoacetyl-CoA])\}_5 + \{-k_{d,3}[Acetyl-CoA]\}, \\ d[Acetoacetyl-CoA]/dt = \{k_{o,2}[Tiolase] [Acetyl-CoA]^2/(K_{m,2} + 2 \cdot [Tiolase] [Acetyl-CoA] + \\ [Acetyl-CoA]^2)\}_2 + \{-k_{o,5}[HMG-CoA\ synthase] [Acetyl-CoA] [Acetoacetyl-CoA]/(K_{m,5} + \\ [HMG-CoA\ synthase] [Acetyl-CoA] + [Acetyl-CoA] [Acetoacetyl-CoA]+ \\ [HMG-CoA\ synthase] [Acetoacetyl-CoA])\}_5 + \{-k_{d,4}[Acetoacetyl-CoA])\}_4 \\ d[HMG-CoA]/dt = \{k_{o,5}[HMG-CoA\ synthase] [Acetyl-CoA] [Acetoacetyl-CoA]/ \\ (K_{m,5} + [HMG-CoA\ \text{synthase}] [Acetyl-CoA] + [Acetyl-CoA] [Acetoacetyl-CoA] \\ + [HMG-CoA\ synthease] [Acetoacetyl-CoA])\}_5 + \{-k_{o,6}[HMG\ reductase] \\ [HMG-CoA]/(K_{m,6} + [HMG\ reductase] + [HMG-CoA])\}_6 \end{cases}$$

Fig. 11.10 : GeneNet model fragment controlling cholesterol biosynthesis.

Consequently the concentration of receptors bound to LDL increases (*e*) and that of receptors unbound to LDL decreases (*d*). Intracellular concentrations of free cholesterol (*a*) and its esters (*c*) increase. Unless a new intervention occurs, the negative feedbacks restore the initial state of the system: the initial cholesterol concentration in the cell is reestablished in approximately 3 hr, and the overall initial state of the system is restored in 10-15 hr.

REGULATION OF ERYTHROCYTE MATURATION

Hematopoietic tissue belongs to the self-renewing systems of the organism that are operated through specific regulatory and self-regulatory mechanisms. Maintenance of a certain number of erythroid cells is one of the necessary conditions for the organism to perform its vital functions. From this standpoint theoretical research into proliferation and differentiation of hematopoietic tissue cells is of both basic and applied biomedical importance. The main stages of erythrocyte maturation are regulated by the gene network presented in figure 11.7.

The hormone erythropoietin interacts with immature erythroid cells (erythroid stem progenitors of CFU-E type) and stimulates their proliferation, as well as syntheses of hemoglobin and the enzymes involved in heme biosynthesis, that is, maturation and differentiation of erythroid progenitors. Low partial pressure of oxygen in venous blood (hypoxia) is another stimulator of erythropoietin synthesis. Interacting with the cell receptor, erythropoietin activates the transcription factor GATA1, a key regulator of erythrocyte differentiation. GATA1 stimulates syntheses of α-and β-globins and the enzymes of heme biosynthesis.

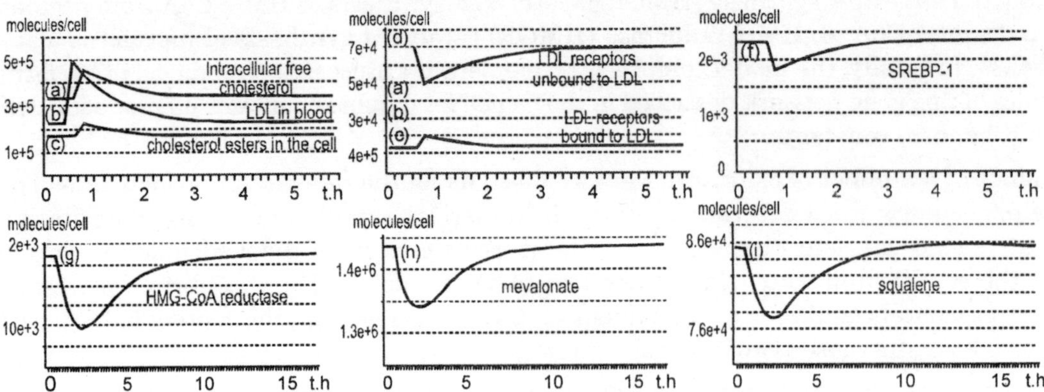

Fig. 11.11 Kinetic changes in the main components of the system regulating cholesterol biosynthesis in the cell (a) in response to a simulated twofold increase (by 30 min) in LDL concentration in blood (b) x axis, time (h); y axis, molecules / cell. The main changes are the following: (d) the number of receptors unbound to LDL decreases; (e) the number of receptors bound to LDL increases; (a) intracellular concentration of free cholesterol and (c) its esters increase; (f) protease (SRP) binds free cholesterol, causing a decrease in SPEBP-1 concentration; (g) productions of the enzymes involved in the intracellular cholesterol synthesis (HMG-CoA reductase) stops; (k, i) production of LDL receptors and intermediate low-molecular-weight components (mevalonic acid, squalene) also stops. The system returns to the initial state until new input of exogenous cholesterol; concentration of free cholesterol in the cell restores approximately in 3 hours; complete restoration requires about 10-15 hours.

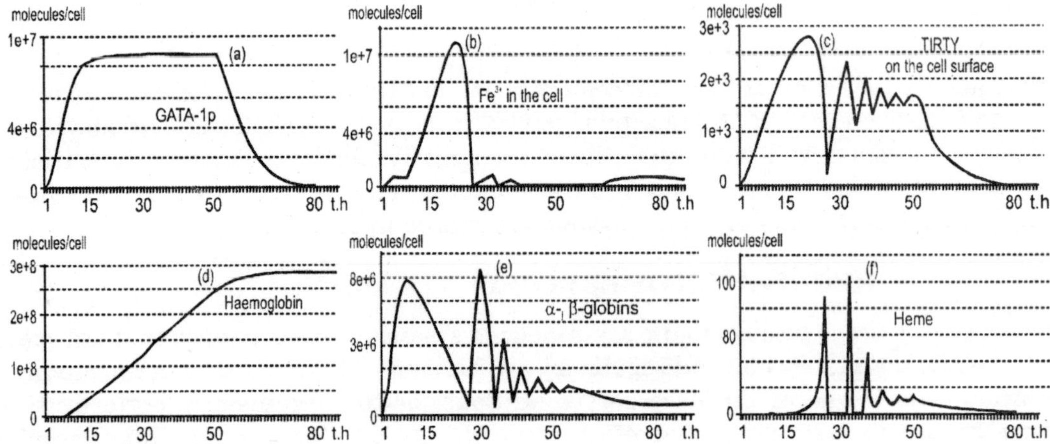

Fig. 11.12 : Dynamics of absolute concentrations of the main components of the erythroid cell differentiation system, calculated with x axis, time (h); y axis, molecules / cell. In (c), TfRTf-transferring receptors bound to transferring. Consecution of events: at beginning of erythropoietin influence on the cell, the precursor initiates the program of erythroid cell differentiation; 0-50 h is the total (transit) time of the function of erythroid cell differentiation in the gene network; at 50 h, the nucleus is lost; and 50-80 h, residual syntheses of the system's components in erythrocyte.

In addition it activates its own gene and the gene of the erythropoietin receptor (positive feedback). Heme, α-globin, and β-globin form hemoglobin, the major component of the mature erythrocyte. This biosystem functions according to a pattern completely different from the system described in the previous section. This gene network is inactive until erythropoietin triggers it to start the irreversible process of erythrocyte maturation, in which positive feedbacks are predominant.

The model of this gene network performance is described as a sequence of events occurring in the maturating cells of the erythroid lineage. The erythropoietin-responsive progenitor cell opens the lineage closed by the mature erythrocyte. The model comprises 119 elementary processes, 68 products, and 178 constants. Direct experimental data were used to specify a number of constants. The rest of the parameters were determined through numerical experiments using quantitative and qualitative characteristics known from the literature. The model developed predicts that production of several components of the system follows oscillatory dynamics. This pattern results from interaction of positive and negative feedbacks, with the former regulation mode.

CONCLUDING REMARKS

A limited set of elementary structures and events underlies the functions of gene networks; however, its elements combine to bring forth a tremendous diversity of existing gene networks and variants of their functions. Obligatory components of each gene network are (*a*) a group of genes expressed in a concerted manner (core of the gene network); (*b*) proteins encoded by these genes, which fulfill structural, transport, catalytic, regulatory, and other functions; (*c*) pathways transducing signals from cell membranes to cell nuclei, which provide for activation or suppression of gene transcription in response to stimuli external to the cell; (*d*) nonproteinaceous components that trigger the gene network function in response to external stimuli (hormones and other signal molecules); (*e*) various metabolites arising during the gene network functions. A characteristic feature of gene network organization is its capacity for self-regulation by means of closed regulatory circuits with negative and positive feedbacks.

These two types of regulatory circuits make it possible to maintain a definite functional state of the gene network or, on the contrary, enable its transition to another function mode under the effect of various factors, including environmental ones. The gene network functions are regulated by hierarchically organized mechanisms whose cores are the key regulatory proteins coordinating the functions of the rest of the genes constituting a particular gene network.

Groups of similar target sites in gene regulatory regions, capable of interacting with key regulatory proteins and thereby providing the cassette-type activation of large gene groups, form the molecular basis of the function of such regulatory circuits. This cassette-type activation of transcription of large gene groups is a characteristic feature of the gene networks described in the GeneNet database. The hierarchical principle of the function of a gene network in a multicellular organism is the most important feature. Figure 11.13 qualitatively illustrates the hierarchy of mammalian gene networks.

The lowest level of this hierarchy corresponds to gene networks controlling basal cell metabolism. Their functions depend on the stage of the cell cycle and can be suppressed or activated by regulatory effects coming from gene networks of higher levels. These effects can change the rates and directions of both metabolic and cell division processes. The highest organization levels correspond to gene networks controlling external signal reception and mental functions. Despite the simplicity of the classification described, it nevertheless allows the groups of gene networks with qualitatively similar functions to be detected, and their interaction and cosubordination to be described.

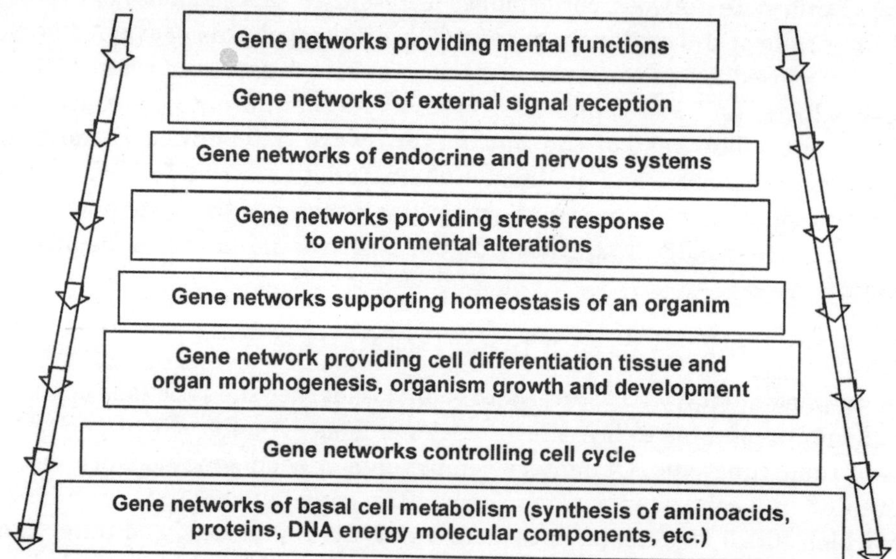

Fig. 11.13. Global gene network of an organism (hierarchical integration).

In this global and hierarchically organized system of gene networks, the controlling signals are directed not only from higher levels to lower levels, but also in the opposite direction; and the signals may have both positive (activating) and negative (suppressing) effects. The extreme complexity of gene networks requires application of mathematical simulation methods for investigating the networks' dynamics. The resulting models will allow the dynamics of both normal gene networks and those exhibiting pathologies, including those connected with gene mutations, to be studied. In addition to the possibilities such models present for basic studies of the gene network operation, they are very important for pharmacology and genetic engineering.

12

Laboratory Detection of Genome

INTRODUCTION

The availability of complete or near-complete catalogues of genes for organisms of increasing complexity has created opportunities for studying numerous aspects of gene function at the genomic level. Gene expression, defined by steady-state levels of cellular mRNA, has emerged as the first aspect of gene function amenable to genome-scale measurement with readily-available technology. It is now possible to carry out massively parallel analysis of gene expression on tens of thousands of genes from a given sample. For model organisms such as *S. cerevisiae,* total genome expression analysis is now routine. For higher eukaryotes, expression measurements that cover a significant proportion of the genome are currently possible, and complete genome expression analysis appears to be an achievable goal. Moreover, this analysis can be repeated on multiple samples, allowing for the statistical analysis of the behaviour of genes across a large number of conditions.

The massive quantities of data generated by this type of analysis have created new bioinformatics challenges, but these obstacles are well worth overcoming as this type of information has already begun to provide new insights into genome function. So far, this promise has been best realized in *S. cerevisiae,* in which whole-genome measurements have been used to examine fundamental processes such as the cell cycle and the roles of specific transcription factors. Although significantly more difficult, similar problems are now being productively approached in mammalian systems. In addition to their impact on fundamental questions in biology, these technologies are being used to probe the complexity of human diseases, particularly cancer. Access to large numbers of measurements allows for the statistical analysis of gene expression across multiple samples.

In the broadest sense, this opens the possibility of identifying patterns of coregulation among genes, which, in turn, reflects underlying regulatory mechanisms and functional interrelationships. Because mammalian genomes are mainly populated by genes of unknown function, it is theoretically possible to assign anonymous genes to pathways or at least to generate hypotheses as to function by identifying the circumstances that alter their expression. Computational techniques for processing gene expression data are rapidly evolving as many investigators attack the problem from diverse perspectives. The following discussion will present an overview of the technologies for generating and processing expression data, with an emphasis on those techniques and databases that are publicly available.

LARGE-SCALE GENOME ANALYSIS

Measurements

Two broad categories of technology have emerged that can provide large-scale expression data. The first is sequence-based, as exemplified by the serial analysis of gene expression (SAGE). An alternative hybridization approach is generically termed microarray hybridization. These two technologies have distinct advantages and disadvantages. In SAGE, a short unique sequence tag is generated from each gene by a PCR-based strategy. Concatemerized tags are sequenced, and the abundance of these tags provides a measurement of the level of gene expression in the starting material. As illustrated below, SAGE tags can be linked to a specific transcript designation in an appropriate database of unique transcripts, such as UniGene. Thus, SAGE is essentially an accelerated technique for cDNA library sequencing.

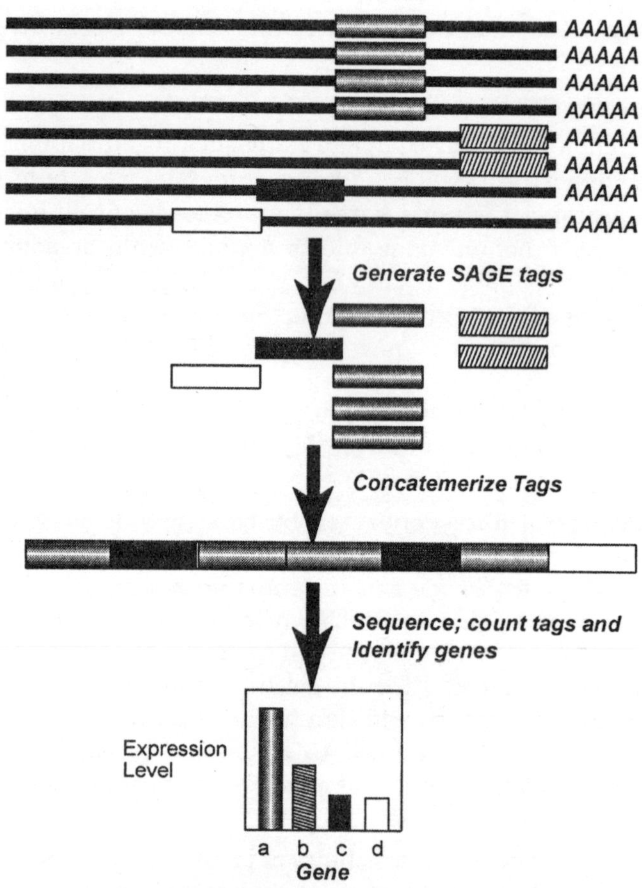

Fig. 12.1 : Serial Analysis of Gene Expression (SAGE) depends on the generation of a tag from the 3′end of an mRNA. Tags are concatemerized and sequenced. These data are compared with a database of tags linked to individual transcripts to generate the frequency of each tag in the library, a measure of the expression level for that gene.

Because it is sequence intensive, SAGE is not well suited to the analysis of large numbers of samples. However, because SAGE does not require a priori knowledge of the pattern of gene expression in a given mRNA source, the same biochemical and bioinformatics procedure can be applied to any sample, given the availability of the appropriate reference database of SAGE tags. By contrast, in the microarray strategy, labeled cDNAs (the "target") are hybridized to an array of DNA elements (the "probes") affixed to a solid support. The array elements can either be synthetic oligonucleotides or larger DNA fragments. High densities are achievable and enable the measurement of over 10,000 genes with either type of microarray.

In contrast to SAGE, microarray analysis can only measure the expression of genes that correspond to the sequences included in the array fabrication process. Therefore complete genome expression analysis by microarrays requires the construction counterbalanced of complete microarrays. This difficulty is counterbalanced by the ease of individual experiments that enable the analysis of numerous samples. Microarray expression databases are being developed that contain information derived from hundreds or even thousands of samples.

Microarray Production

For organisms with completely sequenced and well-annotated genomes such as *S. cerevisiae,* the production of arrays is relatively straightforward. By selecting a primer pair that amplifies each ORF from genomic DNA, a set of PCR products encompassing the complete genome can be produced. Primer pairs for this purpose are commercially available. Likewise, a series of oligonucleotide array elements can be selected from each ORF. More complex genomes, which may not yet be fully annotated, are more problematic. The consensus at this juncture is to utilize a catalog of genes such as UniGene, generated by reduction of EST sequencing data to unique, clustered transcripts.

Clustered ESTs can then be used to predict oligonucleotide sequences for array fabrication or to select a representative EST clone for each cluster; in turn, this is then used to generate a cDNA fragment by PCR for deposition on a microarray. This approach is intrinsically limited by the quality and completeness of the EST database, as well as the intrinsic instability of EST cluster designations that evolve along with the EST sequencing projects.

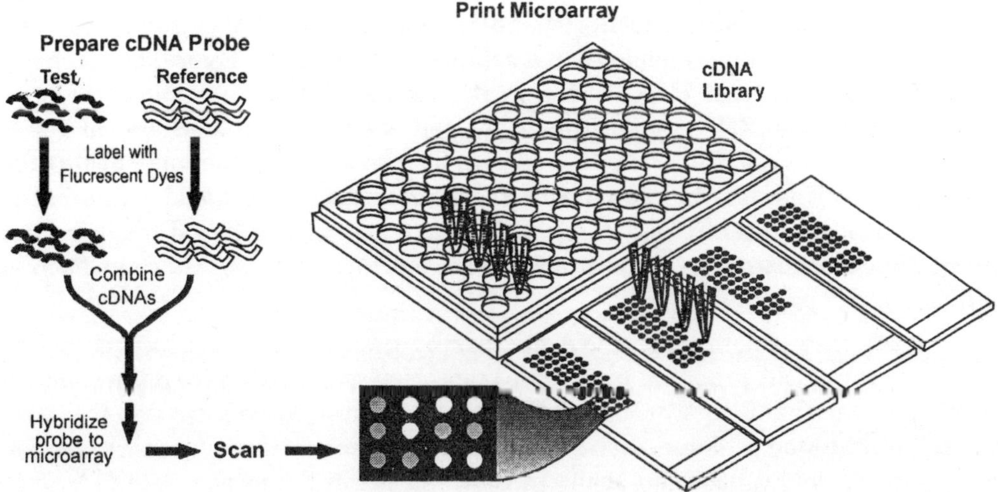

Fig. 12.2 : The process of microarray hybridization using printed DMA probes. A robotic printer deposits DNA in a regular array on a series of glass slides. After they are processed, the slides are hybridized to a mixture of two cDNA pools derived from test and reference samples that have been labeled with spectrally distinct fluorochromes. After stringency washes, the microarray is scanned in a laser-scanning device, and the image is processed to generate numerical data. (See colour plate).

A further practical difficulty unique to the cDNA microarray strategy is brought about by the requirement of this system for a physical clone. Recovery of low-redundancy clones may prove difficult due to problems intrinsic to the handling of arrayed libraries. Ideally, EST-based approaches will be supplemented (or even supplanted) by gene catalogues based on genomic sequence that will ultimately provide stable databases for array construction. ln order for the information generated by expression analysis to be useful, it is necessary for each EST to be linked to as much biological information as possible.

The accessibility and quality of this information depend on the organism. Because the interpretation of results often relies on the expertise of an individual investigator, issues surrounding gene nomenclature and annotation are significant. For example, an investigator may be familiar with a gene by its common literature alias, but it is catalogued under a more cryptic "official" name. ESTs from known genes may fail to be annotated as genes altogether if the available sequence is of insufficient quality to pass the filters required for cluster assembly with known mRNAs. Fortunately, mammalian gene catalogues such as NCBI's LocusLink are being developed that attempt to address these problems.

What is Actually Measured?

A detailed review of array technologies is beyond the scope of this chapter, but the user of expression data must have an understanding of the types of measurement reported by various expression platforms and incorporated in databases. To understand the options available to mine information in expression data, it is essential to understand the measurements generated by the various technologies presently available. Ideally, an expression measurement would be converted into copies of a given mRNA per cell. Unfortunately, no technology measures this value directly. The variety of expression measurement platforms in use creates problems in the cross-comparison of data generated by technologies and in some cases even within a given technology. For example, SAGE measures the abundance of a particular tag.

The validity of this number will depend on the number of tags sequenced. A sample of 1,000 tags will be much less reliable than a sample of 20,000 tags. Hybridization-based systems measure the abundance of cDNAs in a synthetic population of nucleic acids, which is a representation of the mRNA composition of the cell. Variations in biochemical manipulations, hybridization efficiency across array elements, and cross-hybridization may distort the accuracy of measurements. This calls for caution when comparing data generated by different techniques. For example, most synthetic oligonucleotide array assays are based on hybridization of a representation synthesized by amplification from a cDNA template that has been engineered to contain a phage promoter.

In contrast, most fluorescent cDNA microarray assays use direct incorporation of tagged nucleotides in the cDNA prepared by reverse transcription of mRNA from the source of interest. There are few data that directly compare the results of these three assays on the same samples. Moreover, abundance measurements from oligonucleotide array assays are reported as expression levels on a continuous scale. In contrast, printed fluorescent cDNA arrays require the use of a two-colour system incorporating a reference mRNA to compensate for variations in the performance of individual array elements and array slides. Although the intensities of each channel are measured and reported, the most robust measurement is expressed as the normalized ratio of intensities for each gene.

Radioactive hybridization to cDNA microarrays printed on nylon membranes presents additional problems because only a single channel is measured in a given hybridization, and normalization must rely on cross-comparison of experiments. In addition, expression scales vary in an individual way from linearity and have distinct thresholds and saturation levels depending on the technology and instrumentation utilized. In principle, measurements from these various systems could be converted to a common format, but as yet there are no standards for such a conversion. To be certain that computational techniques are appropriately applied, users of array expression databases and experimentalists venturing into microarray research need to be aware of the characteristics of the data generated by the particular experimental platform in use. Aspects of the primary analysis of array data are illustrated for printed cDNA microarrays in the following example.

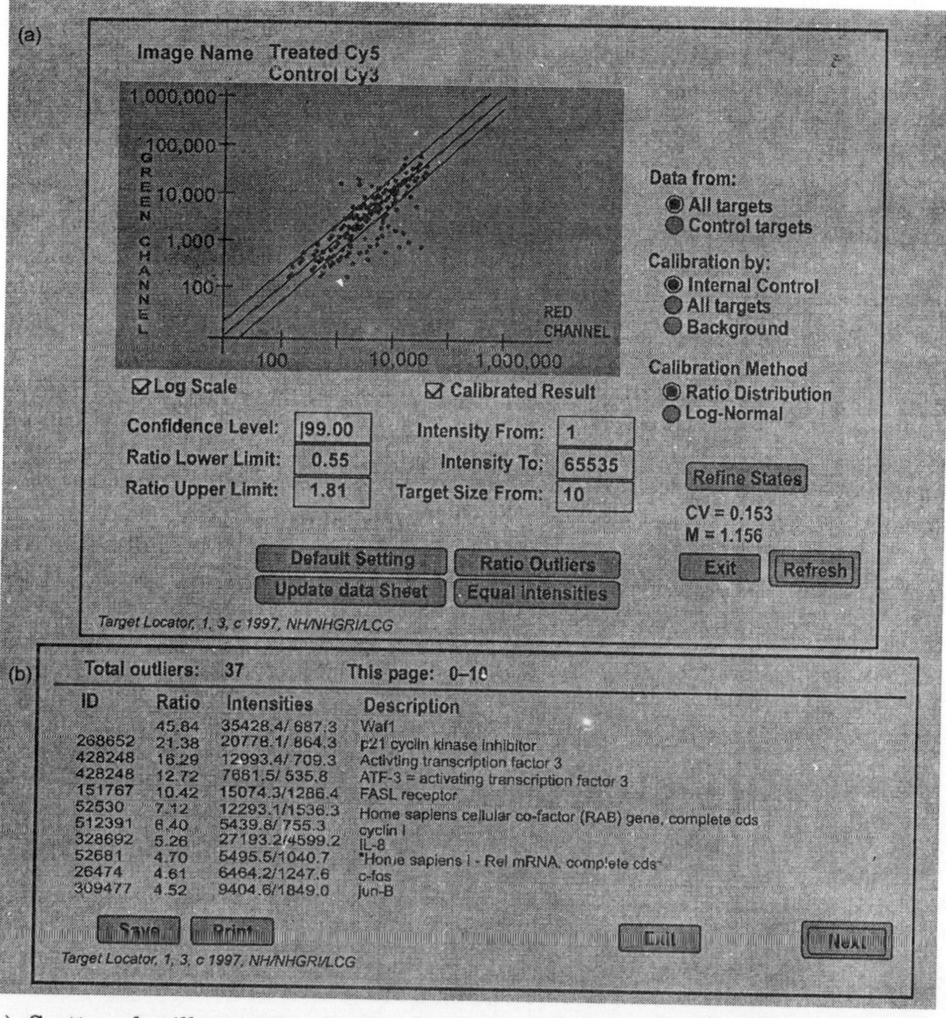

Fig. 12.3(*a*). Scatter plot illustrating the fluorescent intensity measurements in two channels plotted against each other. Note that most measurements fall on the diagonal. Deviation from this pattern at low intensity indicates that the data are skewed due to nonspecific fluorescence in one channel. For analysis, the data should be filtered to exclude these points, (*b*) After data are filtered for intensity and a minimum spot size (to exclude spurious measurements), an outlier list can be displayed. A spreadsheet containing the data from the entire microarray can also be saved.

This methodology achieves accurate ratio measurements of the relative abundance of each mRNA in the test and reference sources by combining the test cDNA pool with a reference cDNA labeled with a spectrally distinct fluorophore. This is accomplished by obtaining images of the hybridized microarray with a two-channel laser-scanning device for analysis with software (such as DeArray, ScanAlyze, CrazyQuant, or proprietary software bundled with scanner instruments) that measures the signal intensity over background, normalizes the two channels, calculates ratios, generates overall array statistics, and outputs a pseudocolor image and data spreadsheet. This process is illustrated here using DeArray, a module of Array Suite. Examining the scatter plot of one channel versus the other is extremely useful. In most hybridizations, the majority of genes will show minimal differences, so the scatter plot would then be centered on the diagonal line.

Deviation from this behaviour, as illustrated in the example at lower intensities, indicates the signal intensity level at which the data should be filtered. In other cases, the scatter plot might simply tend to widen as intensities drop, again indicating that confidence in ratio measurements declines with intensity. A similar scatter plot constructed from single-channel data from two one-colour or radioactive hybridizations is a useful adjunct to the analysis of data generated from these platforms. The analysis of a single microarray experiment is straightforward but not usually very illuminating. The real power of high-throughput gene expression technologies emerges when multiple experiments are subjected to comparison and statistical analysis. To accomplish this goal, expression data must be stored in an appropriate database.

COMPUTATIONAL TOOLS

Public Databases

At the present time, the only centralized, publicly maintained repositories of high-throughput gene expression data, aside from EST sequencing databases, contain SAGE data. A single, central repository for all expression data comparable to GenBank is presently not available. This situation may change if standards for the uniform reporting of expression data are adopted. NCBI maintains a SAGE database called SAGE map, currently containing data from 69 libraries. These data can be searched either for data on individual genes or for lists of genes differentially expressed between libraries or pools of libraries. Given a cDNA sequence (which must include the 3′ end), the user can search for its representation in a number of SAGE libraries. SAGE library data are downloadable, and a submission tool for SAGE laboratories is also available. To search the SAGE map database for data on a given gene using the "virtual Northern tool," the target cDNA sequence is submitted and possible tags are returned.

As in the example, one gene may contain multiple tags. Linked to this tag is a display indicating the relative abundance of this tag in a SAGE library with a proportional gray scale image providing a "virtual Northern" (Fig. 12.5). In addition, the data are linked to the UniGene cluster(*s*) corresponding to this tag. The UniGene designation provides a link to an abundance table of all tags for this particular UniGene entry. Instead of searching for individual genes, SAGE map can be queried using x-Profiler, a tool that carries out a *"virtual subtraction"* to develop lists of tags that are present in one library or group of libraries at a differing frequency from a reference set.

After the user selects the libraries to compare (Fig. 12.7*a*) and the ratio cutoff between the libraries, xProfiler returns a list (Fig. 12.7*b*) of differentially expressed genes in the two sources that can be downloaded for further analysis. It should be noted that SAGE data are subject to error, primarily related to sequencing error in the SAGE libraries and in the sequence databases that are used to define gene clusters. The impact of this error is inversely proportional to the number of times a given tag is counted. Tags that occur only once in the library, therefore, represent the least reliable data. xProfiler can be applied in essentially the same fashion to provide a virtual subtraction of EST sequence data compiled as part of the Cancer Genome Anatomy Project (CGAP).

Another tool available for screening CGAP and dbEST library data, digital differential display (DDD), provides a grayscale output indicating the relative abundance of statistically significant differentially expressed genes. Most public microarray data are accessible only through individual laboratory Web sites, many of which are listed at the end of this chapter. It can be anticipated that unified public expression databases will be developed once issues as to data format and cross-platform comparison are resolved. One of the preliminary efforts currently under development and aimed at the public use and dissemination of gene expression data is the NCBI Gene Expression Omnibus (GEO). This database is intended to house different types of expression data, including oligonucleotide and cDNA microarray data, hybridization filter data, and SAGE data. Although this platform will undoubtedly evolve as more and more gene expression information becomes available, GEO is currently envisioned as having four primary entities:

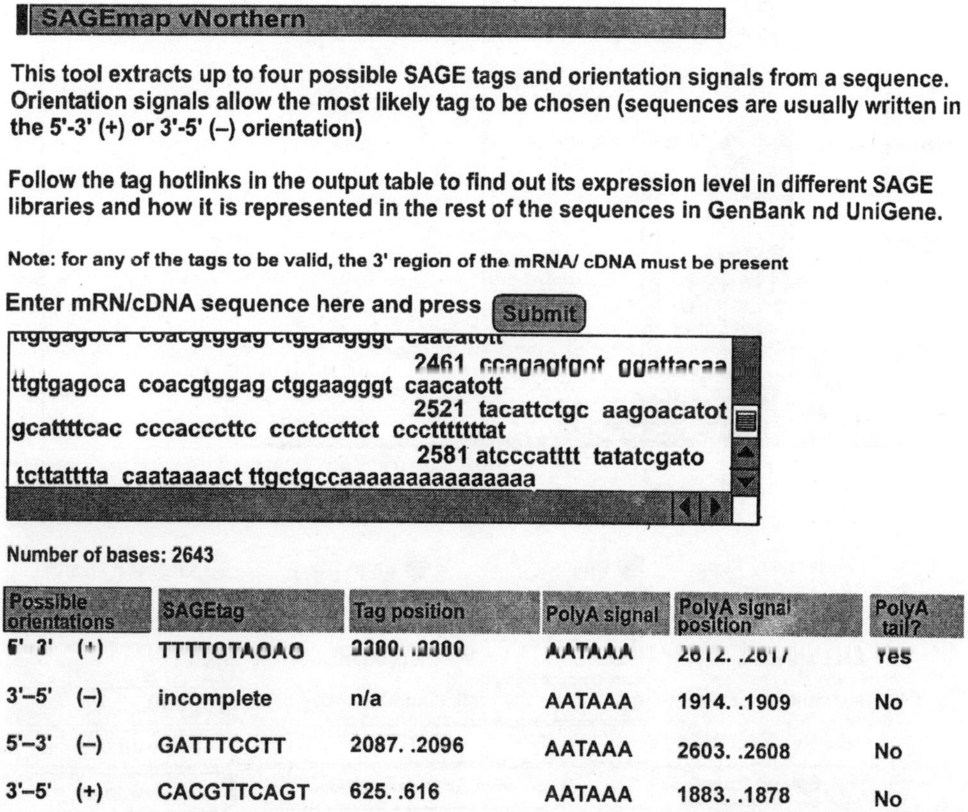

SAGEmap vNorthern

This tool extracts up to four possible SAGE tags and orientation signals from a sequence. Orientation signals allow the most likely tag to be chosen (sequences are usually written in the 5'-3' (+) or 3'-5' (–) orientation)

Follow the tag hotlinks in the output table to find out its expression level in different SAGE libraries and how it is represented in the rest of the sequences in GenBank nd UniGene.

Note: for any of the tags to be valid, the 3' region of the mRNA/ cDNA must be present

Enter mRN/cDNA sequence here and press Submit

```
ttgtgagoca coacgtggag ctggaagggt caacatott
                                 2461 ccagagtgot ggattacaa
ttgtgagoca coacgtggag ctggaagggt caacatott
                                 2521 tacattctgc aagoacatot
gcattttcac cccaccсttc ccctccttct cccttttttat
                                 2581 atcccatttt tatatcgato
tcttatttta caataaaact ttgctgccaaaaaaaaaaaaaaa
```

Number of bases: 2643

Possible orientations	SAGEtag	Tag position	PolyA signal	PolyA signal position	PolyA tail?
5'-3' (+)	TTTTOTAOAO	3300. .3300	AATAAA	2612. .2617	Yes
3'–5' (–)	incomplete	n/a	AATAAA	1914. .1909	No
5'–3' (–)	GATTTCCTT	2087. .2096	AATAAA	2603. .2608	No
3'–5' (+)	CACGTTCAGT	625. .616	AATAAA	1883. .1878	No

Fig. 12.4 : Searching SAGE map for tags representing a given gene ("virtual Northern"). The example illustrates the cDNA sequence of human p53 (TP53). Potential tags and their positions in the p53 mRNA are returned.

1. Submitter, which contains contact and login information on the submitter.
2. Platform, which contains information on the physical reagents used in the actual experiment.
3. Sample, which deals with the mRNA samples in question and the data generated from the actual experiment, and
4. Series, which houses information on collections of samples and the relationship between the samples. The series entity will also contain the results of any data clustering.

Even when public databases for microarray expression data are established, investigators who are setting up microarray laboratories soon learn that data storage is an essential requirement for their research. Sources for freely available software for expression databases are listed at the end of this chapter. One example of such a database is ArrayDB. ArrayDB, which requires either a Sybase or Oracle client, was designed for printed microarray data and allows the storage of experimental data and simple data queries. ArrayDB is designed to store a database of clones used for array fabrication and the data output from individual hybridization experiments. Links to the UniGene database are maintained. Data can be downloaded or viewed through a Web-based Java applet.

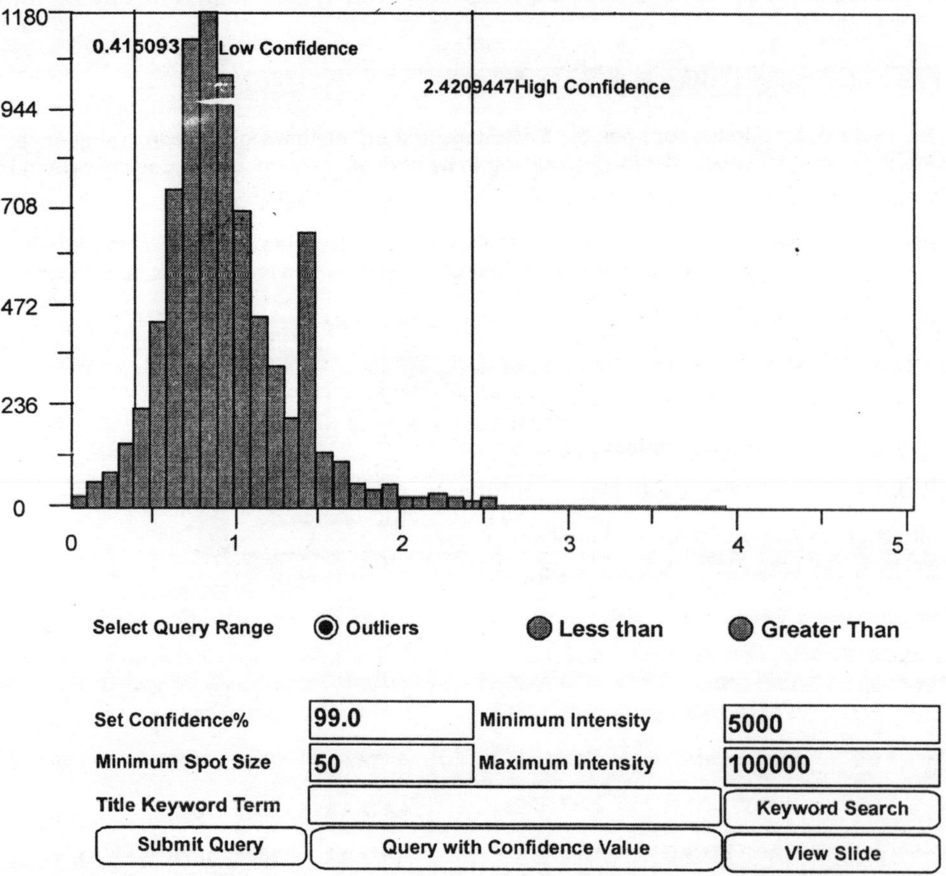

Fig. 12.5 : Ratio histogram of a microarray hybridization retrieved from ArrayDB.

On entry to the database, one selects the experiment to view from a pull-down menu and, on loading, a histogram of ratios appears. This window contains several selectable options for viewing portions of the data. Importantly, the data can be filtered according to intensity and spot size to remove insignificant measurements from subsequent analyses. On submitting a query, a new window opens providing an image of the array and a table of genes that match the query results. The image is useful because it allows confirmation that a given value is not the consequence of hybridization artifact such as a scratch or dye precipitate. The table contains the relevant intensity, spot size, and ratio data for each gene meeting the search criteria. Clicking on the Clone ID field in this view opens a window containing data on the clone printed at that point on the array with links to relevant databases including UniGene, OMIM, GenBank, and GeneCards, which are useful for acquiring functional information about a given gene. Additional features of ArrayDB include the ability to see a list of biological pathways that gene products of interest are involved in; the positions of these gene products within known biological pathways are also shown graphically using image maps from the Kyoto Encyclopedia of Genes and Genomes (KEGG).

Investigators involved in microarray research soon learn the critical importance of rapid access to descriptive information on gene function. Most of the genes of interest in individual experiments are unfamiliar, and an efficient mechanism for obtaining a synopsis of gene function is essential to the interpretation of microarray data. Central storage of microarray data with investigator access via individual work stations is clearly the preferred mode for information management for centers generating significant quantities of microarray data and ultimately will be required for public distribution of expression data. However, it is important to note that projects of significant size can be accommodated in low-cost databases maintained on desktop computers.

As an example of this approach, FileMaker Pro templates for cDNA , microarray data can be downloaded from the NHGRI microarray site. Projects of up to 50 experiments can easily be accommodated. Although, as discussed below, the statistical analysis of expression data with specialized tools is important, the value of simple relational databases should not be underestimated. Even in a database such as FileMaker Pro, it is possible to define patterns of genes expressed consistently across experiments and possible to generate useful graphs displaying numerical data as colour blocks. This type of graphic display is much easier to scan by eye than a large table of numbers.

Reanalyzing a data set using various filters is greately facilitated by placing the entire set of experiments in a searchable database. Various settings can be tested until the output is optimized. Initial processing of data requires the application of significance filters to the raw data set so that only meaningful measurements enter into downstream analysis. This requires the application of a sensitivity threshold to remove genes that have not been measured accurately. The importance of this step was illustrated earlier in this chapter. Individual laboratories will need to establish thresholds applicable to their own data, and investigators should exercise caution when using publicly accessible databases to be certain whether the data has been filtered for a detection threshold. Of course, there is no fixed cutoff that must be applied to all expression data, but, in general, as the threshold is lowered toward the background noise of an assay, the quality of the data will diminish.

Fig. 12.10. Display of microarray results retrieved from FileMaker Pro. This example illustrates the results of a query for genes upregulated in a series of cancer cell lines with ratios coded by a red to green color map.

To maximize the yield of information from an experiment and to simplify data displays by removing uninformative genes, in most cases it is also useful to apply a filter that removes genes that, although measured accurately, do not fluctuate across a series of experiments. Once an appropriately filtered data set has been extracted, the data are ready for analysis. In some instances, simple searches may be sufficient to generate lists of genes that are expressed under a given condition. For example, from a set of experiments in which cells have been treated with a drug or transfected with a gene, simple searching and sorting of the filtered data will yield a list of genes induced or repressed by the treatment. However, the limitations of simple searches are quickly reached with larger data sets involving multiple samples or conditions and computational techniques for data organization are essential.

Cluster Analysis

The goal of cluster analysis is to reveal underlying patterns in data sets that contain hundreds of thousands of measurements and to present this data in a user-friendly manner. Ideally, patterns of similarities and difference among samples are identified and genes are coregulated in distinct patterns. A number of tools for clustering have been developed, which are freely available. In general, these are based on conventional statistical techniques widely used in other contexts and are largely dependent on linear correlation analysis.

It is certain that future research efforts will be directed at the development of increasingly sophisticated tools designed for mining expression data. By applying statistical analysis, definition of subsets in apparently homogeneous tissue samples, development of classifiers for various disease entities, and identification of groups of coregulated genes may be possible. These possibilities are especially intriguing because they may provide a way to assign the large number of anonymous genes in higher eukaryotes to functional groups.

HIERARCHICAL CLUSTERING

Agglomerative hierarchical clustering has established itself as the most frequently applied technique for processing array data for inspection. All pairwise comparisons of expression levels are made between experiments. The resulting matrix of scatter plots can be reduced to a matrix of Pearson correlation coefficients. This is readily displayed in two dimensions as a hierarchical

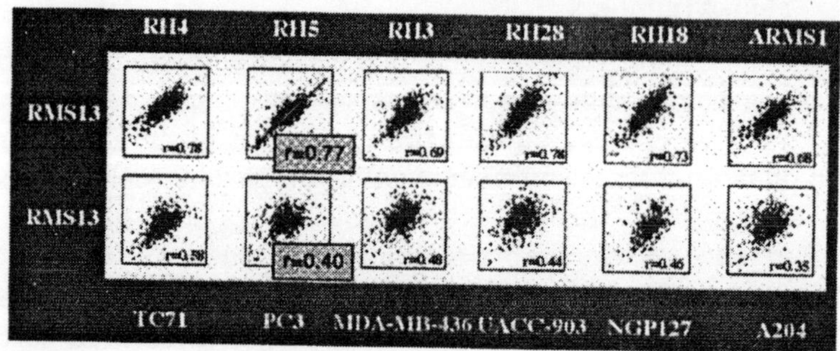

Fig. 12.6 : Portion of a scatter plot matrix illustrating pairwise plots of log ratios across a series of microarray experiments. Each scatter plot is used to calculate a Pearson correlation coefficient between the two sets of measurements.

dendrogram. Both genes and samples can be clustered in this fashion. By colour coding expression levels, a large numerical table can be replaced by a much more compact and easily inspected colour plot. Inspection of the dendrogram and colour plot will reveal samples with closely related patterns of gene expression as well as clusters of genes with similar expression pattern. Software for this type of display is freely available. This approach has been productively applied to yeast and human expression data. It is important to note that there is an arbitrary character to the manner in which a dendrogram is drawn.

Clusters can be rotated about the point of bifurcation affecting the apparent proximity of the edges of a cluster with adjacent clusters. The important information is contained in the cluster contents and their similarity. An alternative approach for displaying the same information is multidimensional scaling (MDS). MDS software is available in standard statistical software such as MATLAB. The distance measure for plotting individual samples is based on $1 - r$ where r is the Pearson correlation coefficient. Thus samples that are closely related will plot closely together, and widely differing samples plot further apart.

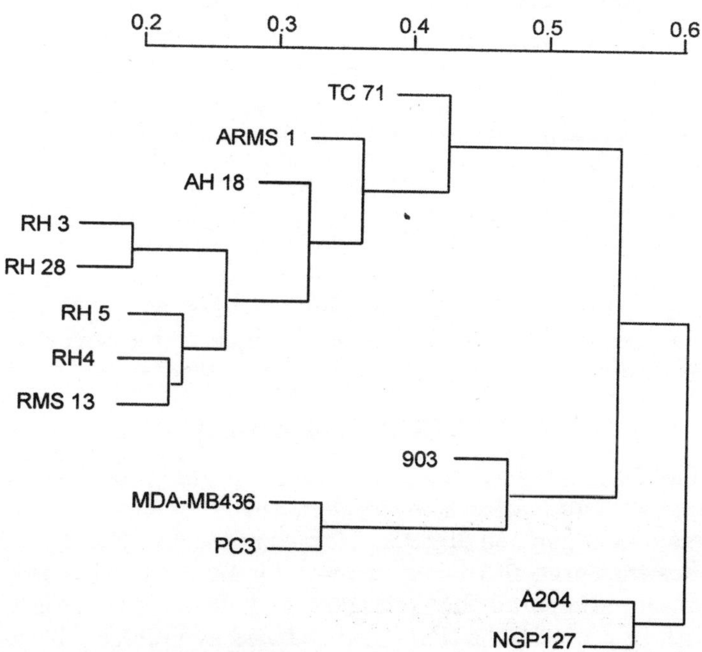

Fig. 12.7 : Hierarchical clustering dendrogram plotted from the Pearson correlation coefficients calculated across a series of experiments (Khan et. al., 1998).

Although MDS does not allow simultaneous display of gene and sample clusters, it has the virtue of retaining a graphical display that places each sample (or gene) plotted in relation to every other. Ideally, an MDS plot, which is a reduction of multidimensional data to a three-dimensional display, should be viewed on a computer screen that allows rotation of the data so that the viewer is not deceived by any single two-dimensional projection. Several other statistical tools based on linear and nonlinear methods have been used to analyze array data, including self-organizing maps, k-means clustering, and principal component analysis.

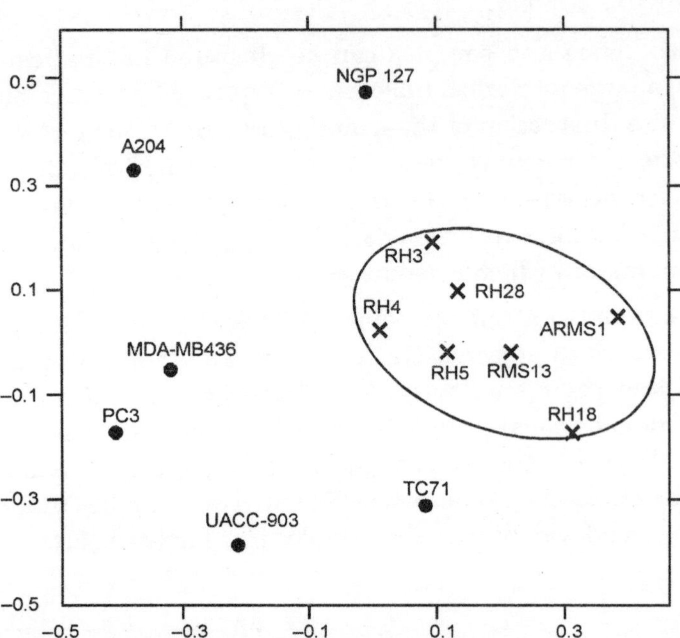

Fig. 12.8. Multidimensional scaling plot of the same data represented in Figure 12.8 : Note that the samples that fall closely together in the dendrogram also plot near to one another in the MDS plot. Ideally, this plot should be viewed in three dimensions on a computer screen so that it can be rotated to allow better inspection of the distances between the data points.

Each of these tools has merit, and software for some of these is publicly available. However, the precise selection of statistical methods for the analysis of array data is still the subject of active investigation, and the limits of these techniques are not well defined. No single method can be recommended at this time to the exclusion of others, and most investigators will base their choice on the availability of software tools and Web-based resources. The range and quality of these tools will certainly improve rapidly in the coming years.

CONCLUDING REMARKS

Existing technologies for high-throughput gene expression analysis and the informatics approaches to analysis of data arising from these methods are already producing an extremely interesting and novel view of genome function. However, it is apparent that there are limitations to current approaches that present the opportunity for significant improvements. At the level of technology, it can be anticipated that microarrays will move progressively closer to whole genome analysis with improved sensitivity and reduced sample requirements. At the level of informatics, several areas of progress can be foreseen. Tools for providing precomputed informative summaries of function for named genes are likely to be developed. Methods of linking gene expression data to genomic sequence would be of great value in dissecting networks of coregulated genes by identifying regulatory motifs shared by clustered genes. Most far-reaching would be methods of defining gene networks in terms of codetermination through the development of computational tools that predict the expression of a gene based on the expression of other genes (Seungchan *et. al.*, 2000). It is hoped that the recognition of gene networks will facilitate the assignment of function to anonymous genes and the definition of pathways. Ultimately, models may be developed that accurately reflect the mesh of self-compensating regulatory networks that maintain ordered genome function.

13

Scanning Genomes

Imprinted genes in mammals are expressed exclusively from one of the parental alleles. This is regulated by parental-allele-specific CpG methylation. For example, H19 is methylated exclusively on the paternal allele, which is repressed, and is expressed exclusively from the maternal allele, which is unmethylated. Therefore, one way to find imprinted genes is searching for parental-allele-specific CpG methylation. Southern analysis using methylation-sensitive restriction enzymes could be used for such a purpose. However, usually only one gene can be analyzed by one Southern analysis. Moreover, Southern analysis requires one DNA probe for each analysis. These facts indicate at least 300 Southern analyses using 300 different probes are required to find only one imprinted gene, because the population of imprinted genes is estimated to be 0.3%. Therefore, this kind of analysis is not appropriate for searching for new imprinted genes, and the development of a new method that can simultaneously analyze thousands of genes was required.

METHODS OF SCANNING

We developed a powerful genome scanning method, termed the restriction landmark genome scanning (RLGS) method for analyzing thousands of genes simultaneously in higher organisms such as mammals. This method is based on end-labeling of genomic DNA at restriction sites and its separation by two-dimensional gel electrophoresis. Before the RLGS method was developed, the detection of genes from higher organisms by end-labeling had been thought to be impossible. There are two reasons for this: (1) The mammalian genome is so complex (for example, the human genome size is 3×10^9 bp and digestion with *Eco*RI will give more than 106 DNA fragments) that DNA fragments digested by restriction enzymes cannot be separated even by two-dimensional electrophoresis. (2) Generally, genomic DNA is cleaved randomly in the preparation step and thus has nonspecific cleaved ends, nicks, and/or gaps. This leads to high background caused by the incorporation of radioisotopes into sites damaged in the labeling process.

To overcome these problems we adopted two solutions: (1) To reduce the complexity, genomic DNA is cut and labeled at rare-cleaving restriction enzyme sites, for which there are thousands of recognition sites in the genome, before cutting with more frequently cutting restriction enzymes that will give more than 10^6 DNA fragments. Using this method, only DNA fragments that have rare restriction sites are detected and used as landmarks in the genome. (2) High background can be avoided by blocking damaged sites with enzymatically incorporated nucleotide analogs, because such analogs prevent exonucleolysis and/or the additional incorporation of nucleotides at blocked ends.

The procedure for the RLGS method is comprised of eight steps.

1. Blocking: Damaged sites in genomic DNA are blocked by nucleotide analogs, 2′-deoxyribonucleoside 5′-[α-thio] triphosphate and 2′3′-dideoxyribonucleoside 5′-triphosphate, to reduce the background.

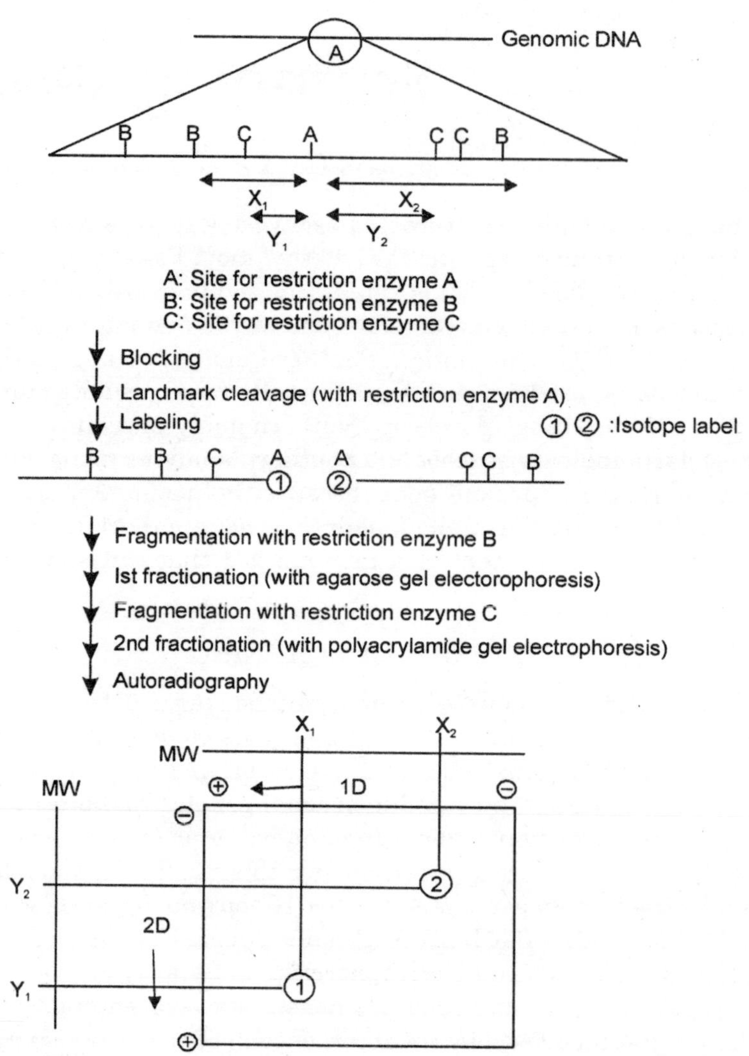

Fig. 13.1 : Procedure for genome scanning by two-dimensional gel electrophoresis. RLGS consists of eight steps: (1) blocking, (2) landmark cleavage (with restriction enzyme A), (3) labeling (4) fragmentation with restriction enzyme B, (5) first fractionation (with agarose gel electrophoresis), (6) fragmentation with restriction enzyme C, (7) second fractionation (with polyacrylamide gel electrophoresis), and (8) autoradiography. X_1 or X_2 and Y_1 or Y_2 represent the distance from a restriction landmark to the neighboring site for restriction enzyme A or B, and that to the site for restriction enzyme C, respectively.

2. Landmark cleavage: Blocked genomic DNA is cleaved with a rare-cutting restriction enzyme (restriction enzyme A; average cleaved fragment size should be more than 100 kbp).
3. Labeling: Cleavage ends of the genomic DNA are labeled with radioisotope.
4. Fragmentation of labeled DNA with restriction enzyme B: Labeled genomic DNA is cleaved with restriction enzyme B (average cleaved fragment size should be between 1 kbp and 100 kbp) to reduce the fragment size to several kilobase pairs, kbp which is appropriate for agarose gel electrophoresis.
5. First fractionation: Genomic DNA fragments are fractionated in one dimension by thin-layer agarose gel electrophoresis.
6. Fragmentation of labeled DNA with restriction enzyme *C*: Fractionated genomic DNA fragments are cleaved in the gel with restriction enzyme *C* (average cleaved fragment size should be less than 10 kbp) to make fragments appropriate for polyacrylamide gel electrophoresis.
7. Second fractionation: Genomic DNA fragments in the agarose gel are fractionated in the second dimension by polyacrylamide gel electrophoresis.
8. Autoradiography: The labeled genomic DNA fragments are detected as spots.

Fig. 13.2 : shows a spot profile of mouse DNA obtained by RLGS. This profile contains about 2000 completely separated spots corresponding to sites for restriction enzyme A.

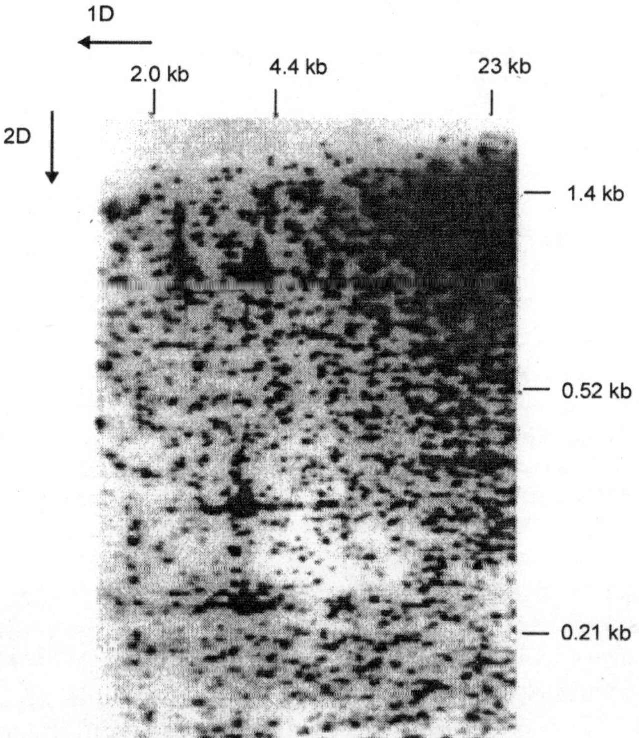

Fig. 13.2 : The RLGS profile of genomic DNA from C57BL/6 mice. NotI, EcoRV, and MboI were used as restriction enzymes A, B, and C, respectively.

RLGS method

Methylation status can be distinguished by the RLGS method when a CpG methylation-sensitive enzyme, such as *Not*I, is used for restriction enzyme *A*. It is expected that the intensity of any spot varies in proportion to the population of methylated DNA molecules. Lack of methylation of both alleles methylation of one allele (middle panel), and methylation of both alleles (right panel) will give full intensity, half-intensity, and no spot, respectively. To identify parental-origin-specific methylation, the spot profiles of F_1 progeny of reciprocal crosses between two inbred strains, for example (DBA2 × C57BL/6)F_1 and (C57BL/6 × DBA2)F_1 should be compared. If there is no parental-origin-specific methylation, both the maternal and the paternal allele can be cleaved with *Not*I and detected as spots, resulting in identical spot profiles of F_1 progeny.

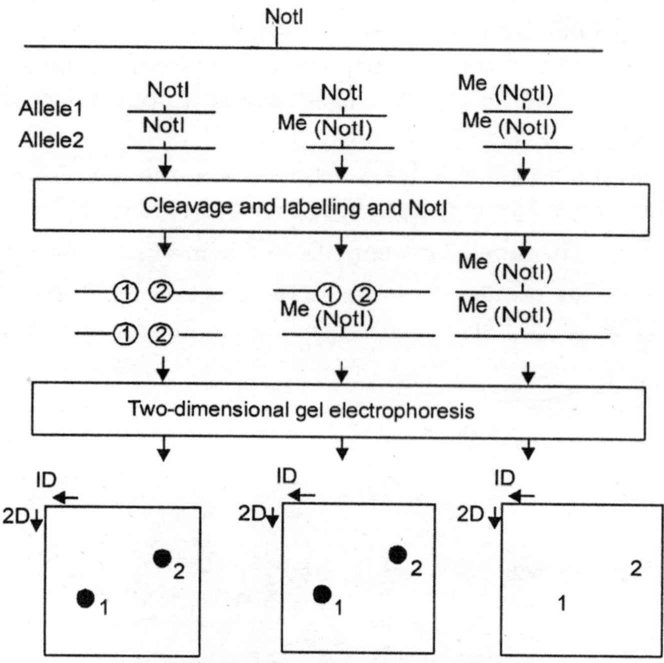

Fig. 13.3 : Detection of methylation differences in genomes by RLGS. Maps show DNA-methylation state at the NotI site, which is used as restriction enzyme A. Three types of methylated state are expected, depending on which allele is methylated, as shown in the lower panels: no allele is methylated (left), one allele is methylated (middle), and both alleles are methylated (right). ^{Me}(NotI) represents the methylated site of this enzyme. As the CpG methylation status of the NotI site affects the cleavage of this site, the RLGS profile results in a change in spot intensity. Each sample would give a full intensity spot (left), half-intensity spot (middle), or no spot (right).

However, if one parental allele is methylated, only one allele can be cleaved with *Not*I and detected as spots, resulting in different spot profiles of F_1 progeny. We found four spots that appeared in one cross while not appearing in another when *Not*I, *Eco*RV, and *Mbo*I were used as restriction enzymes *A*, *B*, and *C*, respectively. For example, spot 2 (specific to DBA2), appeared in one cross (C57BL/6 × DBA2)F_1 while not appearing in the opposite cross (DBA2 × C57BL/6) F_1. Other parental-origin-specific methylation patterns can be sought using other restriction enzymes.

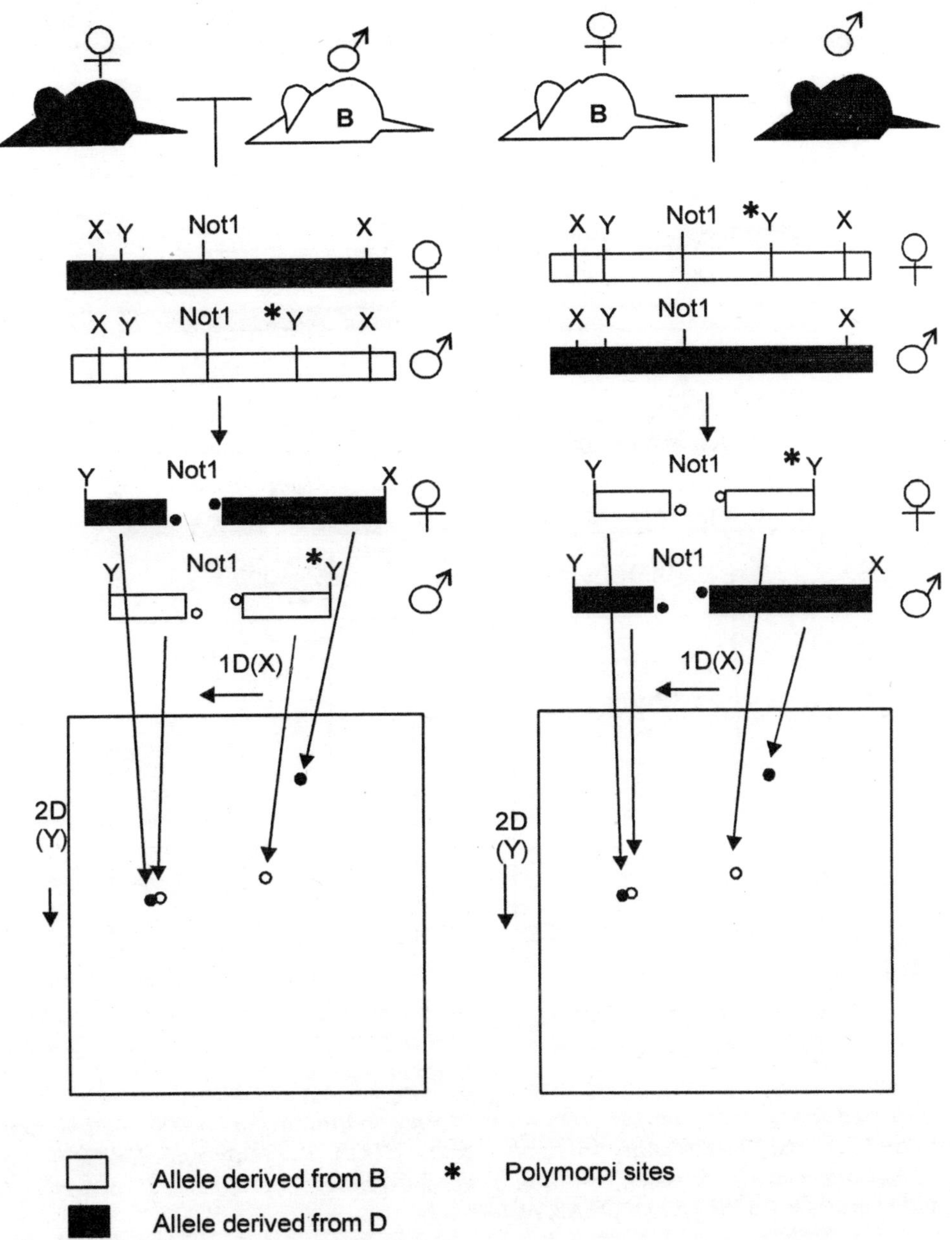

Fig. 13.4 : Expected spot profiles derived from a gene without parental-origin-specific methylation. (DBA2 × C57BL/6)F_1 and (C57BL/6 × DBA2)F_1 are applied to the RLGS. Both the maternal and the paternal allele can be detected as spots, resulting in the identical spot profiles of F_1 progeny of reciprocal crosses. B and D indicate C57BL/6 and DBA2, repectively.

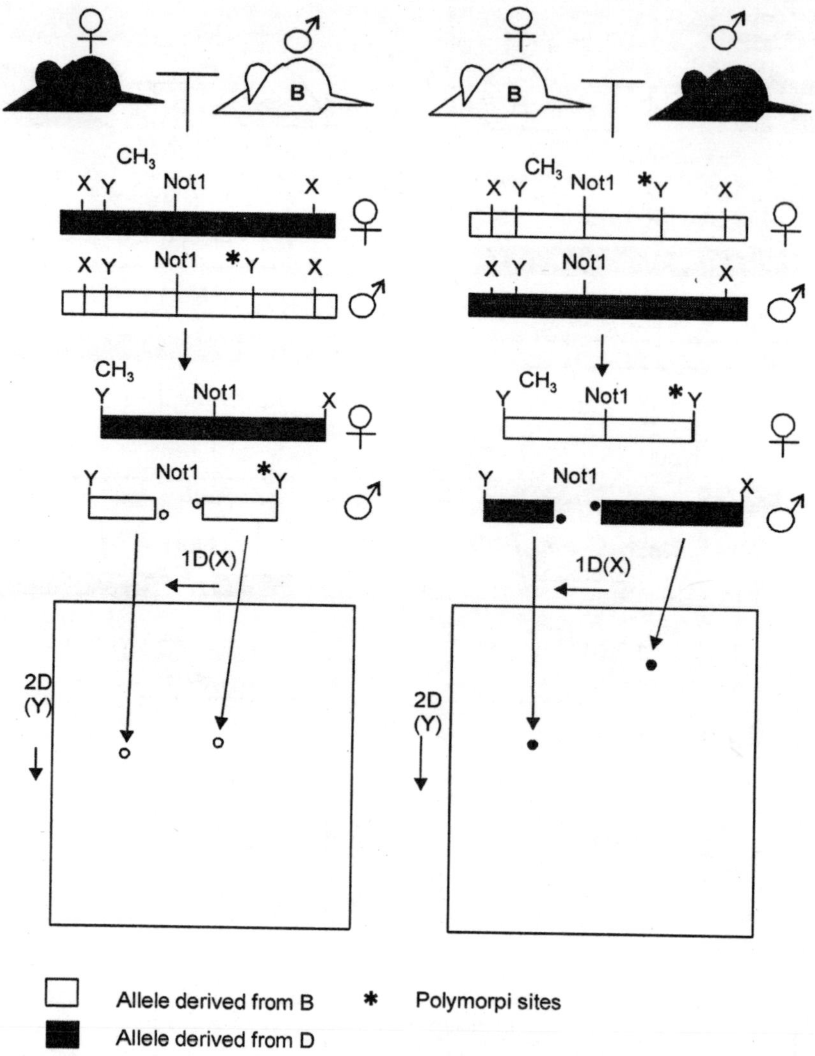

Fig. 13.5 : Expected spot profiles derived from a gene with parental-origin-specific methylation. (DBA2 × C57BL/6)F_1 and (C57BL/6 × DBA2)F_1 are applied to the RLGS. The maternal allele is methylated and only the paternal allele can be detected, resulting in the different sport profiles of F_1 progeny of reciprocal crosses. B and D indicate C57BL/6 and DBA2, respectively.

CLONING OF METHYLATED SPOTS

To clone the spot in which you are interested, a boundary library should be constructed by using restriction enzymes *A* and *B*. For example, when *Not*I, *Eco*RV, and *Mbo*I are used as restriction enzymes *A*, *B*, and *C*, the library is constructed by using *Not*I and *Eco*RV. This library contains only *NotI*I-*Eco*RV fragments and does not contain any *Eco*RV-*Eco*RV fragments. The fragment that must be cloned is concentrated several thousands times in this library. This library is applied to RLGS, except for the blocking and labeling steps, with labeled genomic DNA. The target spot is cut out from the gel and *Not*I-*Mbo*I fragments from this spot are electroeluted for cloning into *Not*I, *Bam*HI sites of the vector.

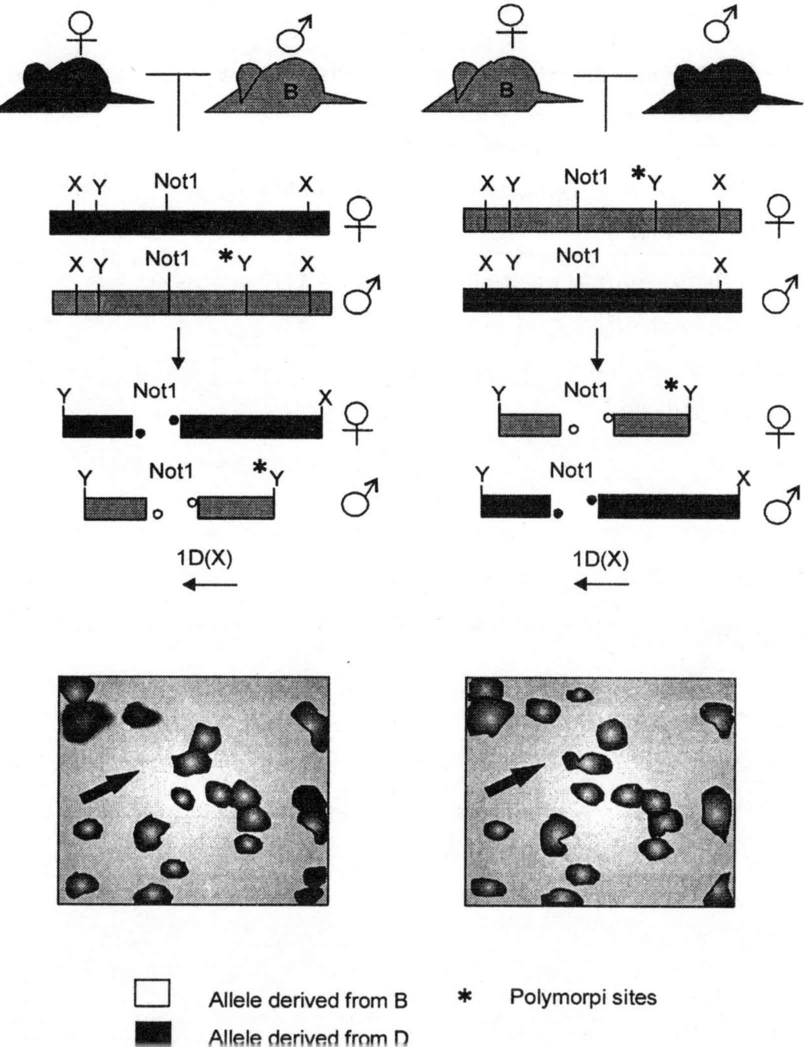

Fig. 13.6 : Detection of the difference in spot intensity in F_1 progeny of reciprocal crosses. (DBA2 × C57BL / 6)F_1 and (C57BL / 6 × DBA2)F_1 are applied to the RLGS. Arrows indicate spot 2. Spot 2 did not appear in (DBA2 × C57BL / 6)F_1 but appeared in (C57BL / 6 ×DBA2)F_1. B and D indicate C57BL / 6 and DBA2, respectively.

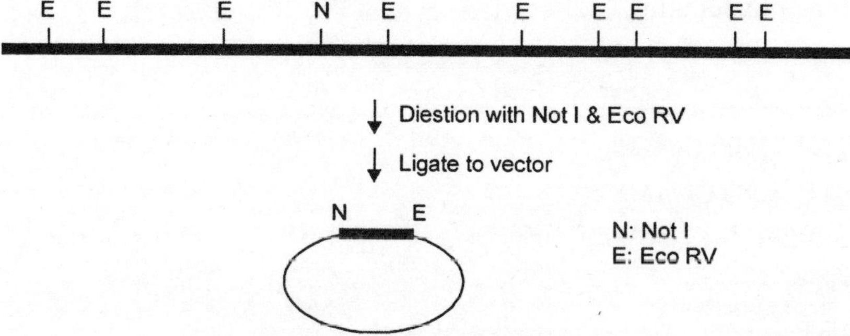

Fig. 13.7 : Construction of a boundary liberary using NotI and EcoRV.

Materials

The materials described here are for a case using *Not*I, *Eco*RV, and *Mbo*I as restriction enzymes *A*, *B*, and *C*.

Extraction of Genomic DNA from the Sample

1. 10 *mM* Tris-HCl (pH 8.0), 0.5 *M* ethylenediaminetetraacetic acid (EDTA), 1% sarkosyl.
2. 1 0 mg/mL Proteinase *K*.
3. Phenol chloroform isoamyl alcohol (*v*/*v*/*v*, 25/24/1).
4. TE: 10 m*M* Tris-HCl (pH 7.4), 1 *mM* EDTA.
5. 1 mg/mL RNase A (boiled for 5 min).
6. Ethanol (70% and 100%).
7. Liquid nitrogen.
8. Aluminum foil.
9. Hammer.
10. Mortar and pestle.

Labeling Genomic DNA

1. 10 × buffer H: 500 *mA* Tris-HCl, pH 7.4, 100 *mM* $MgCl_2$, 1 *M* NaCl, 100 *mM* dithiothreitol.
2. 10 μ*M* dGTP[α*S*].
3. 10 μ*M* dCTP[α*S*].
4. 10 μ*M* ddATP.
5. 10 μ*M* ddTTP.
6. DNA polymerase I.
7. 0.1% BSA.
8. 0.1% Triton *X*-100.
9. *Not*I.
10. Control plasmid: linearized plasmid which has at least one recognition site for restriction enzymes *A* and *B*.
11. 0.25% bromophenol blue, 0.25% xylene cyanol FF. 30% glycerol.
12. 0.8% agarose gel: 0.8% Agarose, 1 × TBE.
13. 1 *M* Dithiothreitol.
14. [α-32P] dGTP (3000 Ci/*mM*).
15. [α32P] dCTP (3000 Ci/*mM*).
16. Sequenase V2.0.
17. 1 *mM* ddGTP.
18. 1 *mM* ddCTP.
19. *Eco*RV.

20. 3 *M* NaOAc.
21. 0.25% bromophenol blue, 0.25% xylene cyanol FF.
22. 100% ethanol.
23. TE.
24. Phenol/chloroform/isoamyl alcohol (25/24/1).

Two-Dimensional Electrophoresis

1. 1 × TAM: 50 *mM* Tris-HCl, pH 7.5, 0.7 *mM* MgAc.
2. 0.8% Seakem GTG (FMC), 1 × TAM.
3. 1 X buffer H: 50 *mM* Tris-HCl, pH 7.4, 10 *mM* MgC_2, 100 *mM* NaCl, 10 *mM* dithiothreitol.
4. TBE: 90 *mM* Tris-borate, 2 *mM* EDTA, pH 8.0.
5. 0.8% SeaKem GTG, 1 × buffer *H*.
6. *Mbo*I.
7. 1-D small plate: 150 × 30 × 5 mm glass plate.
8. 1-D upper plate: 470 × 150 × 5 mm glass plate.

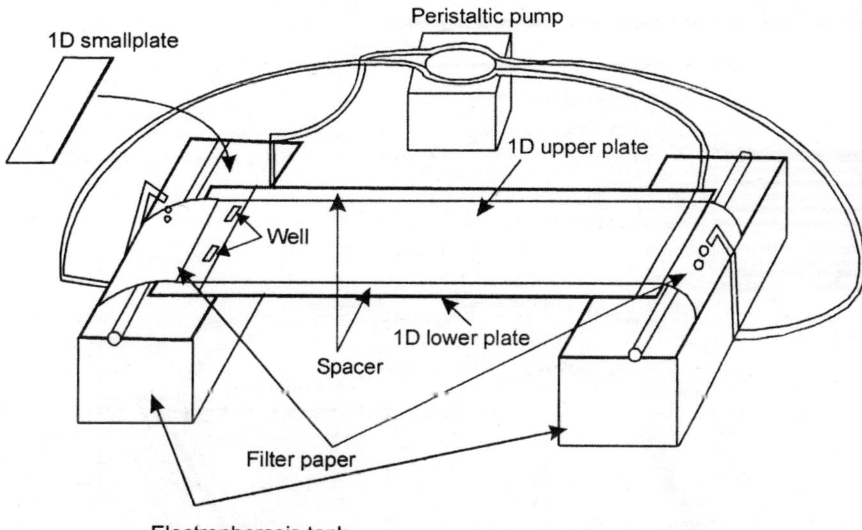

Fig. 13.8 : Apparatus for one-dimensional agarose gel electrophoresis.

9. 1-D lower plate: 500 × 150 × 5 mm glass plate.
10. Well plate: Two plastic sheet (16 × 1 × 0.7 mm) are attached to a glass plate (150 × 30 × 5 mm) for making wells.
11. Upper plate for enzyme digestion: A glass plate (150 × 30 × 5 mm) with two holes (ϕ2 mm) for pouring enzyme solution. Spacers (1.2 mm thick) are attached to four sides of the plate
12 2-D small plate: 415 × 30 × 5 mm glass plate.
13. 2-D upper plate: 460 × 415 × 5 mm glass plate.

14. 2-D lower plate: 490 × 415 × 5 mm glass plate.
15. Electrophoresis tank.
16. Spacer: Thickness is 1 mm.
17. Filter paper: Whatman 3MM.
18. X-ray film: XAR5, Kodak.

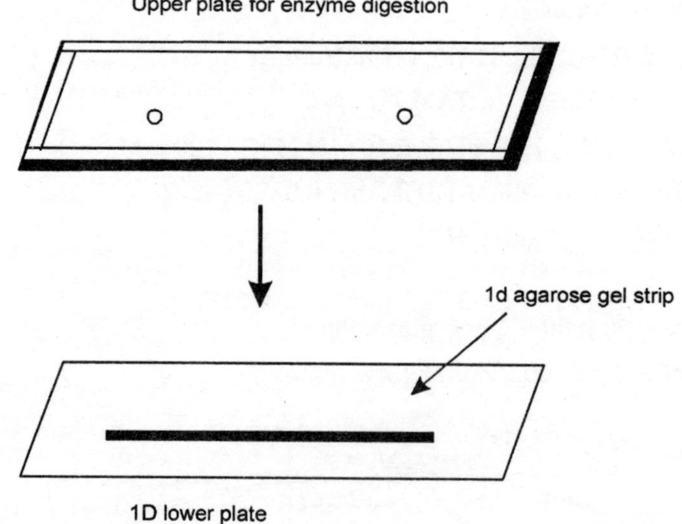

Fig. 13.9 : Apparatus for two-dimensional polyacrylamide gel electrophoresis.

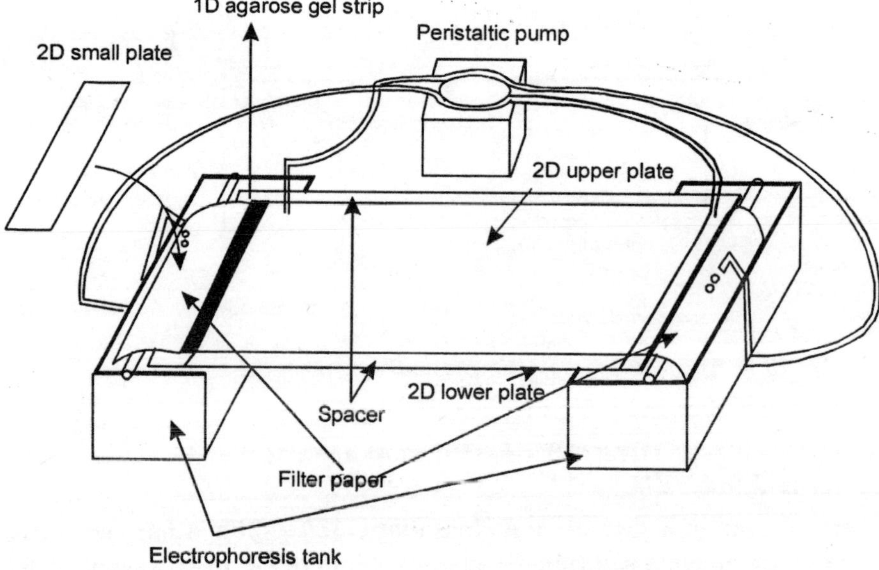

Fig. 13.10 : Apparatus for two-dimensional polyacrylamide gel electrophoresis.

Cloning of Parental-Origin-Specific Methylated Spots

1. *Not*I.
2. *Eco*RV.
3. 10 × buffer *H*.
4. 0.1% BSA.
5. 0.1% Triton X-100.
6. 0.8% Seakem GTG(FMC), TBE.
7. Qiaquick gel extraction kit (Qiagen).
8. Vector plasmid: pBlueScript II.
9. Ligase and ligation buffer.
10. Competent cell for electrotransformation.
11. *Escherichia coli* pulser (Bio-Rad).
12. Agar plate with ampicillin.
13. Rubber policeman.
14. Qiaprep spin (Qiagen).
15. *Bam*HI.
16. L-Broth medium with ampicillin.

Methods

Extraction of Genomic DNA

1. Precool 0.5 g of the sample, a folded aluminum foil, and a mortar and a pestle with liquid nitrogen.
2. Wrap the sample in folded aluminium foil and hammer it flat.
3. Quickly take out the crushed sample from the aluminum foil and transfer it to a mortar and grind it to a powder.
4. Transfer the powdered sample to a tube and add 1 2 mL of 10 *mM* Tris-HCl (pH 8.0), 0.5 MEDTA, 1% sarkosyl, and 1.2 mL of 10 mg/mL proteinase *K*.
5. Mix the solution and incubate it for 30 min at 65°C.
6. Cool the solution on ice. Add an equal volume of phenol/chloroform/isoamyl alcohol (25/24/1) and gently mix the two phases.
7. Separate the two phases by centrifugation at 5000g (3000 rpm) for 10 min at 4°C.
8. Transfer the viscous aqueous phase to a new tube using a wide-bore pipet.
9. Repeat steps 6-8 once more.
10. Dialyze the aqueous phase at 4°C two times against 1 *L* of TE.
11. Transfer the dialysate to a new tube. Add 0.001 vol of 1 mg/mL RNase *A* and incubate for 2 *h* at 37°C.
12. Add 2 vol of ethanol and mix.
13. Centrifuge the solution for 10 min at 4°C.
14. Remove the supernatant.

15. Wash the pellet with 70% ethanol.
16. Centrifuge briefly and remove the supernatant.
17. Dissolve the pellet in 300 μL of TE.

Labeling Genomic DNA

The method described here is for a case using *Not*I, *Eco*RV, and *Mbo*I as restriction enzymes *A, B,* and *C,* respectively.

1. Make the following mixture for the blocking reaction

Prepared genomic DNA	10μg
10 × buffer H	5 μL
10 μM dGTP[αS]	1.7 μL
10 μM dCTP[αS]	1.7 μL
10 μM ddATP	1.7 μL
10 μM ddTTP	1.7 μL
DNA polymerase I	10 units
Distilled water to 50 μL	

2. Incubate the reaction for 30 min at 37°C.
3. Inactivate the enzyme by incubating the reaction at 65°C for 30 min.
4. Add the following reagents for digestion with *Not*I.

10 × bufferH	5 μL
0.1% BSA	10 μL
0.1% Triton X-100	10 μL
*Not*I	100 units
Distilled water to 100 μL	

5. Transfer 5 μL to another tube and add 1 μL control plasmid for monitoring complete digestion.
6. Incubate the sample reaction and the control reaction for 1 *h* at 37°C.
7. Remove 2 μL of the control and add 1 μL of 0.25% bromophenol blue, 0.25% xylene cyanol FF. 30% glycerol. Analyze by electrophoresis through 0.8% agarose gel for checking complete digestion.
8. Add the following reagents for labelling.

1 M Dithiothreitol	1 μL
10 μM ddATP	1.7 μL
10 μM ddTTP	1.7 μL
[α-32P] dGTP (3000/Ci/mM)	5 μL
[α-32P] dCTP (3000Ci/mM)	5 μL
Sequenase V2.0	20 units

9. Incubate the reaction for 30 min at 37°C.

10. Add following reagents for stopping the reaction.

1 mM ddGTP	2 μL
1 mM ddCTP	2 μL

11. Add 100 units of *Eco*RV.
12. Transfer 5 μL to another tube and add 1 μL of control plasmid for monitoring complete digestion.
13. Incubate the sample reaction and the control reaction for 1 *h* at 37°C.
14. Remove 2 μL of the control and add 1 μL of 0.25% bromophenol blue, 0.25% xylene cyanol FF. 30% glycerol. Analyze it by electrophoresis through 0.8% agarose gel, checking for complete digestion.
15. Add an equal volume of phenol/chloroform/isoamyl alcohol (25/24/1) and gently mix the two phases.
16. Separate the two phases by centrifugation 12,000g (15,000 rpm) for 5 min at room temperature.
17. Transfer the aqueous phase to a new tube.
18. Add 0.1 vol of 3 *M* NaOAc and 2 vol of ethanol and mix.
19. Chill the solution at –70°C for 10 min.
20. Centrifuge the solution for 10 min at 4°C.
21. Remove the supernatant.
22. Wash the pellet with 70% ethanol.
23. Centrifuge briefly and remove the supernatant.
24. Dissolve the pellet in 10 μL of TE.
25. Use 1 uL for the estimation of concentration.
26. Transfer 2 μg to a new tube and add 2 μL of 0.25% bromophenol blue, 0.25% xylene cyanol FF and TE to 10 μL for loading.

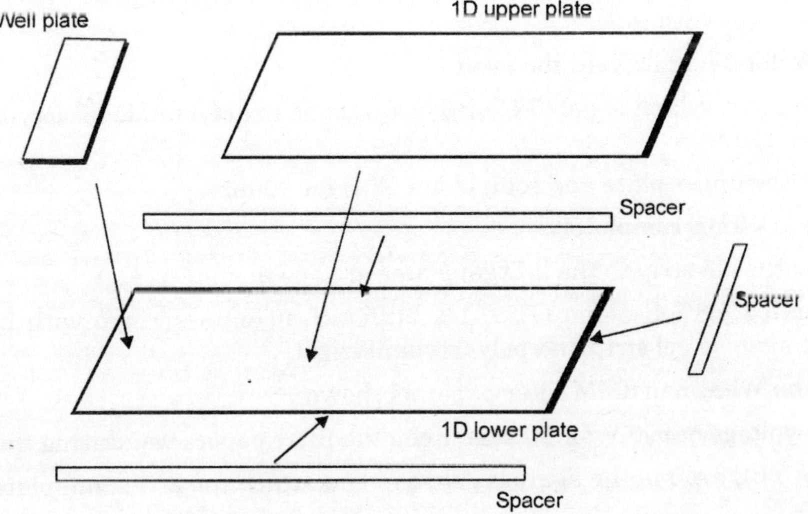

Fig. 13.11 : Preparation of the 1-D plate.

Two-Dimensional Electrophoresis

The method described here is for a case using *Not*I, *Eco*RV, and *Mbo*I as restriction enzymes *A*, *B*, and *C*.

1. Prepare the 1-D glass plates and spacer for pouring the gel and prewarm it at 37°C in a oven.
2. Pour 80 mL of melted 0.8% SeaKem GTG, 1 × TAM, by using a syringe with 20G needle.
3. Lay the glass plates.
4. After the gel is completely set, carefully remove the well plate and mount the gel in an electrophoresis tank filled with 1 × TAM.
5. Cut out the gel 1 cm from the top and mount the Whatman 3MM filter papers as shown.
6. Moisten the filter papers and set up the peristaltic pump as shown.
7. Load the sample in one well and load size markers in the other well.
8. Apply a voltage of 500 V for 10 min. Keep the filter papers wet during this period.
9. Pour 1 × TAM around the wells and attach the 1-D small plate.
10. Apply a voltage of 370 V for 18 h. Keep the filter papers wet during this period by peristaltic pump.
11. Put the glass plates on a graph paper and remove the 1-D small plate and 1-D upper plate.
12. Cut and leave a 10 × 400 mm DNA-containing portion and remove the remaining gel.
13. Soak the gel on the plate for 20 min in 1 × buffer *H*.
14. Remove 1 × buffer *H* completely.
15. Mount the prewarmed upper plate for enzyme digestion and fix using a clip.
16. Pour the following solution from two pores to spread over the gel.

1 × buffer *H*	1.3 μL
*Mbo*I	1500 units

17. Seal the pores with tape.
18. Incubate for 2 *h* at 37°C in the oven.
19. Prepare polyacrylamide gel (5-6% polyacrylamide to acrylamide/bisacrylamide, 29/1) in 2-D glass plate.
20. Remove the upper plate and soak in 1 × TBE for 10 min.
21. Remove 1 × TBE completely.
22. Transfer the gel strip to the 2-D gel plates as shown.
23. Pour melted 0.8% Seakem GTG, 1 × buffer *H*, using a syringe with 24G needle, to fuse the agarose gel strip and polyacrylamide gel.
24. Set up the Whatman 3MM filter paper as shown.
25. Apply a voltage of 900 V for 30 min. Keep the filter papers wet during this period.
26. Pour 1 × TBE around the agarose gel strip and attach the 2-D small plate.

27. Apply a voltage of 900 V for 6 *h*. Keep the filter papers wet during this period using a peristaltic pump.
28. Transfer the gel to Whatman 3MM filter paper and cover with Saran Wrap.
29. Dry the gel under vacuum on a gel dryer set at 80°C.
30. Autoradiograph by exposing the gel to X-ray film (Kodak XAR5) with intensifying screen (Quantalll) at –70°C for 1 wk.

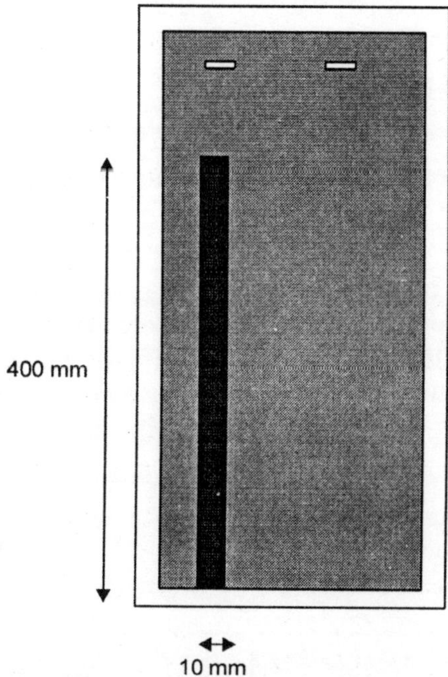

Fig. 13.12 : The agarose gel portion used for digestion with restriction enzyme C is indicated by the black area.

Cloning of Parental-Origin-Specific Methylated Spots

1. Estimate the one-dimensional size of the target spot.
2. Digest 100 μg of genomic DNA with *Not*I and *Eco*RV.
3. Size-fractionate the digested DNA by 0.8% agarose gel electrophoresis.
4. Cut out the portion corresponding to the target spot and recover the DNA with Qiaquick gel extraction kit.
5. The recovered DNA is ligated to the *Not*I-*Eco*RV cleaved vector purified by agarose gel electrophoresis.
6. The ligated mix is applied to electrotransformation.
7. Spread the transformed *E. coli* on to agar plates with ampicillin.
8. Incubate at 37°C for 12 *h*.
9. Recover and mix all colonies from the plate with a rubber policeman.
10. Purify plasmid DNA from the colony mixture with Qiaprep spin.

11. Apply 1 μg of boundary library to the RLGS, except for the blocking and labeling steps, using 1 μg of labeled genomic DNA.
12. Cut out the 2-D gel portion corresponding to the target spot from the gel.
13. Electroelute the *Not*I-*Mbo*I fragment of the target spot.
14. Ligate the eluted DNA to the *Not*I-*Bam*HI cleaved vector purified by agarose gel electrophoresis.
15. Apply the ligated mix to electrotransformation.
16. Spread the transformed *E. coli* onto an agar plate with ampicillin.
17. Incubate at 37°C for 12 *h*.
18. Pick several colonies and culture each colony in L-broth medium with ampicillin.
19 Purify plasmid DNA from each culture and check each DNA.

Notes

1. The method described here is for a case using restriction enzyme A, which can make protruding cohesive 5′ termini. Use 2′-deoxyribonucleoside 5′-[α-thio] triphosphate, which can be incorporated into the cleavage site of restriction enzyme A by fill-in reaction. Use 2′, 3′-dideoxyribonucleoside 5′-triphosphate, which cannot be incorporated into the cleavage site of restriction enzyme *A* by a fill-in reaction.
2. Use restriction enzyme *A* and an appropriate buffer for it.
3. Use 2′, 3′-dideoxyribonucleoside 5′-triphosphate, which cannot be incorporated into the cleavage site of restriction enzyme *A* by fill in reaction. Use labeled 2′-deoxyribonucleoside 5′-triphosphate, which can be incorporated into the cleavage site of restriction enzyme A by a fill-in reaction.
4. Use 2′, 3′-dideoxyribonucleoside 5′-triphosphate, which can be incorporated into the cleavage site of restriction enzyme *A* by a fill-in reaction.
5. Use restriction enzyme *B* and the appropriate buffer for it.
6. Use the buffer appropriate for restriction enzyme *C*.
7. Use restriction enzyme *C* and the appropriate buffer for it.

14

Genomic Imprinting

Genomic imprinting, though most extensively studied in mammals, has long been known to perform an important role in seed development in flowering plants. In this chapter, an overview of what is known to date about genomic imprinting in flowering plants and how this knowledge came into being will be given. Flowering plants (Angiosperms) are unique in that their initial development requires a double fertilization.

After pollination, one sperm nucleus fertilizes the egg cell, giving rise to the embryo. A second sperm nucleus from the same pollen fuses with the two central cell nuclei of the female gametophyte. The triploid cell that results from this fertilization event will develop into a distinct tissue, the endosperm. As in most species both the central cell and the egg originate from a single meoitic product, they are genetically identical. The same holds true for the two sperm in the pollen. As a consequence, embryo and endosperm are genetically identical, their only difference being that the embryo is diploid, containing one maternal and one paternal set of chromosomes, while the endosperm is triploid, with two sets of maternal and one set of paternal chromosomes.

Despite their common ancestry, embryo and endosperm develop along very different pathways. One of the most dramatic consequences of imprinting in mammals is that embryos with a uniparental genetic contribution are inviable. Parthenogenetic mouse embryos, which are derived by the activation of an oocyte and therefore contain only a maternal genome, never develop to term. Androgenetic embryos, derived by nuclear transfer of two male gamete nuclei, and gynogenetic embryos, derived by the fusion of two female gamete nuclei, also result in abortion. The phenotypes of all three types of embryo in mice show a parent-of-origin effect. In parthenogenetic and gynogenetic mice, the embryo proper is relatively well developed, but the extraembryonic tissues are poorly developed or even absent. In androgenones, on the other hand, the extraembryonic tissues are well developed. The underlying reasons for the differences between androgenetic and parthenogenetic/gynogenetic cells have been extensively studied in chimeric embryos.

In chimeras between parthenogenetic and androgenetic cells, the parthenogenetic cells are confined to the embryo while androgenetic cells constitute the bulk of the extraembryonic trophoblast. Both cell types contribute to the yolk sac. Androgenetic cells have a higher proliferation rate than gynogenetic or parthenogenetic cells. While the presence of androgenetic cells in chimeras increases the weight of the embryo, the presence of parthenogenetic cells decreases embryo weight by 30-50%. Thus, even though paternally derived cells are located in the extraembryonic tissues, their presence leads to an increase in embryo weight. In contrast to the situation in mammals, in flowering plants embryos with a genome inherited from either the seed parent or the pollen parent are viable in many species. Parthenogenesis has long been known to occur in, for instance, dandelion (*Taraxacum sp.*) and hawkweed (*Hieracium sp.*) and in buttercup (*Ranunculus sp.*).

In agamospermous *Taraxacum* microspecies, a defective meiosis leads to an oocyte with an unreduced number of chromosomes. In some hawkweed species the megaspore is replaced by a cell of the nucellus or the chalaza. Gynogenesis has been reported for many species, including commercially important plants such as onion (*Allium cepa*), durum wheat (*Triticum tugidum*), sugar beet (*Beta vulgaris*), and pot gerbera (*Gerbera jamesonii*). Similarly, plants with only a paternal genome contribution can be derived via the process of anther culture. By culturing anthers or pollen on appropriate media, pollen cells can develop into haploid embryoids and eventually grow into plants. In most species, androgenetic plants are derived from the vegetative cell of the pollen. This has been reported in, for example, oil seed rape (*Brassica napus*), maize (*Zea mais*), tulip (*Tulipa gesneriana*), wheat (*Triticum aestivum*), and barley (*Hordeum vulgare*).

Androgenesis via the sperm of the pollen is less frequently observed, but has been reported for henbane (*Hyoscamus niger*) and carrot (*Daucus carota*). One has to bear in mind that most flowering plants are hermaphrodite and thus act as both seed and as pollen parents. With regard to this, it can be argued whether embryos derived from the vegetative pollen cell (which is not a gamete insofar as it is never involved in fertilization) can be called true androgenones. However, the fact that plants obtained from haploid or diploid embryos derived from one parent are viable indicates that, in contrast to mammals, genomic imprinting has little or no severe impact on the development of the Angiosperm embryo. The correct genomic constitution of the endosperm, however, is of great importance for seed development.

With a few rare exceptions, restricted mainly to the Asteraceae family, apomictic plants still need pollination and fertilization of the central cell to form a functional endosperm. Strikingly, in many species, shifting the usual balance of the two maternal genomes to one paternal genome in the endosperm results in parent-of-origin effects analogous to those described above for mice. Thus, it seems that though in the early development of the seed of flowering plants both the embryo and the endosperm are of vital importance, genomic imprinting in most species affects the development of the endosperm and not of the embryo. Where there are effects on embryo development, these are thought to be of secondary origin and as a result of imprinting effects on the endosperm.

FUNCTION OF THE ENDOSPERM

In contrast to Gymnosperms, in which the female gametophyte is a source of nutrients for the developing embryo, the ovule in flowering plants contains limited food reserves. In most Angiosperms, the endosperm has taken over this embryo-feeding role. Most probably, in dicotyledons the embryo can develop without nutrient influx from the endosperm until it has reached the globular or the heart stage. From this stage on, the endosperm apparently becomes indispensable as it functions as a sink and a source that acquires resources from maternal tissues for later use by the developing or germinating embryo. In this respect, the endosperm performs an analogous function to the placenta in mammals.

ENDOSPERM DEVELOPMENT

Within the Angiosperms there is a great variety in size, structure, and development of the endosperm. At its maximum size, the endosperm can constitute the bulk of the mature kernel as in, for instance, maize and other cereals. In the other extreme, the endosperm can be very small indeed, as for example in the orchids, where, after initial fertilization of the central cell nuclei little or no proliferation takes place. In general, two types of endosperm can be distinguished. The *persistent* endosperm is maintained into the mature seed and functions as a nutrient store for the seedling during and shortly after germination. This type of endosperm is found in many monocotyledonous (monocot) plant species, for instance, cereals.

Many dicotyledonous (dicot) seeds have *transient* endosperms, which reach their maximum size before the seed reaches maturity and are then consumed by the growing embryo. In mature seeds of such plants, the endosperm is thus reduced to a few cell layers or may be even completely absent. As an example of an transient endosperm, the growth and development of the endosperm of *Arabidopsis thaliana* will be discussed.

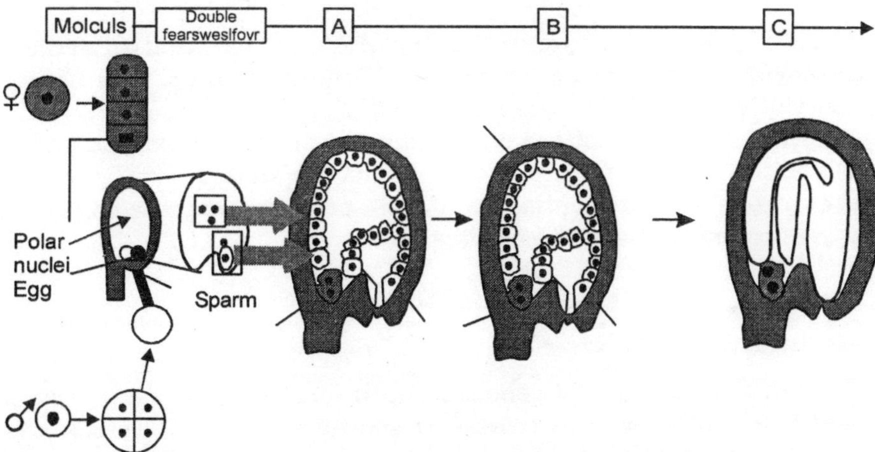

Fig. 14.1 : Double fertilization and seed development in Arabidopsis thaliana. Meiosis in the female germline gives rise to an ovule containing eight genetically identical, haploid nuclei. In the male germline, a pollen is formed containing three genetically identical, haploid nuclei: two generative sperm nuclei and one vegetative nucleus. Fertilization of the egg by a sperm gives rise to a diploid embryo. Fertilization of the two polar nuclei of the central cell by the second sperm results in the formation of a triploid primary endosperm nucleus. Replication and division of this nucleus leads to the formation of a syncytial endosperm lining the inner wall of the seed coat (A). Three different types of endosperm can be distinguished: chalazal, micropylar, and peripheral. When the embryo is in late heart stage, the micropylar and peripheral endosperm cellularise (B). In a mature seed, the embryo has absorbed virtually the whole endosperm (C).

Endosperm Development in *Arabidopsis Thaliana*

After fertilization, the now triploid primary endosperm nucleus replicates and divides. As in many other species, *Arabidopsis* endosperm development is initially free nuclear: the mitotic cycle does not include cytokinesis. Consequently, the central cell forms a syncytium, which can consist of hundreds of individual nuclei. The endosperm differentiates into three distinct types. At the micropylar pole, micropylar peripheral endosperm surrounds the embryo and the suspensor. Central peripheral endosperm constitutes the bulk of the endosperm. A third type of endosperm develops at the chalazal pole. This chalazal endosperm is easily distinguished from both other types because of the high density of its cell mass. The chalazal endosperm is located adjacent to the chalazal proliferating tissue, a maternal tissue that lies adjacent to the site of nutrient unloading from the vascular system. This location suggests that the chalazal endosperm may be involved in translocation of nutrients into the endosperm.

Further indications for a nutrient-importing function chalazal endosperm come from seed ultrastructure studies in several species and radiolabeling of water-insoluble photoassimilates in developing soybean seeds. This is reminiscent of the situation in cereals, where the cells in either the chalazal and/or the micropylar endosperm develop into special transfer cells that have extensive wall ingrowths into the maternal tissue. In these haustoria the surface area available for transport is significantly increased. In *Arabidopsis,* the central cell expands as the peripheral endosperm proliferates.

When the embryo is approximately in late heart stage, the seed reaches its maximum size. Starting at the micropylar pole, the endosperm now cellularizes. First, anticlinal cell walls are formed. The nuclei lining the inner cell wall of the central cell replicate and divide once more and subsequently are separated by periclinal cell walls. This process is repeated by the newly formed second layer of nuclei, until the whole former central cell is filled with cellularized micropylar and peripheral endosperm. Endosperm cells, once formed, no longer replicate. Cellularization thus seems to mark the end of endosperm proliferation. The chalazal endosperm cellularizes later. Exceptions do occur, however: in the pea, the endosperm never cellularizes but is absorbed by the embryo in the free-nuclear state. From the time of endosperm cellularization onwards, the developing embryo absorbs the endosperm, and presumably assimilates the endosperm cells and its contents.

GENOMIC IMPRINTING

Seed Failure in Interploidy Crosses

The first clues to the existence of genomic imprinting in plants came from the study of seed development following interploidy crosses. In some plant species, crosses between diploid and tetraploid plants give rise to viable seed. However, in most plant species, $2x \times 4x$ and $4x \times 2x$ crosses result in seed failure. As this was found to be the case even between diploids and their colchicine-derived autotetraploids, the realization grew that seed failure must be the result of quantitative rather than qualitative differences between the genomes contributed by either parent. In the course of the last century, different hypotheses were proposed to explain seed failure in interploidy crosses. Mintzing suggested that the ploidy balance between maternal: endosperm: embryo tissues was critical for normal seed development. Any deviation from the observed 2:3:2 ratio was predicted to result in seed failure. Although this theory seems to have been widely accepted and referred to over a long time, as early as 1932, Watkins was able to rule out the importance of the ploidy of the maternal tissue.

Working with diploid and autotetraploid *Primula* and *Campanula* species, he came to the conclusion that it is the endosperm: embryo relation that is important. His finding that the relative ploidy of the maternal tissue is of no importance for the success of seed development was confirmed by Howard in 1939. Using diploid and autotetraploid *Brassica oleracea,* he found that seeds from a $4x \times 2x$ cross develop to a comparable size to exceptional $2x \times 2x$ seeds in which the mother plant had produced diploid instead of haploid embryo sacs. While the maternal tissue ploidy differs between the crosses, being 4x for the first and $2x$ for the latter, in both cases the embryo: endosperm ratio is $3x$: $5x$. Cooper and Brink were the first to report that the seed failure in interploidy crosses is due to failure of the endosperm itself. They proposed that embryo death is only a secondary effect of a disturbed endosperm development in interploidy crosses. Cooper and Brink were among the first to make a detailed study of the initial development of embryo and endosperm in crosses between diploid and tetraploid plants. This would eventually shed light on the impact of genomic imprinting in plant development.

Interploidy Crosses

In the early 1950s, Hakansson and other researchers provided detailed descriptions of the phenotypes of seeds produced in reciprocal crosses between diploid and tetraploid plants, in both monocot and dicot species. It became apparent that in many species such seeds show specific parent-of-origin effects dependent on the ploidy of the seed and the pollen parent. These characteristics of $2x \times 4x$ and $4x \times 2x$ crosses were reciprocal in nature. In 2x x 4x seeds, Hakansson emphasised the following peculiarities of the endosperm:

1. Mitotic irregularities were frequently observed. In both rye (*Secale cereal*) and barley (*Hordeum vulgare*), endomitosis and resulting giant nuclei were observed in the endosperm. Mitotic endosperm irregularities were also commonly seen in the offspring of interploidy crosses of the dicot *Galium mollugo* and in interploidy crosses between different species within the genus *Triticum* (61) or within the genus *Avena*.
2. Cellularization of the syncytial endosperm is delayed. In barley and maize a slight delay in the onset of formation of cell walls was observed. In $2x \times 4x$ rye seeds, however, there was no sign of cellularization, or at its best it was very much delayed, even though high numbers of endosperm nuclei were achieved.
3. In some species, abnormal meristematic activity was observed in the mycropylar region of the endosperm. Growth of this part of the endosperm resulted in enclosement of the embryo in endosperm tissue both in barley and in maize. This is reminiscent of the situation in the dicot *Galeopsis pubescens,* where in $2x \times 4x$ seeds the micropylar haustorium is well developed and shows extraovular outgrowth with large nuclei. This may indicate a tendency for enhanced development of those (basal) parts of the endosperm that are in close contact with maternal tissues and presumably play an important role in the import of nutrients into the endosperm.
4. Deposition of starch occurred late compared to $2x \times 2x$ seeds, and usually in small quantities, for example, in barley and in maize. This late onset of starch synthesis and deposition may be correlated with the delay in cellularization observed in these species.

In endosperm of the reciprocal $4x \times 2x$ cross Hakansson noted the following characteristics:

1. Mitotic irregularities were rare or absent.
2. Cellularization occurred early in many species, including maize, barley, and rye. Fagerlind reported in 1937 that "cell walls in a $4x \times 2x$ cross in the dicot *Galium mollugo* were formed earlier than I ever observed in any other *Galium-cross.*"
3. Abnormal meristematic activity was not observed; the volume of the endosperm remains small and mitosis stops early.
4. Starch deposition began early.

Although all interploidy crosses described above resulted in seed failure at some stage during development, the early stages of endosperm development displayed a strong indication of a parent-of-origin effect pattern. High ploidy x low ploidy crosses showed a tendency to form a small endosperm with a low cell cycle activity that cellularized early, while seed of the reciprocal cross had a tendency toward a bigger endosperm in which a longer and more active mitotic stage was correlated with a later onset of cell wall formation. As for the cause of these phenotypes, Hakansson and others still believed this had be sought in the different ploidy ratios of embryo and endosperm.

Parental Ratios

It was not until the 1960s that the emphasis finally shifted from the ploidy ratio between different tissues toward the balance of paternal and maternal genomes within the endosperm itself. In 1962, von Wangenheim published his results on the development of endosperm resulting from interploidy crosses in the evening primrose, *Oenothera*. *Oenothera* is an exception within the Angiosperms in that its central cell has a single haploid nucleus rather than the usual two.

As a consequence, after pollination and double fertilization, both embryo and endosperm have the same number of chromosomes. Thus, *Oenothera* provided a unique means of testing the importance of ploidy ratio of maternal, embryo, and endosperm to the development of the seed. Both a $4x \times 2x$ and a $2x \times 4x$ cross resulted in a $3n$ embryo and a $3n$ endosperm. In the former cross both "zygotes" consisted of two sets of maternal and one set of paternal and in the latter of one maternal and two paternal sets of chromosomes. Despite both reciprocal crosses having the same ploidy and the same endosperm: embryo ratio, the phenotypes of the developing seeds and endosperms were strikingly different. In line with the results of interploidy crosses in cereals described above, $2x \times 4x$ crosses in *Oenothera* resulted in seeds with large, late-cellularizing endosperms, in whose nuclei endopolyploidy was frequently observed. In many seeds the chalazal endosperm was overproliferated. On the other hand, $4x \times 2x$ seeds developed small, early-cellularizing endosperms with a small chalazal endosperm.

Von Wangenheim in excluding a role for the ploidy level of the maternal tissue, concluded that the different parental contributions to the endosperm were responsible for the opposite phenotypes in the reciprocal crosses. Since the paternal: maternal ratio represented the only difference between the genomes of a $4x \times 2x$ and a $2x \times 4x$ seed von Wangenheim proposed that the origin of the differences had to be extrachromosomal. Extrapolating a theory proposed by Kihara and Nishiyama, he first speculated that pollen, in addition to its genome, contributed a certain unknown active compound that inhibited the formation of cell walls as well as differentiation of parts of the endosperm. Only after a series of mitoses and the accompanying growth of the endosperm syncytium would this compound have become so diluted that cellularization finally could take place.

In order for this theory to be consistent, the concentration of this compound had to be related to both the ploidy of the pollen nucleus and that of the recipient central cell nucleus. However, calculations of how the concentration of this hypothetical compound would decrease in the course of endosperm development in both $2x \times 4x$ and $4x \times 2x$ seeds could not explain the reciprocal outcomes observed in both crosses. A second theory, which states that it is the embryo sac that carries an extrachromosomal compound that stimulates the formation of cell walls, proved even harder to defend. Von Wangenheim finally concluded that "the observed phenomena are more likely to be caused by an extrachromosomal compound which is capable of autoreduplication. Von Wangenheim's observation that it was the ratio of paternal to maternal genomes in the endosperm that is critical for endosperm development was confirmed in 1966 by Nishiyama and Inomata. Working with diploid and autotetraploid *Brassica,* they concluded that a maternal:paternal ratio of 2:1 is required for normal development. Deviation from this ratio led to endosperm dysfunction and hence to seed failure. A more restricted theory was proposed by Sarkar and Coe who from their work with maize suggested that the triploidy of the endosperm itself is of vital importance for proper endosperm development. A definitive answer to the question of what is required for normal endosperm development was provided by the *indeterminate gametophyte* (*ig*) mutation in maize.

Maize plants carrying this mutation are aberrant in the formation of polar nuclei. This may result in the production of female gametophytes with extra polar nuclei in the central cell. Kermicle was able to obtain a number of viable triploid plants from a cross between a diploid *ig/ig* and a tetraploid *wt* pollen donor. He explained this by assuming that those plants resulted from seeds that combined a triploid (1 maternal:2 paternal) embryo with a hexaploid (4 maternal: 2 paternal) endosperm, in which the required 2:1 balance was restored. This hypothesis was confirmed by Lin in 1984.

Using diploid *ig/ig* females and both diploid and tetraploid pollen donors, he obtained diploid as well as triploid embryos, which were combined with a range of different endosperm karyotypes. Both triploid and hexaploid endosperms could yield normal seeds, but only if the maternal: paternal contribution was either 2:1 or 4:2. Endosperms with a ratio of 5:1 or 2:2 invariably failed to produce viable seeds. Strikingly, tetraploid endosperms with an aberrant constitution of 3 maternal: 1 paternal genomes were also found to give rise to viable seeds. Although such seeds were smaller than seeds with a 2:1 or a 4:2 endosperm, their survival indicates that although a 2:1 dosage of maternal and paternal genomes in the endosperm is required for normal seed development, at least in certain species small deviations from this ratio are tolerated.

In this regard it is noteworthy that whereas in maize an endosperm with one extra set of maternal genomes (3:1) is viable, the reciprocal endosperm, that is, with an extra set of paternal genomes (2:2), is not. This is reminiscent of the outcome of interploidy crosses of other species: there are several reports of interploidy crosses that fail to give viable seed when the pollen parent has the highest ploidy, even when the reciprocal cross may be successful.

Arabidopsis Thaliana

In *Arabidopsis sp.,* Redei analysed seed set following interploidy crosses. He found that $4x \times 2x$ crosses produced a high proportion of good seed, but the reciprocal $4x \times 2x$ cross resulted in a low percentage of viable seeds. However, the first study in which seed development following interploidy crosses was analyzed in detail was by Scott *et. al.* They found that in contrast to the findings of Redei and unlike the closely related *Brassica* species in *Arabidopsis* crosses between diploid and tetraploid plants in either direction gave rise to a high percentage of viable seeds. This offered the opportunity to study seed development in such crosses in detail from the moment of fertilization until the seed reaches maturity. Although viable seeds were produced in both $4x \times 2x$ and $2x \times 4x$ crosses these seeds were very different from each other, as well as from balanced seeds produced by self-pollinated $2x$, $4x$, or 6x plants, in final size, weight and development of the different parts of the endosperm. Seeds of a $4x \times 2x$ cross are significantly smaller and lighter than $2x \times 2x$ seeds, while a $2x \times 4x$ cross gives rise to bigger and heavier mature seeds.

Using Feulgen staining combined with confocal microscopy, the development of embryo and endosperm in the different crosses was analyzed. In $4x \times 2x$ crosses, both endosperm mitosis and embryo differentiation were found to be slower than in balanced crosses. Endosperm cellularization began slightly early compared to balanced crosses. As a result of this, cellularization led to a small endosperm with a limited number of large cells. The chalazal endosperm remained small and binucleate and often began to collapse after 5 DAP (days after pollination), when the embryo is still only at the heart stage. Seeds with a paternal excess showed complementary phenotypes to those with a maternal excess, that is, endosperm hyperplasia. In $2x \times 4x$ crosses, embryos developed at about the same rate as in balanced crosses.

The central peripheral endosperm, however, underwent accelerated mitosis, giving rise to a large number of peripheral endosperm nuclei. The endosperm also cellularized late, so that when cytokinesis finally occured there were many, small endosperm cells. The chalazal endosperm was often enlarged and vacuolated. Chalazal nodules, peripheral endosperm protoplasts that developed characteristics of the chalazal endosperm could also be very large. Although crosses between diploid and tetraploid plants yield viable seed interploidy crosses between diploid and hexaploid nearly always resulted in seed failure. For the first few days, embryogenesis in $2x \times 6x$ crosses proceeded at about the same rate as in $2x \times 4x$ crosses, but never passed the heart stage.

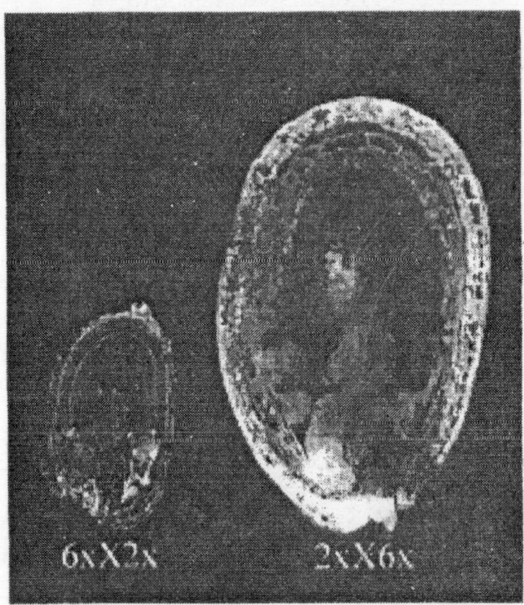

Fig. 14.2 : The effect of imbalanced crosses on seed development in Arabidopsis thaliana. Left, a seed from a $6x \times 2x$ cross at DAP 5. Characteristic for this extreme maternal excess phenotype are the absence of chalazal endosperm and a small seed with few peripheral endosperm nuclei. The peripheral endosperm has cellularized prematurely, resulting in a few, large cells. Right, a seed from a $2x \times 6x$ cross at DAP 5, with a large, uncellularized peripheral endosperm containing many nuclei and a massive chalazal endosperm with a number of chalazal nodules. In both crosses the embryo is at the globular stage. Seeds from both crosses fail to germinate.

Hyperplasia of the endosperm was very dramatic. The central peripheral endosperm divided rapidly without cellularizing. Both micropylar and chalazal endosperm, as well as chalazal nodules, became hugely overgrown and vacuolate, eventually engulfing the embryo and filling the seed. Seeds from $6x \times 2x$ crosses had a similar but more extreme hypoplasic phenotype than those from $4x \times 2x$ crosses. The peripheral endosperm cellularized early, and embryos aborted by the globular to heart stage. No distinct micropylar peripheral endosperm was seen, and the chalazal endosperm disappeared by 5 DAP. The results from the interploidy crosses in *Arabidopsis* are very much in line with the earlier observations on initial seed development in other species, discussed above.

Seeds with double the number of paternal relative to maternal genomes show accelerated mitosis and delayed cellularization of the endosperm, have a well-developed chalazal endosperm, and are abnormally large at maturity. In contrast, those containing a double dose of maternal genomes exhibit reduced endosperm mitosis and precocious cellularization. They develop small chalazal endosperms and are abnormally small at maturity. In all cases, embryo development appears roughly normal (though slow when there is a maternal excess). As others working predominantly on maize had previously, Scott et al. concluded that the observed parent-of-origin effects were best explained by assuming that endosperm development requires the activity of imprinted genes.

Imprinting in Flowering Plants

The results of the interploidy crosses in *Arabidopsis* and other species are in accordance with the parental conflict theory of Haig and Westoby. This model predicts that maternally and paternally derived alleles will be selected to have opposite effects on endosperm (and, hence, ultimately embryo) growth. Adding paternal genomes to the seed is expected to provide extra doses of the uniparentally expressed alleles that increase seed size, while extra maternal genomes are predicted to provide an excess of alleles that limit seed size.

The *Arabidopsis* results provide support for the parental conflict theory, as *Arabidopsis* seeds with double the normal dose of paternal genomes produce large endosperms and embryos, whereas those containing a double dose of maternal genomes have the opposite effect. There is an apparent contradiction in the assumed presence of genomic imprinting in *Arabidopsis thaliana*. As this species is an inbreeding plant that in the wild reproduces almost solely by self-fertilization, it unites mother and father in a single individual. As in this case one can hardly speak of a "parental conflict," it seems there should be no need for uniparentally expressed genes. However, it is probable that *A. thaliana,* like other inbreeding plants, evolved from outcrossers. Therefore it is suggested that A *thaliana* retains a parental imprinting system inherited from outcrossing ancestors, which has partially broken down as inbreeding has become predominant, allowing development of viable seeds with a limited degree of maternal or paternal excess. The advantage of using *Arabidopsis* in studies of imprinting and parent-of-origin effects lies in the fact that interploidy crosses in this species do lead to viable seeds.

It demonstrates that, by influencing the development of the transient endosperm, eventually the embryo and thus the whole seed are affected by altering the ratio of maternal to paternal genomes in the endosperm. In other words, extra paternal genomes, which are expected to contribute active copies of maternally imprinted genes, stimulate endosperm growth and development, leading to a bigger endosperm with most probably a larger capacity for storing nutrients. As a result, the embryo in such a seed will reach a larger size at maturity.

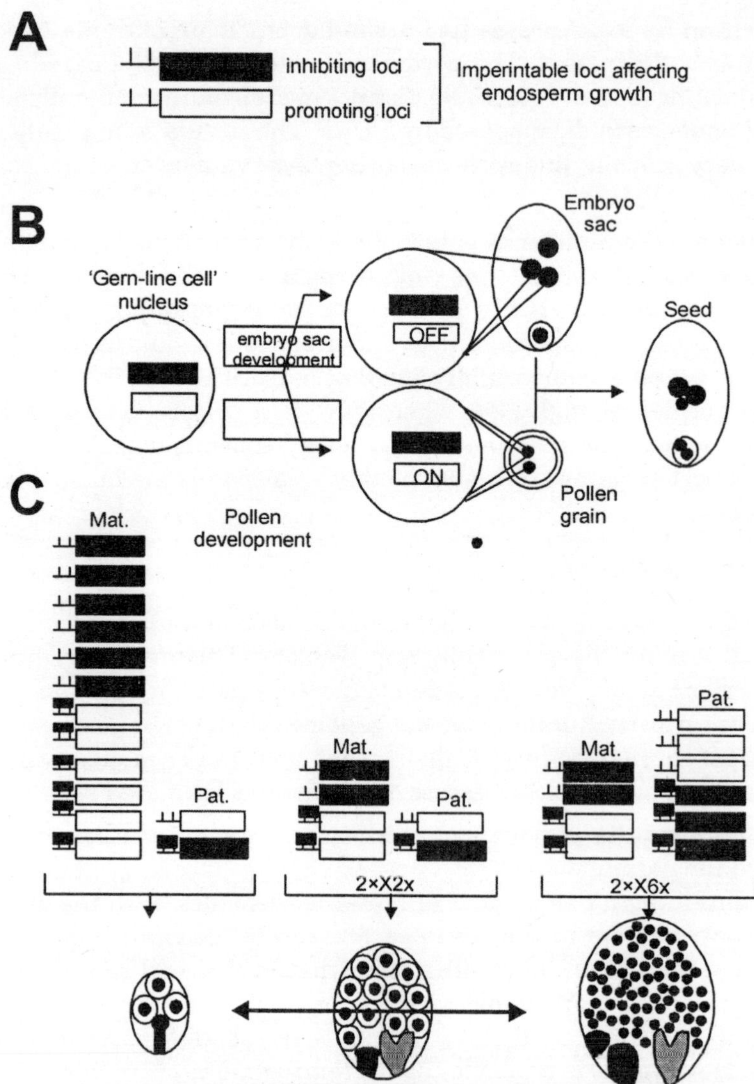

Fig. 14. 3 : (*Opposite page*) Model of the effect of parental genome dosage on seed development in flowering plants. (*A*) In this model it is assumed that some loci affecting endosperm growth are imprintable, while some are nonimprintable. Each of these types can be subdivided into loci that promote endosperm growth (*white*) and those that inhibit (*black*). Here we represent only the imprintable loci (vertical lines on upstream regions are sites for imprint associated methylation). A crucial aspect of the model is that imprintable growth-promoting genes are maternally inactivated, while imprintable growth inhibiting genes are paternally inactivated. (*B*) Each "germline" nucleus contains both classes of loci. Germ cells in flowering plants are derived from somatic cells late in development. There is no sequestered germline, and inactivation via imprinting is presumed to occur during gametogenesis (the polar nuclei as well as the egg are considered as female gametes here). In this model, polar nuclei transmit inactivated growth-promoting genes (*white*) as well as active growth-inhibiting genes (*black*), while sperm transmit the reciprocal. Both polar nuclei and sperm also carry potentially active growth-promoting and inhibiting genes that are not imprintable (not shown). Double fertilization yields a triploid endosperm containing two maternal genomes and one paternal genome (as well as a diploid embryo). (*C*) Crosses between individuals of different ploidies generate unbalanced endosperm, which results in abnormal seed development. In the model, this is explained in terms of the relative numbers of active imprintable growth-promoting and inhibiting genes in the endosperm. Paternal excess generates endosperms with a high *active positive: negative* ratio (>1:2) and consequently increased vigor because sperm genomes contribute active growth-promoting genes but only inactive growth-inhibiting genes. Maternal excess has the reciprocal effect.

MECHANISMS OF IMPRINTING

Little is known about the parental imprinting mechanism in plants, although there is evidence that, as in mammals, DNA methylation is involved. In maize endosperm, imprinted *zein* genes are expressed only when inherited from the seed parent, and these loci are methylated at fewer sites on maternally then paternally derived chromosomes. Differential methylation also corresponds with parent-specific expression of the *R* locus. Recently, Adams *et. al.*, analyzed seed development in crosses in which the genome of one or either parent had undergone extensive demethylation. Transgenic *Arabidopsis* lines expressing the *Methyltransferase I* antisense gene (*METI* a/s) have only 15% of the methylation level of a wild-type plant. If DNA methylation is indeed essential to the imprinting mechanism in *Arabidopsis,* and if the antisense transgene prevents imprinting-specific methylation, one would predict that hypomethylated plants produce gametes in which imprinted alleles have lost most or all of their silencing.

Indeed in crosses using *METI* a/s plants as one of the parents, both seed weight and the development of the endosperm showed that hypomethylation closely phenocopies the effect of interploidy crosses. *METI* a/s × 2*x* seeds had a strong paternal excess phenotype with high seed weight, many endosperm nuclei, delayed endosperm cellularization, and overgrown chalazal endosperm, although both parents were diploid, and the seed was nourished by a hypomethylated mother that suffers a variety of defects in vegetative and floral development. This behaviour is consistent with a model in which hypomethylation of the maternal genome in *METI* a/s plants has prevented silencing of endosperm-promoting genes that would normally be expressed only from the paternal genome.

Meanwhile, the wild-type paternal genome contributes its normal complement of silenced endosperm-inhibiting genes and active endosperm-promoting genes. The net effect, according to the model, is that endosperm has an excess of imprinted genes that behave as if they were inherited from the father, thus phenocopying an excess of paternal genomes. As predicted, seed resulting irom a 2*x* × *METI* a/s cross phenocopied the maternal excess obtained in a 4*x* × 2*x* cross. Hypomethylated pollen gave rise to small seeds with fewer peripheral endosperm nuclei and a small chalazal endosperm. Cellularization in such endosperms was early compared to wild-type seeds. This phenotype can be explained by assuming that in the pollen donor, the activity of the *METI* a/s gene has erased genomic imprints from, or prevented silencing of, paternally imprinted genes, thereby reactivating these genes.

As paternally imprinted genes are predicted to have an endosperm-inhibiting effect, the endosperm genome after fertilization will have an extra set of active endosperm-inhibiting genes. In other words, demethylation of pollen genomes "maternalizes" such genomes, whereas demethylation in female gametes leads to "*paternaliazation*". Crosses with demethylated plants that were hemizygous for the METI a/s gene (hemi-Met) demonstrated that the determining factor is the demethylated status of the gamete genome and not the presence of the transgene. Seed development in offspring of crosses with a hemi-Met as a parent showed the expected, above-described phenotypes, independently of the presence or absence of the *METI* a/s transgene in the seed.

It is not known in which way methylation may lead to gene silencing in flowering plants. In a number of plant species, transgene silencing has been found to be correlated with a change in both methylation level and in chromatin structure. In *Petunia hybrida,* the maize *Al* transgene is expressed in some lines but silent in others. In *Al*-expressing lines the transgene is hypomethylated and sensitive to DNase I and nuclease S7 digestion. In lines where the transgene is silent, however, the locus is hypermethylated and significantly less sensitive to digestion, suggesting a more condensed chromatin structure on the transgenic locus. Likewise, in *Arabidopsis,* transgene silencing of the *HPT* gene is correlated with both increased methylation and an increased resistance to DNase I and micrococcal nuclease, again indicating a change in chromatin structure. These data suggest that, as in animals in plants methylation, chromatin structure and gene silencing may well be linked. This need not be always the case, however. Activation of the pea *rbcS* gene involves a change in chromatin structure, but not in methylation status. Recently, an *Arabidopsis* gene has been cloned that plays an important role in the maintenance of transcriptional gene silencing.

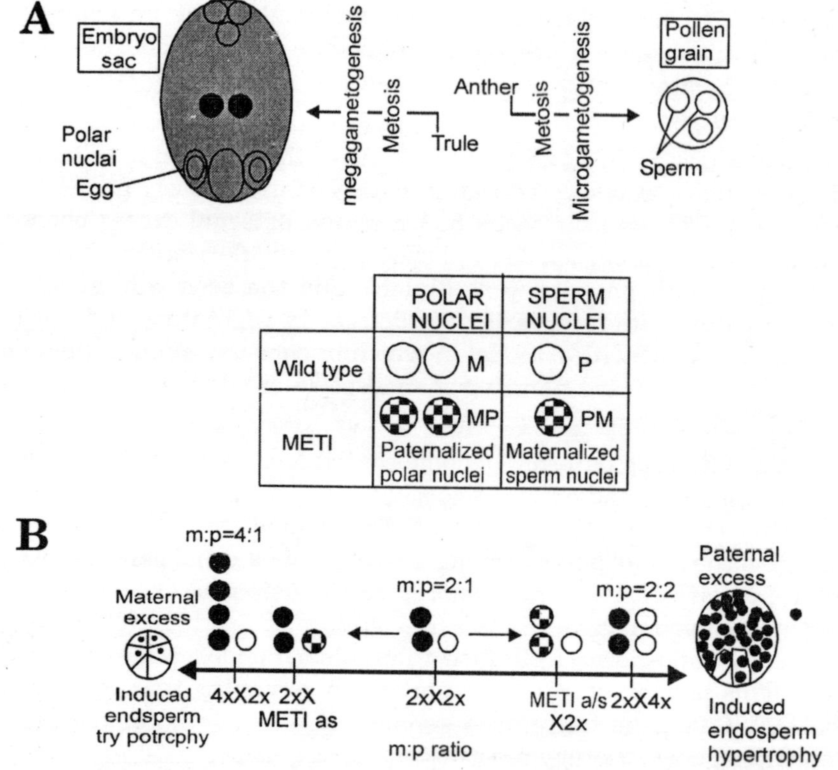

Fig. 14.4 : Model of the effect of global DNA hypomethylation on parental imprinting in Arabidopsis thaliana. (A) Normally, endosperms contain two copies of the maternal genome, contributed by the polar nuclei, and one copy of the paternal genome, contributed by the sperm. In the maternally inherited genomes, endosperm-promoting genes are expected to be imprinted and hence silenced. In paternally inherited genomes, these genes will be expressed, but endosperm-inhibiting genes are expected to be imprinted. When maternal genomes are contributed by a MET I a/s parent, the endosperm-promoting genes are expected to be largely derepressed, producing a "paternalized" genome. Similarly, a MET I a/s pollen parent is expected to contribute a "maternalized" genome. (B) Interploidy crosses (e.g., 4x × 2x or 2x × 4x) result in seeds with extra maternal or paternal genomes, and therefore extra doses of active maternal or paternal alleles of imprinted loci. Maternal or paternal excess has dramatic and complementary effects on seed development. A diploid MET I a/s parent does not contribute extra genomes but appears to contribute extra doses of active endosperm-promoting or-inhibiting genes, resulting in phenotypes similar to those produced by parental genomic imbalance.

A mutation in this *Morpheus 'Molecule* (*MOM*) gene leads to reactivation of transcriptionally silent transgenic loci. Interestingly, the methylation level of at least one of these loci does not change: even after nine generations, the reactivated *HPT* transgenic locus remains hypermethylated.

As in addition no change in methylation status of a 180-bp CEN repeat could be detected, it is likely that in a *mom* mutant background reactivation of silenced genes and methylation patterns are inherited independently. What is the function of *MOM* in gene silencing? The MOM protein may act downstream of methylation. It may for instance be involved in the link between methylation and the actual transcriptional silencing. Alternatively, it is also possible that *MOM* regulates gene expression in a separate, methylation-independent way. It will be interesting to determine what effect the *mom* mutation has on the chromatin structure of affected genes, and to what extent *MOM* regulates the expression of imprinted genes.

IMPRINTS IN ARABIDOPSIS ARE NOT ESSENTIAN FOR DEVELOPMENT

The seed development of plants with a hypomethylated parent suggests the importance of methylation and the activity of the *Methyltmnsferase I* gene in *Arabidopsis* in establishing imprinting-associated methylation. As inbreeding of demethylated plants yields viable seeds, one can also conclude that genomic imprinting is not a prerequisite for seed development in this species. Apparently, removal of all or most genomic imprints from both parental genomes does not prevent embryo or endosperm development. Jaenisch proposed that removal of imprints or of imprinted genes themselves should have few developmental consequences, as they exist in "'paired sets' of genes involved in the same pathway" (*e.g.*, of growth promoters and inhibitors, as predicted by Haig and colleaguesp.

This has been difficult to test in mammals as embryos with reduced methylation die during gestation. In contrast to mammals, in flowering plants one would expect to observe consequences for seed development if imprints are erased in both parents. It is the endosperm genome that is mainly subject to imprinting, and as described earlier, this genome consists of two maternally inherited and one paternally inherited genome copies. As a consequence of this, paternally imprinted genes are expressed from two copies of the maternally inherited genome. Maternally imprinted genes, on the other hand, are expressed from only one allele, inherited via the pollen donor. Thus, in the endosperm genome the bias of two sets of maternally expressed, endosperm-inhibiting genes to one of paternally expressed, endosperm-promoting genes is the normal, *"balanced"* situation that results in a wild-type endosperm.

Overall removal of imprints in both parents would shift this bias from a 2 : 1 ratio to a 3 : 3 ratio: all three copies of the endosperm genome are now allowing expression from formerly imprinted genes. This is predicted to mimic the effect of a $2x \times 4x$ cross, in which the endosperm genome consists of two maternal and two paternal copies of the genome. In other words, a *"paternal excess"* phenotype with a large endosperm containing many peripheral endosperm nuclei that cellularizes late and forms a large chalazal endosperm would be expected. What is seen, however, is that in *Arabidopsis,* seeds resulting from a *METI* a/s × *METI* a/s cross more closely resemble wild-type seeds than those produced by crosses between one *METI* a/s and one wild-type plant. They do, however, contain small chalazal endosperms and weigh less than wild-type seeds, both features indicating a maternal excess.

One aspect of the phenotype, however, is consistent with this prediction: the number of peripheral endosperm nuclei in *METI* a/s × *METI* a/s crosses is higher than in $2x \times 2x$ and about sixfold greater than in $4x \times 2x$ crosses. It is not known why sell-fertilization of hypomethylated plants leads to seeds that show a combination of maternal and paternal excess phenotypes, but several factors may contribute. First, demethylation is not complete in *METI* a/s plants, perhaps in part because of other methyltransferases in the *Arabidopsis* genome that are not affected by the *METI* a/s transgene. Partial demethylation could affect gamete genomes or individual sequences unequally.

In addition, some genes may even become hypermethylated in a *METI* a/s background, like the (nonimprinted) *SUPERMAN* locus (94). Finally, due to the complicated regulation of imprinted genes, global DNA hypomethylation in mouse can repress as well as activate imprinted alleles. It is conceivable that the same occurs in plants, although it seems more likely that the overwhelming effect of hypomethylation is to activate normally silent imprinted alleles.

IMPRINTED GENES

In contrast to the situation in mammals, where several well-characterized examples of imprinted genes are known, in flowering plants to date only a small number of genes are known or even suspected to be imprinted. In maize, four loci show a strong parent of origin effect and hence are likely to be imprinted. These are the *R* gene, which encodes a transcription factor active in the regulation of anthocyanin biosynthesis in the aleuron layer of the endosperm, the *delta zein-regulator* (*dzr*), a storage protein regulator a *zein* gene, and an *alpha tubulin* gene.

None of these genes seems to be involved in seed development. Recently, three mutants have been isolated in *Arabidopsis thaliana* that shed more light on the backgrounds of genomic imprinting in this species. All three genes, *MEDEA/FIS1/EMB173 F 152* (*Fertilization Independent Seed-I*), *FIS2* (*Fertilization Independent Seed 2*) and *FIE* (*Fertilization Independent Endosperm*)*/FIS3* are involved in the regulation of the development of the endosperm. *FIS2* contains a putative zinc-finger motif and three putative nuclear localization signals, suggesting that the FIS2 protein might be a transcription factor

MEDEA, FIS2, and FIE in Endosperm Development

Some mutant alleles *of FIE, MEA,* and *FIS2* confer a degree of autonomous endosperm development; in other words, the central cell proliferates and develops into an endosperm-like structure even in the absence of pollination and fertilization. Without being stimulated by fertilization, the central cell nucleus starts replicating and dividing and the central cell develops into a premature endosperm in the absence of embryo development. In addition to endosperm development, both the seed coat and the silique wall, which constitute maternal tissue, proliferate.

Autonomous endosperm development is limited: differentiation into either chalazal or micropylar endosperm does not take place. In an unpollinated mutant *fie* ovule, endosperm development arrests in the free nuclear stage. In contrast, the autonomous endosperm in a mutant *mea orfis2* ovule will cellularize. One of the functions of *MEA, FIS2,* and *FIE* must be to control the fertilization-dependent block of endosperm development in the female gametophyte. Mutant seeds cannot be rescued by fertilization with pollen carrying the wild-type allele.

One of the possible explanations would be to assume that the genes are paternally imprinted, in which case the wild-type allele carried by the pollen would be silenced. A closer look at what happens after fertilization of a mutant *mea or fie* ovule by wild-type pollen provides an insight into the processes regulating seed development. Pollination of an ovule carrying the putative loss-of-function *mea-3* allele with wild-type pollen leads to a seed in which the embryo arrests at the heart stage. In such seeds, the peripheral endosperm overproliferates without cellularization, resulting in endosperm with approximately 150% the number of nuclei produced by a wild-type seed.

Fertilization of a *fie* mutant ovule with wild-type pollen leads to a comparable phenotype. A *fie-I/FIE* heterozygote as seed parent pollinated by a wild-type pollen donor produces viable and shriveled seeds in a 1 : 1 ratio. In the plump seeds, which carry a wild-type maternal *FIE* allele, embryos reach maturity, and endosperm development appears to be identical to that in wild-type seeds. In shriveled seeds, inferred to be the result of the fertilization *of a fie* mutant ovule, the embryo becomes vacuolate and does not develop past the late heart-early torpedo stage, the endosperm fails to cellularize, and the chalazal endosperm undergoes massive overproliferation. Both *the fie* and *mea* phenotypes are reminiscent of the lethal paternal excess phenotype obtained in a 2*x* × 6*x* cross in *Arabidopsis*. If *MEA* and *FIE* are assumed to be involved in inhibiting proliferation of the central cell and endosperm, this would explain why these undergo extra divisions in both unpollinated and pollinated *fie-1* and *mea-3* mutants.

According to this model, the seed phenotypes of pollinated fie^{-1} and *mea-3* mutants on one hand and wild-type 2*x* × 6*x* seeds on the other all result from failure to inhibit endosperm proliferation. In the case *of mea-3 and fie-1,* the loss of wild-type MEA and FIE protein causes derepression of gene activity promoting endosperm growth. In the case of the interploidy cross, an overdose of paternal genomes contributes to the seed extra active alleles of genes that promote endosperm growth. The analogy between the *fie* and *mea* mutants and the (lethal) paternal excess in a 2xX6x seed can be extended. If either mutant leads to a lethal *"paternalizing"* effect on seed development, then introducing factors that *"maternalize"* seed development are predicted to shift back the paternal excess phenotype, possibly to such an extent that seed viability could be restored.

In the case *of MEA,* it has been found that *mea* mutant ovules can be rescued by pollination, provided that the resulting seed is homozygous for a loss-of-function mutant allele of the *Decrease in DNA Methylation* 1 gene *(DDM1).* In homozygous *ddml* mutants, the over all cytosine methylation has been reduced by 70% (However, *ddml* mutations do not affect methyl-transferase activity and *DDM1* has recently been found to be a member of the SWI2/SNF2 family of chromatin remodeling proteins. It is possible that hypomethylation or chromatin remodeling or both have activated the silenced paternal copy of the *MEA* gene. Alternatively, it is also possible that it is the altering of the methylation status of the sperm genome itself that rescues the maternal *mea* mutant, possibly by activating genes that function downstream of *MEA.* Ovules carrying *a fie-1* mutant allele can be rescued by hypomethylated pollen from a plant expressing the *METI* a/s transgene.

Interestingly, rescue of *fie* seeds was only obtained by using demethylated pollen carrying the wild-type *FIE* allele. It was concluded that an active wild-type paternal *FIE* allele is needed for the *METI* a/s-mediated rescue. Most probably, *FIE* is a paternally imprinted gene that is silent in wild-type pollen. In a hypomethylated background of the pollen genome, the normally silenced *FIE* allele apparently becomes reactivated.

The rescue of a maternal fie mutant by such pollen can be explained in two ways. It is possible that *FIE* is crucial for endosperm development. The paternally inherited, active copy of the *FIE* gene, probably in concert with the over all "maternalizing" of the paternal genome, inhibits endosperm development to such an extent that the lethal paternal overexcess phenotype caused by the maternal *fie* mutation is attenuated to a state in which the seed can survive. It is also possible, however, that the maternalization of the sperm genome is due entirely to the overall demethylation of the pollen genome, which is predicted to lead to removal of imprints from endosperm-inhibiting genes.

Indeed, as described earlier, demethylation of the genome of a pollen parent has been shown to have such an effect. An active *FIE* allele may be needed for proper embryo development. In this case, the failure of demethylatedfie pollen to rescue a mutant *fie* ovule would not be due to an endosperm, but to an embryo defect. With regard to this, it is interesting to note that while *mea/mea* and *fis2/fis2* homozygous plants do occur sporadically, despite extensive searching *fie/fie* plants have not been recorded. Rescued *fie-1* mutant seeds showed many features of paternal genomic excess, including large size and weight, and overproliferation and delayed cellularization of endosperm. Vielle-Calzada *et. al.*, also reported that rescued *mea/MEA* seeds were enlarged, containing large embryos and sometimes excess persistent and partially cellularized endosperm. These phe-notypes can be interpreted as indicating that reactivation of paternally silenced alleles—which in effect maternalizes the sperm genome—can compensate partially but not completely for the paternalizing effect of maternal *fie-1* or *mea-3* mutations.

The activity of wild- type *FIE* and *ME A* will probably inhibit (over)expression of endosperm-promoting genes, which would otherwise result in a lethal paternal excess phenotype. This endosperm-inhibiting activity might come too late to fully restore the mutant phenotype caused by the loss of *FIE* or *MEA* function in the female gametophyte, though. It is also possible, however, that the partial rescue is due to a gene dosage effect of the Polycomb proteins. The single copy of activated *FIE* or *MEA* from the paternally inherited genome might not be as effective as the two copies normally transmitted by the diploid central cell.

Uniparental Expression

Interestingly, *mea, fis2,* and *fie* are all maternal effect mutations: the aberrant phenotype is observed only when the mutant allele is inherited via the mother. A heterozygote seed in which the mutant allele is inherited via the father is indistinguishable from a homozygote wild-type seed. This could be explained by assuming that these three genes are subject to paternal imprinting and hence are expressed only from the maternal copy of the genome. There are several indications that for *MEA* as well as *FIS2,* in the endosperm genome the paternally inherited allele is silenced throughout the initial stages of seed development. Expression of β-glucoronidase (GUS) driven by either the *FIE* or the *FIS2* promoter could only be detected in the endosperm when the construct was inherited via the seed parent and not when it was inherited via the pollen parent. Vielle-Calzada *et. al.*, analyzed the expression of *MEA* in embryo and endosperm using *in situ* hybridization. Using a MEA-specific probe, they showed the presence of two nuclear dots in the polar nuclei both before and after fertilization.

Nuclear dots have been observed in mammals and insects. In *Drosophila,* they are indicative of the presence of nascent transcripts of actively expressed genes. The presence of only two nuclear dots after fertilization suggests that the paternal *MEA* allele remains silent after fertilization and that the observed dots correspond to the two maternal *MEA* alleles. No dots could be observed in the nucleus of the egg. Using parent plants with distinguishable *MEA* alleles, the parent specific expression of the *MEA* gene was analyzed. With reverse transcriptase polymerase chain reaction (RT-PCR) analysis on cDNA isolated from complete siliques at 54 h after pollination, when the embryos are in the midglobular stage, only expression of the maternally inherited allele could be detected. The authors conclude that in the early stages of seed development the paternally inherited *MEA* alleles are silent in both embryo and endosperm. Silencing of the paternal allele in the endosperm, but not the embryo, was confirmed by Kinoshita *et al.* They managed to dissect *Arabidopsis* seeds under a stereomicroscope into embryo and endosperm plus seed coat fractions at 4, 6, 7, and 8 DAP, corresponding to the heart, torpedo, walking stick, and early maturation stages of embryo development.

Pure endosperm fractions were obtained from seeds of 7 DAP. Using ecotypes that have distmguishable *MFA* alleles, the expression of both parental alleles in reciprocal crosses was analysed by RT-PCR. Expression of the paternal allele in whole-seed samples could be detected in all stages under study. Paternal and maternal *MEA* mRNA could be detected in embryos at 6, 7, and 8 DAP. In contrast, only expression from the maternal allele could be detected in endosperm plus seed coat material at 6 and 7 DAP, and in isolated endosperm at 7 DAP. From these observations, it can be concluded that during these stages the paternally inherited *MEA* allele is specifically silenced in the endosperm, but not in the embryo. The *MEA* gene thus provides the first example of a plant gene where an observed parent-of-origin effect on seed morphogenesis can be linked to and explained by genomic imprinting at the molecular level.

If the expression pattern of the *MEA* gene is representative of other paternally imprinted genes in flowering plants, it at least partially explains why the effects of interploidy and interspecific crosses are observed only in the endosperm and not in the embryo. Apparently the uniparental silencing of imprinted genes, at least during some stages of seed development, takes place only in the endosperm and not in the embryo. Interestingly, Vielle-Calzada *et. al.*, recently published data that suggest that the whole paternal genome in both embryo and endosperm may be silenced during the first 3 or 4 d after fertilization in *Arabidopsis thaliana*.

By screening a library of enhancer detector and gene trap lines expressing the *GUS* gene, they identified 19 transposants that show *GUS* expression in the developing seed, either in embryo, endosperm, or both. In reciprocal crosses between these transposants and wild-type plants, it became apparent that when the transposants were used as seed donors, *GUS* expression was detected from very early stages of seed development onwards. In contrast, when the transposants were used as pollen donors to fertilize wild-type ovules, *GUS* expression could only be detected in the seeds from up to 80 *h* after pollination, when the embryo is in the globular stage. Having shown that in at least one of these genes the absence of expression from the paternal allele is not related to transgene silencing, the authors concluded that in all genes under study, the paternal allele is silenced.

As the genes tested are distributed throughout the genome and represent a wide variety of functions, they proposed that most, if not all, of the paternal genome is silenced during early seed development in both embryo and endosperm. This overall silencing early during seed development may explain why Vielle-Calzada *et. al.*, could not detect expression of the *FIE* gene in the embryo at 54 h after pollination, whereas Kinoshita *et. al.*, detected embryonic expression from 4 DAP onwards. The reason for this delayed transcriptional activation of the paternally inherited genome is not clear, nor is the mechanism behind it.

The authors suggest that the silencing of the paternal genome probably occurs during sperm cell differentiation and may be related to either a tight packaging of the DNA into heterochromatin, or an alternative methylation level of sperm DNA. As a consequence of the overall silencing of the paternal genome in the initial stages of seed development, one would have to assume that at 3 or 4 DAP the barriers preventing expression from the paternal genome are removed. An exception would have to be made for the paternally imprinted genes in the endosperm genome. Alternatively, these would have to become silenced immediately after the activation of the paternal genome.

GENOMIC IMPRINTING

Both the *FIE* and the *MEA* gene encode *Polycomb group proteins*. In *Drosophila* the best-understood function of Polycomb proteins is to maintain transcriptional repression of homeotic genes through many rounds of cell division by forming complexes that modulate chromatin configuration or prevent access of transcription factors. Both *MEA* and *FIE* are expressed in flowers before fertilization and in developing siliques afterwards. The MEA protein contains a SET domain (so called because it was initially found in the *Drosophila* genes Suppressor of variegation, £nhancer of Zeste, and Trithorax), and shows homology to the SET domain polycomb protein Enhancer of Zeste [E(z)] in *Drosophila*. The FIE protein contains several WD40 motifs, and shows highest homology to the extra sex combs (esc) protein in *Drosophila* and the Embryonic ectoderm development (Bed) protein in mice and humans. The WD40 motif is thought to promote protein-protein interactions. The presence of multiple WD motifs allows the protein to be bind to multiple other proteins at the same time. WD polycomb proteins have been reported to be active in repressing expression of insect and mammalian genes during embryo development.

In *Drosophila* mouse and human it has been reported that WD polycomb group proteins interact with a SET domain polycomb group protein. ln *Drosophila* and in mammals, polycomb group proteins function to maintain a repressed state of homeotic genes. Polycomb group proteins function in complexes that bind to chromatin and downregulate gene expression through epigenetic silencing. The polycomb group proteins are thought not to initially repress expression, but to maintain a repressed state of already silenced target genes. The protein complexes formed by polycomb group proteins and other proteins interact with Polycomb-response elements (PRE) in the DNA. Silencing of the gene containing the PRE and presumably other genes in the vicinity as well is thought to take place though packaging the DNA in a higher level of condensation. Given the observation that loss of function mutations in *FIE, MEA,* or *FIS2* gives rise to highly comparable phenotypes, it is possible that the proteins of these genes also form a complex.

The absence of any one of them may lead to disruption or inactivation of the whole complex. In *Drosophila,* the WD40 polycomb protein esc is thought to form a complex that comprises other polycomb group proteins plus the zinc-finger transcription factor Hunchback. Interaction of this complex with the transcription machinery at the *Ultrabitho-rax* locus leads to silencing of the *Ubx* gene, most likely by packaging the DNA into a condensed chromatin form. As the WD polycomb group protein FIE is an *Arabidopsis* homolog of esc, it is not unlikely that it will perform a similar function in the plant's genome. In this case, FIE may form a complex with FIS2, a putative zinc-finger transcription factor, and the SET-polycomb protein MEA. In early *Arabidopsis* seed development, this complex would then negatively regulate the expression of endosperm-promoting genes from the central cell and endosperm genome.

FIE AND DNA METHYLATION

In plants that are heterozygous for the loss of function *fie-1* allele, proliferation of the central cell without pollination is seen in half of the ovules. This phenotype has been interpreted as showing that wild-type *FIE* represses endosperm development before fertilization. However, much of the endosperm developmental programme infie-1 mutants does not take place. For example, there is no regional specification of micropylar and chalazal endosperm, and cellularization of the peripheral endosperm does not take place. Therefore there is a block to complete endosperm development in unpollinated *fie-1* mutants.

Aside from a wild-type *FIE* allele, several components of normal seed development are missing from unpollinated *fie-1* ovules: pollination and fertilization themselves, and therefore gene expression they might trigger; and a paternally transmitted genome, which is not equivalent to maternal genomes because of genomic imprinting. Recently it was shown that demethylating *fie-1/FIE* heterozygotes using the *METI* a/s construct allowed autonomous endosperm to develop much further than previously reported. In half of the ovules of a hemizygous METI/a/s, unpollinated *fie-1/FIE* plant—presumably those carrying wild-type *FIE* alleles—the central cells did not proliferate.

This indicates that hypomethylation on its own does not promote fertilization-independent seed development. The other half of the ovules did show autonomous endosperm development. Those ovules fell into two classes. In type 1, the central cell usually underwent more rounds of mitosis than in normally methylated *fie-1* mutants, and the endosperm cellularized, but there was still little or no regional differentiation. In the type 2 seedlike structures, autonomous endosperm development went much further. These seeds resembled sexually produced seeds, containing more peripheral endosperm nuclei, which cellularized later than in type 1, and large chalazal and micropylar endosperms. In neither of the types were embryos found. The presence of two types of autonomous endosperm may be caused by the presence or absence of the *METI* a/s transgene in the ovule, or it may reflect a less direct effect of the transgene in the mother plant; for example, hypomethylation might affect different DNA sequences in different embryo sacs. In which way does hypomethylation relieve the partial endosperm block in *fie-1* mutants? One explanation would be that some genes necessary for full endosperm development are maternally imprinted, and that this is mediated by means of methylation. Demethylation would lead to activation of these genes, which would in effect supply the missing paternal genome.

As demethylation of the maternal genome alone does not lead to autonomous endosperm development, clearly not all components of this pathway are regulated by methylation. In normally methylated plants, fertilization of *a fie-1* ovule leads to a strong paternal excess. This, in concert with *FIE* encoding a polycomb group protein, suggests that *FIE* itself represses endosperm-promoting genes in the maternal gametophyte. So, it is possible that there are two (possibly overlapping) pathways of controlling endosperm-promoting genes in female gametophytes: (1) some genes are maternally imprinted by means of methylation; (2) another set of genes is controlled by the expression of the maternal *FIE* gene. Alternatively, *FIE* and DNA methylation could participate in repressive complexes at the same loci. *The fie-l* mutation alone may not be sufficient to completely release the expression of all these genes i

FIS COMPLEX CONFERS

Earlier we described how pollination of a mutant *fie* ovule with wild-type pollen results in seed abortion. The phenotype of the abortive seeds closely resembles that of a $2x \times 6x$ or a *METI* a/s $\times$ $4x$ seed: the embryo does not develop past the late heart-early torpedo stage, the chalazal endosperm undergoes massive overproliferation, and the endosperm fails to cellularize. One explanation for this observation is that seed abortion in *a fie* $\times$ $2x$ cross results directly from massive paternal excess in the endosperm. Since the abortive endosperms resemble those generated by a $2x \times 6x$ cross, we speculate that the m: p ratio must be close to 2:3. Consistent with this proposition is the fact that the lethal phenotype encountered in *fie* $\times$ $2x$ crosses can be rescued by providing *the fie* mutant ovule with demethylated instead of wild-type pollen, provided this carries a wild-type *FIE* allele.

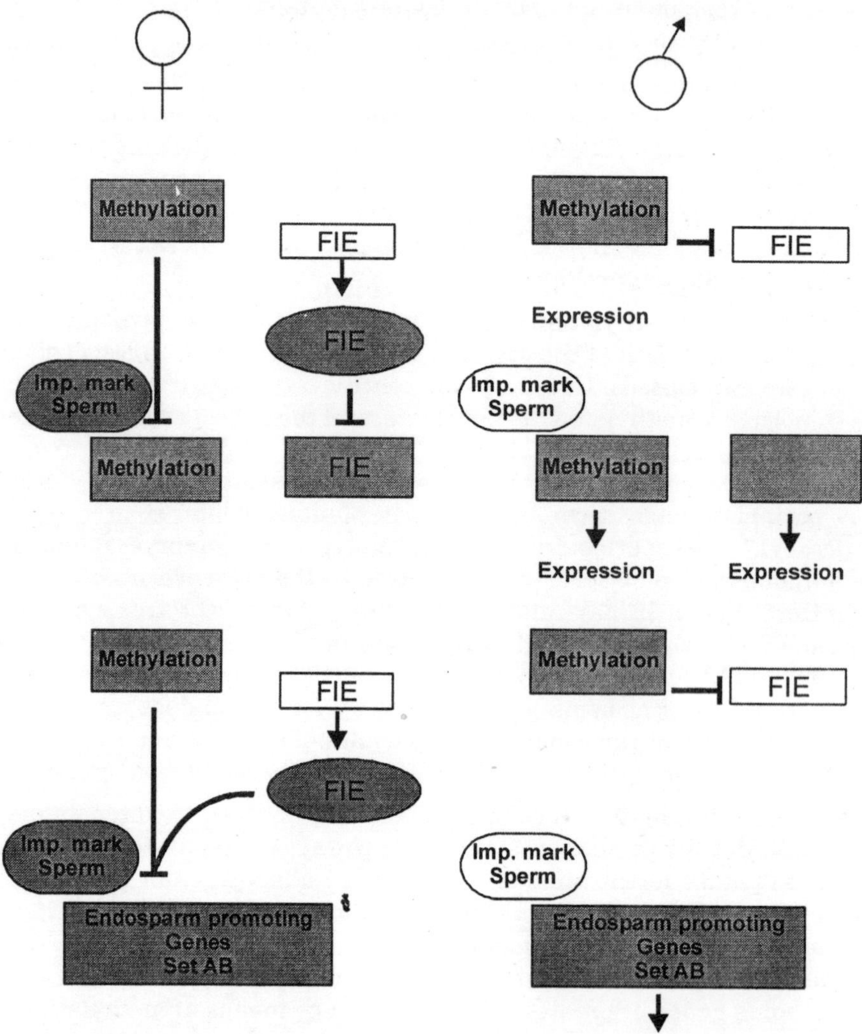

Fig. 14.5 : Hypothetical models for the role of FIE in the regulation of endosperm development in the ovule. (A) The FIE protein represses expression of a set of endosperm-promoting genes (Set B). Other endosperm-promoting genes are down-regulated by methylation (Set A). In the pollen genome, both sets of endosperm-promoting genes are expressed. As a (hypothetical) imprinting mark on the genes of Set A is absent in the male germline, the genes are no target for methylation and hence can be expressed. The FIE gene is paternally imprinted (probably involving methylation) In the absence of active FIE protein, the endosperm-promoting genes in Set B can be expressed as well. (B) Methylation and the FIE protein are components of the imprinting mechanism that represses expression from the endosperm promoting genes (represented as Set AB) Possibly, the FIE protein (presumably in a complex with FIS1, FIS 2, and other proteins) maintains transcriptional repression of genes that are uniparentally methylated.

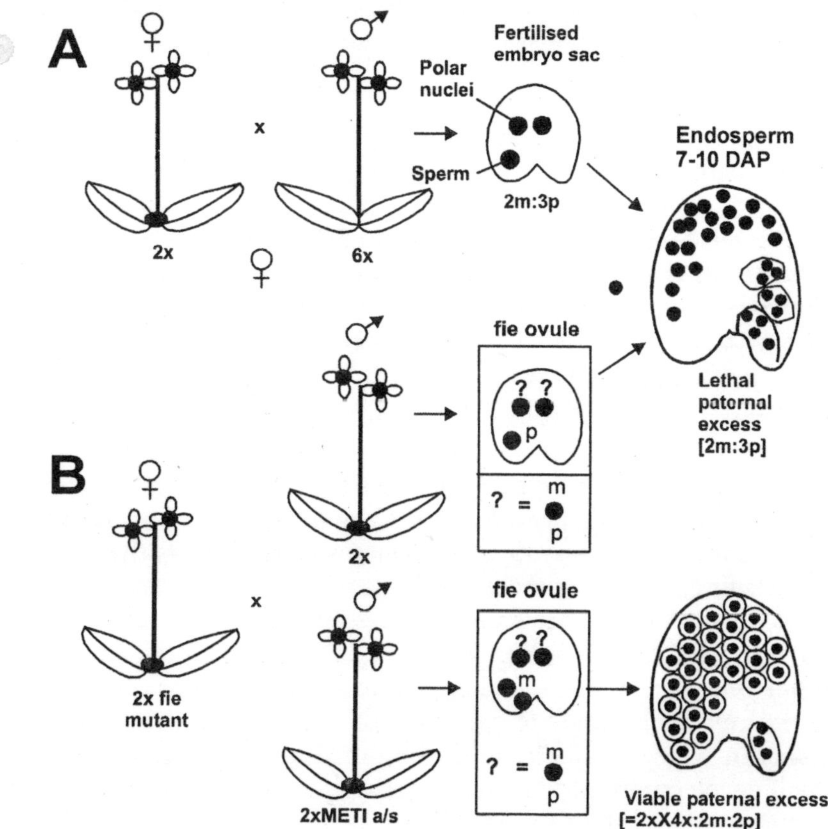

Fig. 14.6 : Fertilized fie mutant ovules abort with a paternal excess phenotype. (A) A seed resulting from a 2x × 6x cross expresses two sets of maternally inherited endosperm-inhibiting genes (2m) for every three sets of paternally inherited endosperm-promoting genes (2m: 3p). This leads to the development of a lethal paternal excess phenotype in the endosperm: massive, overgrown chalazal endosperm, many peripheral endosperm nuclei, and no cellularization. (B) In a fie mutant ovule, the imprinting status of the polar nuclei is not known. Fertilization with wild-type pollen (and, thus, adding one set of active endosperm-promoting genes to the endosperm), however, phenocopies a 2x × 6x cross (2m:3p) and results in a lethal paternal excess phenotype. Fertilization of a fie mutant ovule with demethylated pollen (expressing both a set of endosperm-promoting and a set of endosperm-inhibiting genes from which the imprint has been lifted) phenocopies a 2x × 4x cross (2m:2p) and results in a mild (viable) paternal excess phenotype. It is therefore likely that the genomes of the polar nuclei in a fie mutant ovule are not imprinted and express both the endosperm-inhibiting and the endosperm-promoting genes (1m: Ip).

The reactivation of paternally imprinted, endosperm-inhibiting genes in such pollen is predicted to restore the lethal 2m:3p ratio to a viable 3m:3p ratio, giving rise to seed developing with a viable paternal excess phenotype, which resemble those generated by a 2*x* × 4*x* cross. We argued that *FIE* as well as *FISI/MEDEA* and *FIS2* may be involved in the regulation of the expression of imprinted genes. These proteins, possibly in a complex with other proteins (from now on referred to as the FIS complex), are predicted to prevent autonomous . endosperm development in the absence of fertilization and inhibit endosperm-promoting genes after fertilization. Another way to interpret *the fie* autonomous and fertilized phenotype is to propose that *FIE* participates in establishing or maintaining the gender of the female gametes (polar nuclei).

Gametes are morphologically and biochemically different in ways that reflect their different roles as vehicles in the fertilization process. However, nuclear transfer experiments in mammals and the outcomes of interploidy crosses in flowering plants reveal that gamete genomes in these organisms are nonequivalent. The basis of this genomic nonequivalence appears to be due to genomic imprinting. Therefore, in the absence of imprinting, would the gamete genomes of flowering plants, or indeed mammals, have a gender of their own? Or perhaps more correctly, a memory of the gender of the individual that produced them? As discussed earlier viable parthenogenetic, gynogenetic, and androgenetic embryos can be formed in flowering plants. This is in accordance with the hypothesis that the embryo genome is not subject to imprinting.

lnactivation of the FIS complex and thus removal or pievention of maintenance of imprints in the female gametophyte could be interpreted as additionally erasing imprinting from the polar nuclei. The formation of autonomous endosperm in *a fie* mutant suggests that maternally imprinted genes have been reactivated, thereby generating a genetic 2m:2p ratio in the central cell. In other words, by removing or preventing the application of imprints the genome has returned to, or remained in, the same neutral or hermaphrodite state as the progenitor somatic cells (*i.e.*, 1m: Ip). Therefore, gametic gender can be seen as a consequence of imprinting, and whether a gamete is "female" or "male" as depending on which class of genes is imprinted: female gametes have the set of endosperm-promoting genes imprinted, while production of male gametes implies imprinting of the endosperm-inhibiting genes. Logically, therefore, in the absence of gender, any combination of gamete genomes could in principle lead to a viable embryo. In organisms in which genomic imprinting of the genome is thought to be absent (viz., species in which development of the offspring is not dependent on nutrient influx from a maternal tissue), it should therefore be possible to obtain viable uniparental embryos. In contrast to the situation in mammals, where androgenetic and gynogenetic embryos abort with opposite phenotypes, nonimprinted uniparental embryos are not only predicted to be viable but also to show no parent-of-origin effects in their development. To date, natural and induced uniparental embryos have been reported for a number of species representing several vertebrate and invertebrate taxa.

Uninseminated turkey eggs can develop parthenogenetically into viable turkeys. Naturally occurring parthenogenesis in reptiles has been recorded for several lizard species. In the amphibian species *Xenopus laevis,* both gynogenetic and androgenetic individuals can be experimentally induced and grown into adulthood. The first unisexual vertebrate discovered was a fresh-water fish, the Amazon molly (*Poecilia formosd*). The Amazon molly is an "all female" species that reproduces solely by gynogenesis. During reproduction, the entire female genome is transmitted to the next generation. Sperm of related, sexual species is needed to activate embryogenesis in the ova of the gynogenetic female, but the paternal genome is incorporated nor expressed. Androgenetic and gynogenetic embryos can be induced in many species of fish, including commercially interesting species such as salmon and trout and the much-studied zebrafish.

In zebrafish, the development of induced haploid and diploid andro- and gynogenetic embryos has been compared. Development of haploid zebrafish embryos arrests in an early stage. Haploid androgenotes are indistinguishable in appearance from haploid gynogenotes, indicating that the failure to develop into adults is not due to parent-of-origin effects. Diploid andro- and gynogenotes, obtained by inhibiting the first mitotic division of the haploid embryo, complete embryogenesis and develop into adults. Although parent-of-origin transgene methylation has been observed in zebrafish it appears that even if endogenous genes are imprinted, this has no detectable effects on the development of uniparental organisms. In insects, parthenogenesis is found in many species, for example, aphids and grasshoppers.

Natural androgenesis has only been reported for the Sicilian stick insect but both androgenesis and gynogenesis (leading to viable adults) can be induced in *Drosophila.* All tested members of the order Hymenoptera, which comprises bees, wasps, and ants, are haplodiploid: the sexes differ in ploidy. Whereas the females are diploid and develop from fertilized eggs, male bees, wasps, and ants are haploid and arise gynogenetically from unfertilized eggs. In wasps, it is the diploid status itself and not the contribution of the male genome that determines the female gender. Infection of the parthenogenesis-inducing *Wolbachia* bacterium disrupts meiosis in female hosts and leads to the formation of diploid eggs, which develop into female wasps. Haplodiploidy occurs also in other insect orders and in Arthropods other than insects, such as ticks and mites. Genomic imprinting has been observed in insects. In fact, the term "imprinting" in an epigenetic context was first used to describe the maternally inherited control of sex-specific elimination of *X* chromosomes in the fly *Sciara coprophila.*

As in zebrafish, the occurrence of androgenetic, gynogenetic, and parthenogenetic insects and their development into adulthood suggests that, unlike in mammals and flowering plants, genomic imprints do not play a pivotal role in early development of insect embryos. The widespread occurrence of natural uniparental embryogenesis and the possibilty to induce androgenesis and/or gynogenesis in species representing four of the five vertebrate classes and many invertebrate taxa can be explained by assuming that in these taxa the gamete genomes do not have a gender. Any combination of gamete genomes (or, at least in some species, any haploid gamete genome) can result in the formation and development of a viable embryo.

Where the development of gynogenetic or parthenogenetic and androgenetic embryos has been studied within a species, no parent-of-origin effects have been found, again indicating an apparent absence of differences between the male and female gamete genomes. In the gametes of mammals and in the endosperm genome of flowering plants, it is highly likely that it is the establishments of genomic imprints that determines the gender of the gamete genome. With genomic imprints present and sets of genes uniparentally silenced, gamete genomes of different parental origin have become complementary and hence only a combination of maternal and paternal genomes can be successful. Because of their suggested role in maintaining genomic imprints in the polar nuclei, *FIS1, FIS2,* and *FIE* in *Arabidopsis* may thus give the central cell genome her maternal identity. One can therefore speculate about the existence of genes that impose and/or maintain maleness onto the sperm genomes.

Paternally imprinted genes are predicted to inhibit endosperm development. At the moment, we can only speculate about the nature of such genes. They may be genes involved in import of sugars in the endosperm, or cell cycle-inhibiting genes controlling the rate of proliferation in the endosperm. It is possible that the expression of such imprinted genes is controlled by paternally expressed upstream regulatory genes (possibly polycomb genes). Analogous to *MEA, FIS2,* and *FIE* on the female side, the activity of such regulatory genes would lead to parent-of-origin-specific silencing of genes and thereby impose a male gender onto the genome.

Plants carrying loss of function mutations in *FIS1/MEDEA* have been accused of being bad mothers and hence the gene was named after Medea, who sacrificed her children to revenge their father's infidelity. By first imposing a female identity on her genome, we think that *FIS1* is more like Medea's illustrious predecessor, Eve.

APPLICATIONS OF GENOMIC IMPRINTING

It will be obvious that large parts of the knowledge of genomic imprinting in plants are still lacking. We are just beginning to identify some of the imprinted genes and determine their function in endosperm development in Angiosperms. Though the importance of DNA methylation for the establishment and/or maintenance of genomic imprints is becoming clear now, the way in which alleles are marked for parent-of-origin-specific methylation is not known, nor is the exact role of methylation. Nonetheless, it has already become clear that the development of embryo and endosperm and the involvement of genomic imprinting is an area of research that is not only scientifically highly interesting, but also offers several opportunities at the level of bioengineering. First, as has been shown by both interploidy crosses and crosses with demethylated parents in *Arabidopsis,* differential genomic imprints in the parental genomes influence endosperm size and development and thereby eventually embryo and seed size. Many of the world's most important crop species, such as wheat, rye, rice, barley, and maize, have a persistent endosperm.

In contrast to *Arabidopsis,* in such plants the endosperm is still present in the mature seed and can make up to 90% of the seed weight.In cereals, where the mature seed is the crop product, endosperm therefore constitutes the bulk of the harvest. Opportunities to influence (increase) endosperm size could thus lead to an enhanced yield. As described above, changing the methylation level of one of the parental gamete genomes would be one of the ways to achieve this. -When more imprinted genes, and more components of the imprinting machinery in flowering plants, are characterized, additional and more sophisticated ways to influence the expression of imprinted genes are likely to become possible. For instance, transgenic plants overexpressing one or more maternally imprinted, endosperm-promoting genes could be generated. When such plants are used as pollen donors, this will make the pollen even more vigorous concerning the promotion of endosperm growth.

Depending on the nature of silencing of maternally imprinted genes, it may even be possible to have such genes expressed from the maternally inherited copy of the endosperm genome as well. If the mark that eventually leads to imprinting and silencing in the maternal copy is located in the promoter sequence, then expression of the gene from a different promoter should be possible. In the mouse *Igf2r* gene, the information necessary for both *de novo* methylation of the imprinted paternal allele and an allele-discrimination signal are located on a 113-bp sequence in intron. Deletion or mutation of these signals abolishes *de novo ft* methylation of a normally methylated site in the imprinted allele, as well as differential methylation of the parental alleles in the embryo.

Acquiring comparable data on the sequences involved in the regulation of imprinting of plant genes will make the manipulation of parental imprinting more likely. lnfluencing the extent of genomic imprinting in plants will also be very useful in the establishment of interspecific crosses. Many related plant species have developed interspecific cross barriers that appear to be of an epigenetic rather than a genetic character. In 1942, Stephens noted that a cross between 4*x* Asiatic cotton and 2*x* American cotton managed to produce viable seeds despite their being of a different ploidy. As this was in conflict with the then-accepted required 2:3:2 ratio for maternal, endosperm, and embryo tissue, he suggested that within each species, a single set of chromosomes has a certain strength that may differ from that in related species, whether of the same or of a different ploidy.

More examples of interspecific interploidy crosses in different species followed, with similar results. Howard, by comparing the seed weight following different interspecific and interploidy crosses, was able to calculate that the relative "seed strengths" of the diploid *Nasturtium officinale* and its allotetraploid *N. uniseriatum* were 1 and 1.41 instead of the expected 1 and 2. Perhaps the most striking examples were found in the genus *Solanum*. Seed set in the interploidy cross between the two tetraploid species *S. acaule* and *S. tuberosum* invariably fails due to endosperm breakdown. However, when an autotetraploid of 5. *acaule* was used, the cross could be made, notwithstanding the fact that the endosperm now consisted of 8 maternal: 2 paternal sets of chromosomes. Johnston *et. al.*, accounted for these and similar observations with their endosperm balance number (EBN) hypothesis.

According to this theory, in the case where there is a difference in "genome strength" between the parents, an effective ploidy ratio—which is not necessarily the same as the absolute ploidy ratio—has to be reached within the endosperm. The endosperm balance number hypothesis has been shown to be of highest value in the genus *Solanum*. Johnston and Hanneman assigned the diploid species *S. chacoense* a random EBN of 2. When crossing this species and its autotetraploid (EBN 4) into other *Solanum* species, only those crosses that gave rise to an EBN balance of 2:1 were successful. In this way, a number of $2x$ and $4x$ *Solanum* species could be assigned an endosperm balance number of 1, 2, or 4. The EBN can be interpreted in terms of genomic imprinting.

Species with a high EBN are likely to be highly imprinted species, that is, species in which uniparental imprinting has a strong effect on the development of the seed. This can be explained either by assuming that a large number of genes in such a species is imprinted, or because imprinted genes have a particularly strong effect on seed development. If differences in EBN indeed can be accounted for by differences in gamete genome imprinting, then changing the level of imprinting should allow interspecific crosses between species with a different EBN without the need for polyploidization of one of these species. A third and probably most promising implication of engineering the level of genomic imprinting in flowering plants is the possible creation of apomictic plants.

Many crop plants are produced as F_1 hybrids between inbred strains. Because of the heterosis effect, F_1 hybrids are more vigorous then either parent. The obvious drawback is that each F_1 generation has to be produced by cross-fertilization of the two parent strains. Introducing apomixis into the desired F_1 hybrid strain would allow this strain with its desired genetic background to be propagated into future generations without meiotic recombination and segregation. One of the problems involved in trying to obtain apomictic plants is the dependency on double fertilization. For the successful sexual reproduction of a plant, both a functional embryo and a functional endosperm are needed. In an ideal, true apomict, seeds would develop without pollination, implying that both embryo and endosperm development would have to be initiated and carried on without fertilization.

Most natural apomicts have successfully overcome the fertilization-dependent barrier of embryo development. A $2n$ embryo, derived from either the fusion of two haploid maternal gametes or from an aberrant $2n$ maternal gamete, is formed and develops within the seed. However, with a few exceptions, in all these seeds the endosperm is still dependent on fertilization of the polar nuclei by a sperm, and in most cases the requirement of a 2:1 maternal: paternal ratio remains. Screening for mutants that allow both endosperm and embryo development in the absence of fertilization has not yet been successful. In *Arbaidopsis thaliana* it is now possible to obtain seeds in which the endosperm develops autonomously in the absence of fertilization to such an extent that at least morphologically it cannot be distinguished from a sexual endosperm.

Combining the mutant *fie-1* allele and hypomethylation leads to the development of such an endosperm in mutant ovules. If these, purely maternally derived endosperms are functional, that is, if they are capable of importing and storing enough nutrients and expressing the genes needed for breakdown and conversion of those nutrients when embryo growth and development demands it, then in principle they could support the development of an (apomictic) embryo. If so, the autonomous endosperm mutants may be the ideal background for mutagenizing and screening for autonomous embryo development in *Arabidopsis.* If this proves to be successful, then characterization of the genes and pathways involved in autonomous development of either tissue may open the way to the introduction of apomixis in other (crop) species.

Index

W

Y

❑❑❑